高职高专“工作过程导向”新理念教材 计算机系列

Java Web程序设计与项目实践

陈建国 主编

张晓云 代英明 副主编

清華大學出版社

北 京

内 容 简 介

本书以"用户管理系统"项目为案例，采用项目导向、任务驱动、案例讲解的方法，可以使读者更直接、深入地掌握 Java Web 编程的核心知识点。

本书主要包括 Web 应用程序概述、JSP 基础、JSP 内置对象、JavaBean 技术、JSP 的数据访问、Servlet 技术、EL 和 JSTL、JSP 应用开发等内容。本书共 8 章，包含 8 个项目，28 个任务，提供了 150 多个教学案例程序。通过这些项目，将 Java Web 程序设计中常见的开发技术融入其中。

本书内容翔实，实例丰富，非常适合作为零基础学习人员的学习用书和大中专院校的教材，也可供相关培训机构的师生和软件开发人员参考。

图书在版编目(CIP)数据

Java Web 程序设计与项目实践/陈建国主编. —北京：清华大学出版社，2022.10
高职高专"工作过程导向"新理念教材. 计算机系列
ISBN 978-7-302-59049-1

Ⅰ. ①J… Ⅱ. ①陈… Ⅲ. ①JAVA 语言－程序设计－高等职业教育－教材 Ⅳ. ①TP312

中国版本图书馆 CIP 数据核字(2021)第 178879 号

责任编辑：孟毅新
封面设计：傅瑞学
责任校对：李 梅
责任印制：宋 林

出版发行：清华大学出版社
网 址：http://www.tup.com.cn，http://www.wqbook.com
地 址：北京清华大学学研大厦 A 座 **邮 编**：100084
社 总 机：010-83470000 **邮 购**：010-62786544
投稿与读者服务：010-62776969，c-service@tup.tsinghua.edu.cn
质量反馈：010-62772015，zhiliang@tup.tsinghua.edu.cn
课件下载：http://www.tup.com.cn，010-83470410
印 装 者：三河市铭诚印务有限公司
经 销：全国新华书店
开 本：185mm×260mm **印 张**：26 **字 数**：598 千字
版 次：2022 年 10 月第 1 版 **印 次**：2022 年 10 月第 1 次印刷
定 价：89.00 元

产品编号：085394-01

前言

在学习本课程之前，首先要了解与本课程密切相关的前修知识。读者要先学习 Java 基础知识，了解并掌握使用 Java 语言进行代码编写的基本语法及程序逻辑，掌握如何遵循面向对象思想的原则设计并开发程序；能使用 HTML 语言制作静态网页，了解网页的基本结构，并具备网页布局设计及编写的能力；能利用客户端脚本 JavaScript 语言实现与用户的动态交互。本书的主要知识点有：Java 基本语法、HTML 制作静态网页、面向对象程序设计、数据库基础、JavaScript 编程。

本书以"用户管理系统"项目为案例，采用项目导向、任务驱动、案例讲解的方法，使读者更直接、深入地掌握 Java Web 编程的核心知识点。

本书主要包括 Web 应用程序概述、JSP 基础、JSP 内置对象、JavaBean 技术、JSP 的数据访问、Servlet 技术、EL 和 JSTL、JSP 应用开发等内容。

本书内容的组织采用由浅入深、循序渐进的方法，根据软件开发模式将知识结构分为三大模块并对应相应的项目任务，按照软件开发的分层结构的思想，将项目的实现按照业务逻辑层、数据访问层、控制层、表示层分解到各章中。

(1) JSP+JDBC 开发模式。读者要掌握 JSP 的基础语法、JSP 的内置对象、JDBC 等相关的知识点，完成静态网页设计、表单设计、用户登录、用户注册、聊天室、网站计数器、购物车、验证码等网页设计，并实现与数据库进行简单的交互处理。

(2) JSP+DAO 开发模式。读者要掌握 JavaBean 等相关知识，理解数据封装、业务封装等基本概念，完成后台业务的处理，初步理解三层结构的含义。

(3) MVC+DAO 开发模式。读者要掌握 EL、JSTL、标签、Servlet 程序开发等相关知识，理解 MVC 结构的设计方法，掌握如何利用 Servlet、过滤器、监听器实现对控制层的程序设计，完成特殊显示、登录验证等设计。

本书具有如下特点。

(1) 采用项目导向、任务驱动、案例讲解的方法，使读者更直接、深入地掌握 Java Web 编程的核心知识点。

(2) 将 Java 开发核心设计模式(MVC、DAO、工厂模式等)贯穿全书,并应用于项目。

(3) 从源代码分析出发,使学生知其然也知其所以然。

(4) 用一个完整的教学案例"用户管理系统"贯穿教学的始终,将项目分解为相应的任务,由任务引入相关知识,通过案例理解相关的知识点。在最后一章讲解 Java Web 常见的应用开发,用一个完整的在线消息管理系统,按照分层模式开发的方式,详细介绍其开发的整个过程。

(5) 本书采用提出问题、分析问题、引申知识点、案例讲解、任务实施、项目案例实践的顺序进行编写,符合"生疑—思疑—释疑、再生疑—再思疑—再释疑"的学习思维过程,能快速地让读者入门并精通这门技术。

(6) 本书每章有习题、知识要点、关键词,从多角度帮助读者掌握基础知识。

本书由陈建国任主编,张晓云、代英明任副主编。在本书编写的过程中,编者根据多年的工程实践和教学经验,参考了其他同行先进的思想体系,力求内容简洁、准确,既结合工程项目,又能让初学者易于接受。由于编者水平有限,书中难免有不足之处,殷切希望广大读者批评指正。

编　者

2022 年 5 月

目录

第 1 章　Web 应用程序概述

在掌握 Java 基础知识、理解面向对象编程的思想以后，就可以开始逐步学习开发 Web 应用程序。本章将一步一步地深入讲解进行 Web 应用开发所使用的技术结构——B/S 结构，以及 B/S 架构所具有的优势。

【技能目标】 掌握 Tomcat 服务器的发布与运行方法；掌握静态页面的设计。

【知识目标】 B/S 结构的基本概念；B/S 结构与 C/S 结构的区别；静态网页的设计；Tomcat 服务器的发布与运行方法。

【关键词】 超文本(hypertext)　传输(transfer)　协议(protocol)　资源(resource)　浏览器(browser)　服务器(server)　客户(client)　部署(deploy)

1.1　Web 相关概念

在信息管理系统的应用程序中有两种模式，一种模式是在客户端安装相应的应用程序；另一种模式则不需要在客户端安装应用程序，直接利用浏览器访问服务器就可以了，这就是所谓的 C/S 结构和 B/S 结构。

1.1.1　C/S 结构与 B/S 结构

C/S 结构与 B/S 结构如图 1-1 所示。

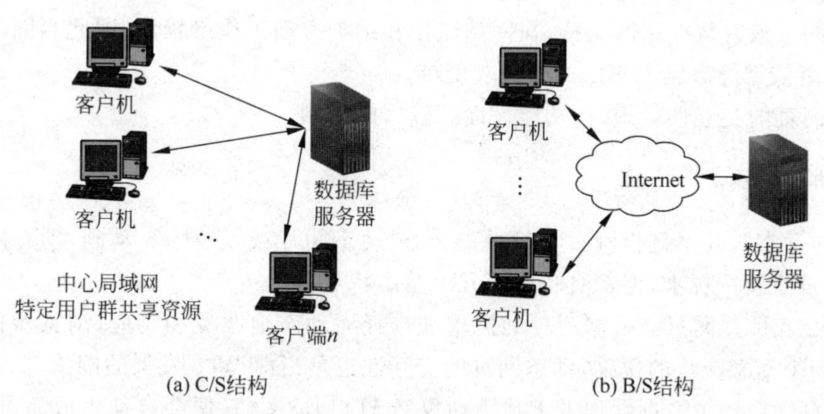

图 1-1　C/S 结构与 B/S 结构

C/S(client/server，客户/服务器)结构下的计算工作分别由服务器和客户机完成。服务器主要负责管理数据库，为多个客户程序管理数据，对数据库进行检索和排序等工作。客户机主要负责与用户的交互，收集用户信息，通过网络向服务器请求数据库、电子表格等信息的处理工作。在 C/S 结构下，资源明显不对等，是一种“胖客户机(fat client)”或“瘦服务器(thin server)” 结构。

B/S(browser/server,浏览器/服务器)结构下,客户端不需要开发任何用户界面,而统一采用如 IE 之类的浏览器,通过 Web 浏览器向 Web 服务器提出请求,由 Web 服务器对数据库进行操作,并将结果逐级传回客户端。

B/S 结构简化了客户机的工作,客户机上只需配置少量的客户端软件。服务器将担负更多的工作,对数据库的访问和应用程序的执行在服务器上完成。浏览器发出请求,而其余如数据请求、加工、结果返回以及动态网页生成等工作全部由 Web 服务器完成。

1.1.2 静态网页与动态网页

网页一般又称 HTML 文档,是一种可以在 WWW 上传输、能被浏览器识别和翻译成页面并显示出来的文件。网页是构成网站的基本元素,是承载各种网站应用的平台。通常看到的网页,大都是以. htm 或. html 为扩展名的文件,除此之外网页文件还有以. cgi、. asp、. php 和. jsp 为扩展名的。目前网页根据生成方式,大致可以分为静态网页和动态网页两种。

1. 静态网页

静态网站是最初的建站方式,浏览者所看到的每个页面是建站者上传到服务器上的一个 HTML 文件(静态网页)。这种网站每增加、删除、修改一个页面,都必须重新对服务器上的文件进行一次下载上传。其特点如下。

(1) 网页内容不会发生变化,除非网页设计者修改了网页的内容。

(2) 不能实现和浏览网页的用户之间的交互。信息流向是单向的,即从服务器到浏览器。服务器不能根据用户的选择调整返回给用户内容。

(3) 网页的内容相对稳定,因此容易被搜索引擎检索。

(4) 网页没有数据库的支持,在网站制作和维护方面工作量较大,因此当网站信息量很大时完全依靠静态网页制作方式比较困难。

(5) 网页的交互性较差,在功能方面有较大的限制。

2. 动态网页

所谓"动态",并不是指网页上简单的 GIF 动态图片或是 Flash 动画,动态网站的概念现在还没有统一标准,但都具备以下几个基本特点。

(1) 交互性。网页会根据用户的需求和选择而动态地改变和响应,浏览器作为客户端,成为一个动态交流的桥梁,动态网页的交互性也是当前 Web 发展的潮流。

(2) 自动更新。站点管理者无须手动更新 HTML 文档,便会自动生成新页面,可以大大节省工作量。

(3) 因时因人而变。浏览者在不同时间、不同用户访问同一网址时会出现不同页面。

3. 静态网页与动态网页的区别

动态与静态最根本的区别是网页在服务器端运行状态的不同,在服务器端运行的程序、网页、组件,属于动态内容,它们会随不同客户、不同时间,返回不同的网页,例如

ASP、PHP、JSP、ASP. NET、CGI 等。运行于客户端的程序、网页、插件、组件，属于静态内容，例如 HTML、Flash、JavaScript、VBScript 等，它们是不变的。

静态网页和动态网页各有特点，网站采用动态网页还是静态网页主要取决于网站的功能需求和网站内容的多少，如果网站功能比较简单，内容更新量不是很大，采用纯静态网页的方式会更简单；反之一般要采用动态网页技术来实现。

静态网页是网站建设的基础，静态网页和动态网页之间也并不矛盾，为了网站适应搜索引擎检索的需要，即使采用动态网站技术，也可以将网页内容转化为静态网页发布。

动态网站也可以采用静动结合的原则，适合采用动态网页的地方用动态网页，如果必须使用静态网页，则可以考虑用静态网页的方法来实现。在同一个网站上，动态网页内容和静态网页内容同时存在也是很常见的事情。

4. 动态网页实现的手段

动态网页大多是由网页编程语言写成的网页程序生成的，访问者浏览的只是其生成的客户端代码，而且动态网页要实现其功能大多还必须与数据库相连。目前比较常见的互动式网页编程语言有 ASP、PHP、JSP、ASP. NET。

(1) HTML 网页适用于所有环境，它本身也相当简单。

(2) ASP 及 ASP. NET 网页主流环境为 Windows Server 的 IIS＋Access/SQL Server。

(3) PHP 网页主流环境为：Linux/UNIX＋Apache＋MySQL＋PHP4＋Dreamweaver。

(4) JSP 网页环境为：JDK＋Tomcat＋Eclipse＋MyEclipse。

1.1.3　Web 运行环境

在了解如何开发 Web 应用程序之前，要首先了解一下这些应用程序的运行平台和环境，包括 Web 访问基本原理、HTTP 协议、Web 服务器以及 Web 浏览器。

1. Web 访问基本原理

图 1-2 显示了浏览器访问 Web 服务器的整个过程。

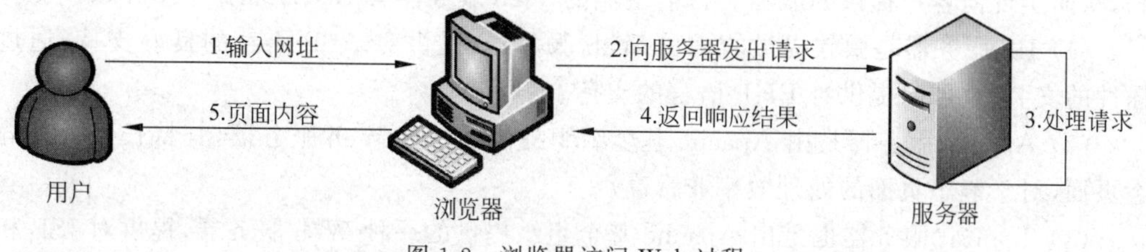

图 1-2　浏览器访问 Web 过程

(1) 用户在浏览器中输入网站的 URL。

(2) 浏览器寻找到指定的主机之后，向 Web 服务器发出请求(request)。

(3) Web 服务器接收请求并做出相应的处理，生成处理结果。

(4) 服务器把响应的结果返回给浏览器。

(5) 浏览器接收到响应结果后，在浏览器中显示相应的内容。

2. HTTP 协议

超文本传送协议(hypertext transfer protocol,HTTP)是一种用于分布式、协作式和超媒体信息系统的应用层协议。HTTP 是一个客户端终端(用户)和服务器端(网站)请求和应答的标准(TCP),是万维网数据通信的基础。

HTTP 协议定义客户端如何从 Web 服务器请求页面,以及服务器如何把页面传送给客户端。HTTP 采用了请求/响应模型。客户端向服务器发送一个请求报文,服务器以一个状态行作为响应。HTTP 请求/响应的步骤如下。

(1) 客户端连接到 Web 服务器。一个 HTTP 客户端,通常是浏览器,与 Web 服务器的 HTTP 端口(默认为 80)建立一个 TCP 套接字连接。

(2) 发送 HTTP 请求。通过 TCP 套接字,客户端向 Web 服务器发送一个文本的请求报文,一个请求报文由请求行、请求头部、空行和请求数据 4 部分组成。

(3) 服务器接收请求并返回 HTTP 响应。Web 服务器解析请求,定位请求的资源。服务器将资源副本写入 TCP 套接字,由客户端读取。一个响应由状态行、响应头部、空行和响应数据四部分组成。

(4) 释放 TCP 连接。若连接模式为 close,则服务器主动关闭 TCP 连接,客户端被动关闭连接,释放 TCP 连接;若连接模式为 keepalive,则该连接会保持一段时间,在该时间内可以继续接收请求。

(5) 客户端浏览器解析 HTML 内容。客户端浏览器首先解析状态行,查看表明请求是否成功的状态代码。然后解析每一个响应头部,响应头部告知以下为若干字节的 HTML 文档和文档的字符集。客户端浏览器读取响应数据 HTML,根据 HTML 的语法对其进行格式化,并在浏览器窗口中显示。

3. Web 服务器

Web 服务器一般指网站服务器,与通信相关的处理都是由服务器软件负责,开发人员只需要把功能代码部署在 Web 服务器中,客户端就可以通过浏览器访问这些功能代码,从而实现向客户提供的服务。目前主流的 Web 服务器有 IIS、Apache、Tomcat 等。

(1) IIS 服务器是微软提供的一种 Web 服务器,提供对 ASP 语言的良好支持,通过插件的安装,也可以提供对 PHP 语言的支持。

(2) Apache 服务器是由 Apache 基金组织提供的一种 Web 服务器,其特长是处理静态页面,对于静态页面的处理效率非常高。

(3) Tomcat 服务器也是由 Apache 基金组织提供的一种 Web 服务器,提供对 JSP 和 Servlet 的支持,通过插件的安装,Tomcat 是一个小型的轻量级 Web 服务器,是开发和调试 JSP 程序的首选。

(4) JBoss 服务器是一个开源的重量级的 Java Web 服务器,在 JBoss 中,提供对 J2EE 各种规范的良好支持,而且 JBoss 通过了 Sun 公司的 J2EE 认证,是 Sun 公司认可的 J2EE 容器。

(5) WebLogic 是 BEA 公司的产品,支持 J2EE 规范,而且不断地完善以适应新的开发要求。WebLogic Server 支持企业级、分布式的 Web 应用，支持包括 JSP、Servlet、EJB 在内的 J2EE 体系，并提供必要的应用服务，如事务处理，支持集群技术。WebLogic Server 功能特别强大，配置操作简单、界面友好，在电子商务应用中被大量采用。

（6）WebSphere 是 IBM 公司的产品，支持 J2EE 规范。IBM 的 WebSphere Application Server 可运行于 Sun Solaris 等多种操作系统平台上；除可以使用 Servlet 和 JSP 之外，还可以补充用 EJB 编写的业务逻辑。几种技术结合起来，以开放的 Java 标准为基础，提供一种完整的编程模型，以实施各种 Web 站点。

（7）Nginx（Engine x）是一个高性能的 HTTP 和反向代理 Web 服务器，也提供 IMAP/POP3/SMTP 服务。其特点是占用内存少，并发能力强，事实上 Nginx 的并发能力在同类型的网页服务器中表现较好。

4. Web 浏览器

网页浏览器用于显示网页服务器或档案系统内的文件，并实现用户与这些文件进行交互操作。目前，有很多 Web 浏览器，但是比较普及和流行的为 Microsoft Internet Explorer（IE）、Mozilla Firefox 和 Google Chrome，其他的浏览器还有傲游浏览器（Maxthon）、腾讯 TT 浏览器、Opera 等。

1.2　Java Web 开发环境的安装与配置

开发 Java Web 应用程序需要搭建 Java Web 的开发和运行环境。

（1）JDK（Java development kit）：它是 Sun 官方的 Java 开发和运行环境。

（2）Eclipse 和 MyEclipse：它是 Java Web 集成开发环境（integrated develop environment，IDE）。

（3）Tomcat：它是开源 Web 应用服务器。

1.2.1　开发工具包 JDK

要运行 Java 的程序，首先要安装 Java 运行环境，即 JRE（Java runtime environment）。作为开发人员，需要安装 Java 的开发环境 JDK，JDK 包含开发 Java 程序的所有工具。

1. JDK 下载与安装

JDK 可以在 Sun 公司的主页下载，直接进入官网 https://www.oracle.com/，找到下载页后，选择相应的版本直接下载即可。通常 32 位的系统只支持 32 位的 JDK，64 位系统可以兼容 32 位和 64 位的 JDK。

下载完成后就可以运行 JDK 安装程序进行安装，安装过程中所有选项保持默认值即可。

2. JDK 配置

安装结束后进行环境变量的配置，基本步骤如下。

(1) 依次单击“控制面板”→“高级系统设置”→“高级”，或右击“我的电脑”再依次单击“属性”→“高级系统设置”→“高级”，如图 1-3 所示。

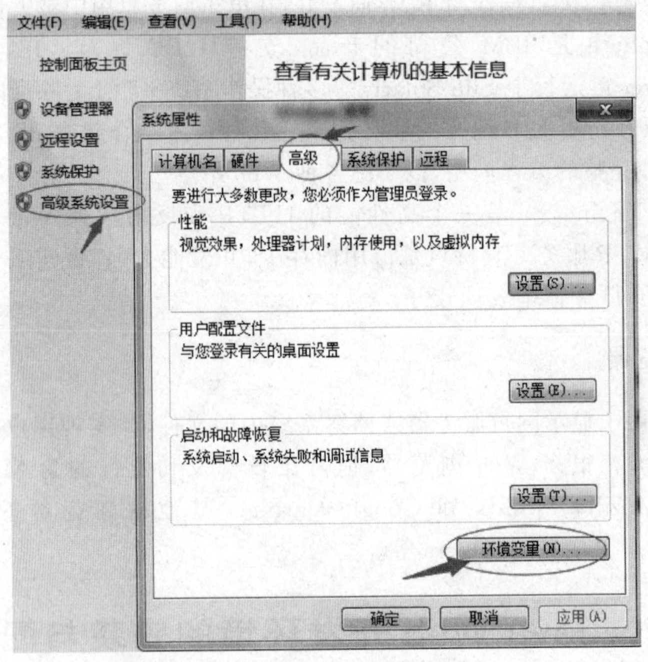

图 1-3　单击“环境变量”按钮

(2) 单击“环境变量”按钮，在系统变量下新建变量 JAVA_HOME，变量值指向 JDK 安装的文件夹，如图 1-4 所示。

(3) 选中 Path，单击“编辑”按钮，直接在末尾添加%JAVA_HOME%\bin;%JAVA_HOME%\jre\bin，如图 1-5 所示。

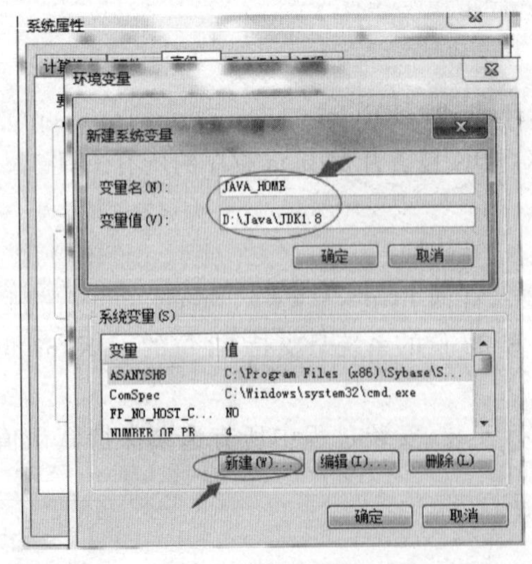

图 1-4　新建环境变量

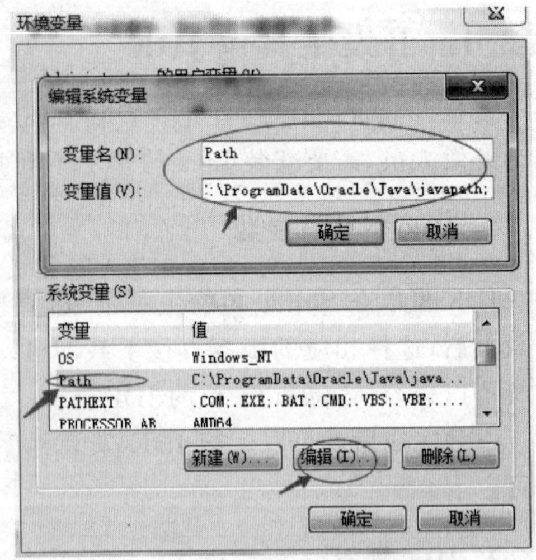

图 1-5　编辑系统变量 Path

(4) 测试 JDK 环境配置是否成功。

按 Win+R 键在“运行”中输入 cmd,单击“确认”按钮,输入 java -version(java 后空一格)按 Enter 键。如果出现 JDK 版本信息,即 JDK 环境配置成功。如果出现“java 不是内部命令”说明配置失败,如图 1-6 所示。

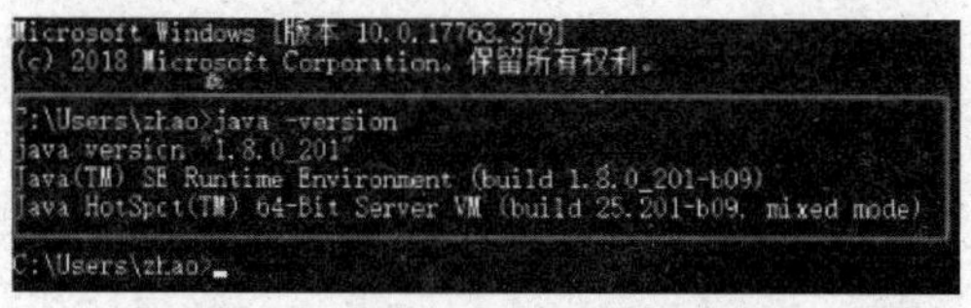

图 1-6 测试 JDK 环境配置

配置不成功的原因通常是 JAVA_HOME 变量值错误,编辑 Path 时,新建变量输入的值不对,或者输入完成后,未单击“确认”按钮,而是直接关闭。

注意:如果以后要安装诸如 Eclipse、Borland JBuilder、JCreator、IntelliJ IDEA 等集成开发环境,应该在 IDE 中编译运行一个简单的程序进行测试,以保证 IDE 可以识别 JDK 的位置。

1.2.2 Tomcat 服务器

1. Tomcat 的下载与安装

进入 Tomcat 官方安装包下载页面 https://tomcat.apache.org/,在 Download 目录下,找到并下载合适版本的 Tomcat。将下载的压缩包直接解压到 D 盘根目录,然后按照安装提示完成安装。安装时要注意 JDK 的安装路径。

2. Tomcat 的配置

1) 配置 Tomcat 的环境变量

安装好 Tomcat 之后,依次单击“计算机”→“属性”→“高级系统设置”→“高级”→“环境变量”,打开环境变量设置对话框。

(1) 新建变量名 CATALINA_BASE,变量值为 D:\apache-tomcat。

(2) 新建变量名 CATALINA_HOME,变量值为 D:\apache-tomcat。

(3) 为 Path 添加变量值%CATALINA_HOME%\lib;%CATALINA_HOME%\bin,注意要用分号把 Path 的各个变量分开。

Tomcat 的环境配置是否成功:在命令行中,输入 startup,按 Enter 键,启动 Tomcat。设置成功,则能正常启动。

2) 修改 Tomcat 的 JDK 目录

打开 tomcat/bin/catalina.bat 文件,在最后一个 rem 后面增加 set JAVA_HOME=C:\Program Files\Java\jdk1.8.0。

3) 启动内存参数的配置

打开 tomcat/bin/catalina.bat 文件(如果是 Linux 系统则是 catalina.sh),在 rem 的

后面增加 set JAVA_OPTS= -Xms256m -Xmx256m -XX:MaxPermSize=64m。

4）Tomcat 的端口配置

Tomcat 默认使用 8080 端口，可以通过 server.xml 文件修改 Tomcat 的端口号。将 port 定义的内容修改即可，如下面将端口号修改为 80 端口。

```
< Connector port = "80" protocol = "HTTP/1.1" connectionTimeout = "20000" redirectPort =
"8443" />
```

这样以后直接输入"http://localhost"即可进行访问，不用再输入端口号。

5）配置虚拟目录

在 Tomcat 服务器的配置中，最重要的就是配置虚拟目录，每个虚拟目录保存了一个完整的 Web 项目。设置虚拟目录为 site，通过 http://localhost:8080/site 访问物理路径 D:\site 目录中的内容。设置过程如下。

（1）复制 Tomcat\webapps\ROOT\WEB-INF 文件夹到 D:\site 目录中。

（2）打开 Tomcat\conf\server.xml 文件，在< Host >和</Host >之间加入以下内容。

```
< Context path = "" docBase = "ROOT" debug = "0" reloadable = "true"></Context >
< Context path = "/site" docBase = "d:\site" reloadable = "true"/>
```

其中，path="/site" 是虚拟目录的名称；docBase="d:\site"为物理路径。

（3）打开 Tomcat\conf\web.xml 文件，找到以下内容。

```
< init - param >
  < param - name > listings </param - name >
  < param - value > false </param - value >
</init - param >
```

把 false 改成 true 后保存，重启 Tomcat，就可以应用 http://localhost:8080/site 虚拟目录了。其浏览效果如图 1-7 所示。当系统正式运行时，将< param-value >的值设置为 false。

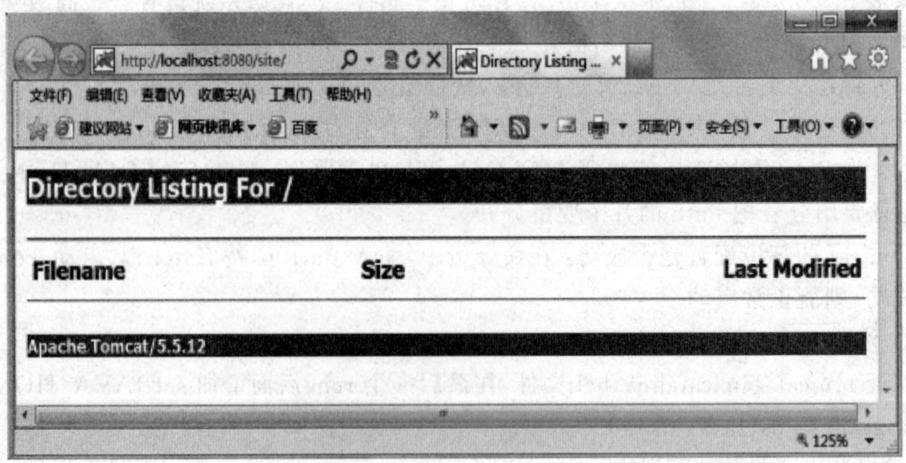

图 1-7　浏览虚拟目录

因为默认情况下,Tomcat 启动过程中配置虚拟目录的时候会从 webapps 目录下查找 webContent 应用。这样配置好了,即使以后从一台服务器移植到另一台服务器,不做任何修改也能运行。

6）解决 GET 方式下 URL 乱码问题

打开 tomcat/conf/server.xml,在最后增加以下代码。

```
<Connector port="80" maxHttpHeaderSize="8192"
URIEncoding="UTF-8" useBodyEncodingForURI="true"/>
```

其中的 UTF-8 可根据需要自己修改,比如也可以是 GBK。

7）配置虚拟主机文件

```
tomcat/conf/server.xml
<!-- 默认的主机 -->
<Host name="localhost" appBase="webapps" .unpackWARs="true"
autoDeploy="true" xmlValidation="false" xmlNamespaceAware="false">
<Context path="" docBase="ROOT" debug="0" reloadable="true"></Context>
</host>
<!-- 以下是新增的虚拟主机 -->
<Host name="" appBase="webapps" unpackWARs="true" autoDeploy="true" xmlValidation=
"false" xmlNamespaceAware="false">
<Context path="" docBase="d:\" debug="0" reloadable="true"/>
<!-- 虚拟目录 -->
<Context path="/count" docBase="d:\counter.java2000.net" debug="0" reloadable="true"/>
</Host>
<Host name="java2000.net" appBase="webapps" unpackWARs="true" autoDeploy="true"
xmlValidation="false" xmlNamespaceAware="false">
<Context path="" docBase="d:\ " debug="0" reloadable="true"/>
<Context path="/count" docBase="d:\counter.java2000. net" debug="0" reloadable="true"/>
</Host>
```

3. Tomcat 的安全设置

默认安装 Tomcat 时,Tomcat 作为一个系统服务运行。如果没有将其作为系统服务运行,通常会将其以 Administrators 权限运行。这两种方式都允许 Java 运行时访问 Windows 系统下任意文件夹中的任何文件,对系统的安全有极大的威胁。

为了保证系统的安全,应以 System 权限启动。根据权限最小的安全原则,降低脚本所获取的操作本地系统权限。操作步骤如下。

1）新建一个账户

通过操作系统创建一个普通用户账户,设置用户密码,并设置"密码永不过期"被选中。

2）修改 Tomcat 安装目录的访问权限

对 WebApps 目录设置只读权限,如果某些 Web 应用程序需要写权限,则单独为其授予。

3）Tomcat 作为系统服务运行

打开“控制面板”，选择“管理工具”选项卡，然后选择“服务”，找到 Tomcat，比如 Apache Tomcat. exe 等，打开“属性”，选择“登录”，选择“以……登录”(Log ON Using)选项。输入新建的用户名，输入密码，重启机器。

4. Tomcat 服务器的启动

Tomcat 的启动和停止脚本存在于 bin 目录下。其中，各脚本用途如下。

catalina：Tomcat 的主要脚本，它会执行 Java 命令以调 Tomcat 的启动与停止类。

configtest：Tomcat 的配置项检测脚本。

digest：生成 Tomcat 密码的加密摘要值，用于生成加密过的密码。

service：该脚本以 Windows 服务的方式安装和卸载 Tomcat。

setclasspath：这是唯一用于系统内部，以设定 Tomcat 的 classpath 及许多其他环境变量的脚本。

shutdown：运行 catalina. bat stop 可以停止 Tomcat 运行。

startup：运行 catalina. bat start 可以启动 Tomcat。

tool-wrapper：用于 digest 脚本系统内部。这是最常用的 Tomcat 命令行工具，用于封装可用于设置环境变量的脚本，并调用 classpath 中设置的完全符合限定的主要方法。

version：这是运行 Catalina 的版本，会输出 Tomcat 的版本信息。

执行 catalina. bat 时，必须附带一个选项，常用的有 start、run 及 stop。catalina 脚本启动选项含义如下。

-help：输出命令行选项的摘要表。

-nonaming：在 Tomcat 中停用 JNDI。

-security：启用 catalina. policy 文件。

debug：以调试模式启动 Tomcat。

embedded：在嵌入模式中测试 Tomcat，应用程序服务器的开发者通常使用此选项。

jpda start：以 jpda 的调试方式启动 Tomcat。

run：启动 Tomcat，但不会重定向标准输出与错误。

start：启动 Tomcat，并将标准输出与错误送至 tomcat 的日志文件。

stop：停止 Tomcat。

version：输出 Tomcat 的版本信息。

catalina. bat version：打印环境变量和版本信息。

当以 start 选项调用 catalina 时，它会启动 Tomcat，并将标准输出与错误流导出到 $TOMCAT_HOME/logs/catalina. out 文件中。选项 run 会让 Tomcat 保留当前的标准输出与错误流(如控制台窗口)。

如果使用 catalina 及 start 选项，或调用 startup 脚本而非使用参数 run，那么会在控制台上看到前几行 Using ……其余的输出信息则被重定向到 catalina. out 的日志文件中。

shutdown 脚本会调用加 stop 选项的 catalina，它会让 Tomcat 连接于 server 元素中设定的默认端口，并送出停止信息。

在 DOS 窗口下运行 Tomcat 的步骤如下。

输入 runas /user：*TOMCATUSER* CMD 命令。其中，*TOMCATUSER* 为计算机系统用户名，在询问 *TOMCATUSER* 用户的密码时输入设置的密码。这将打开一个新的 DOS 窗口，如图 1-8 所示。

图 1-8　DOS 环境下启动 Tomcat

在新开的 DOS 窗口中，切换到 Tomcat 的 bin 目录。输入 catalina run 命令，然后关闭第一个 DOS 窗口。

在 Windows 环境下启动 Tomcat 服务器的步骤如下。

直接运行 Tomcat Tomcat 的 bin 目录下的 Tomcat9w 程序，如图 1-9 所示，选择"启动"或者"停止"操作。可以在其他选项卡中设置启动参数，如修改用户登录，如图 1-10 所示。

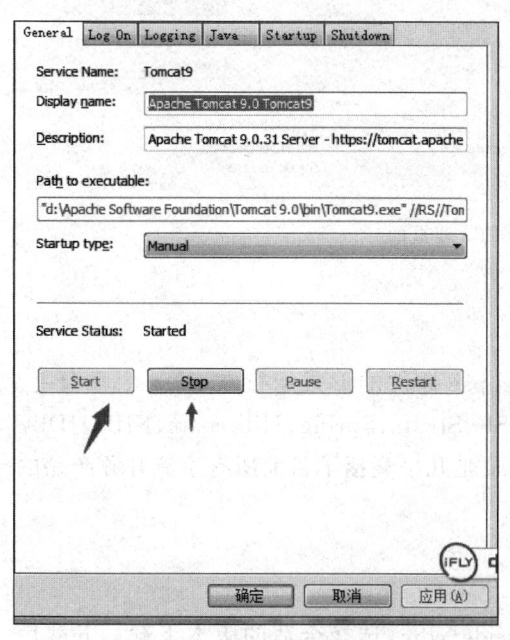

图 1-9　启动或停止 Tomcat

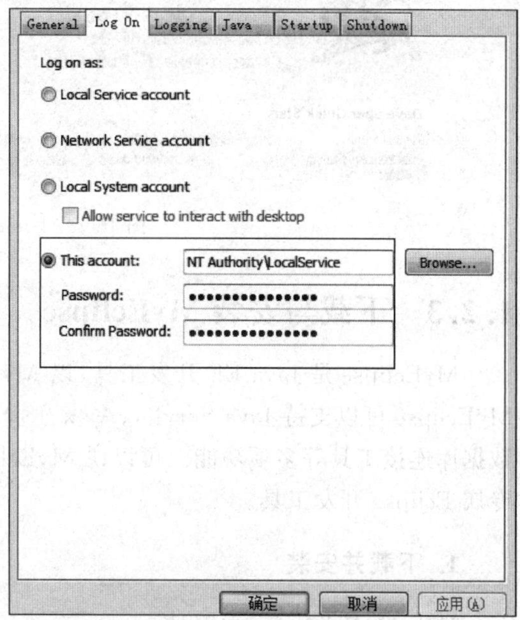

图 1-10　设置启动参数

5. Tomcat 服务器的目录结构

Tomcat 服务器的目录结构及功能如表 1-1 所示。

表 1-1 Tomcat 服务器的目录结构及功能

目　录	功　能
/bin	存放 Windows 或 Linux 用于启动和停止 Tomcat 的脚本文件
/conf	存放 Tomcat 服务器的各种配置文件,其中最重要的是 server. xml
/server/lib	存放 Tomcat 服务器所需的各种 JAR 文件
/server/webapps	存放 Tomcat 自带的 admin 应用程序和 manager 应用程序
/commom/lib	存放 Tomcat 服务器以及所有 Web 应用都可以访问的 JAR 文件
/work Tomcat	把由 JSP 生成的 Servlet 放于此目录下
/webapps	当发布 Web 应用时,默认情况下会将 Web 应用的文件存放于此目录中

6. Tomcat 服务启动检测

服务器启动后,打开浏览器,在浏览器中输入 http://localhost:8080 或者 http://127.0.0.1:8080,即可看到 Tomcat 的欢迎页面,表明安装成功,如图 1-11 所示。

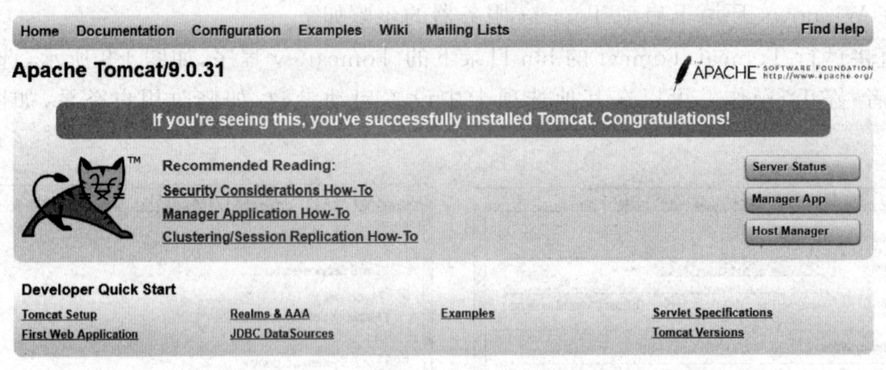

图 1-11　Tomcat 的欢迎页面

1.2.3　下载与安装 MyEclipse

MyEclipse 是 Java EE 开发工具,以 MyEclipse 插件的形式运行在 Eelipse 平台上。MyEclipse 可以支持 Java Servlet、Ajax、JSP、JSF、Struts、Spring、Hibernate、EJB3、JDBC 数据库连接工具等多项功能。可以说 MyEclipse 是几乎囊括了目前所有主流开源产品的专属 Eclipse 开发工具。

1. 下载并安装

进入 MyEelipse 官方网站 www. myeclipseide. com,选择合适的版本下载。下载后按照提示一步一步地执行安装程序,在安装过程中,可以根据自己对界面的喜好,在安装完成时选择相应的主题。

2. 环境配置

启动 MyEclipse,在使用前可以对常用的环境参数进行设置。

1）编码设置

选择 Window→Preferences，然后在窗口左边中选择 General→Content Types，在右边选择要修改的文件类型，可根据习惯将 Class、Text、XML、JSP、Java、Properties File、Spring Properties File 等编码方式进行修改，比如都设为 UTF-8，如图 1-12 所示。

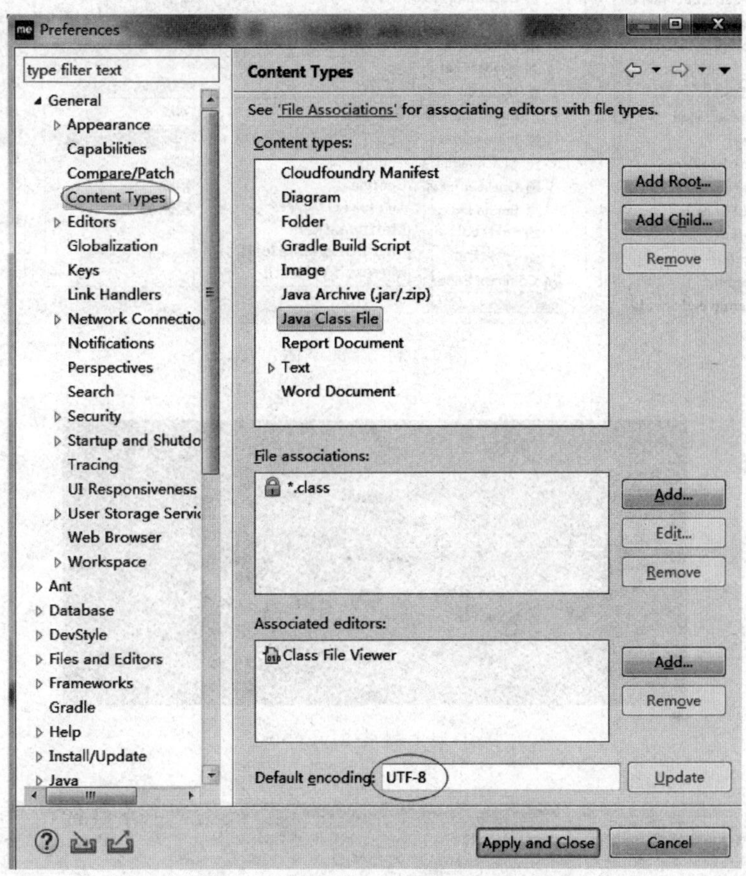

图 1-12　编码方式设置

2）设置字体

选择 Window→Preferences，然后在窗口左边中选择 General→Appearance→Colors and Fonts，在右边选中 Basic，选择要修改字体的项目，如 Text Font，单击 Edit 按钮，定义字体，如图 1-13 所示。

3. 启动 MyEclipse

安装完毕就可以从"开始"菜单来启动 MyEclipse。启动的时候会让用户选择工作文件夹，如图 1-14 所示。

单击 OK 按钮，进入 MyEclipse 工作台。第一次启动 MyEclipse 会出现欢迎画面 Welcome，关掉它将进入 MyEclipse 工作界面，如图 1-15 所示。

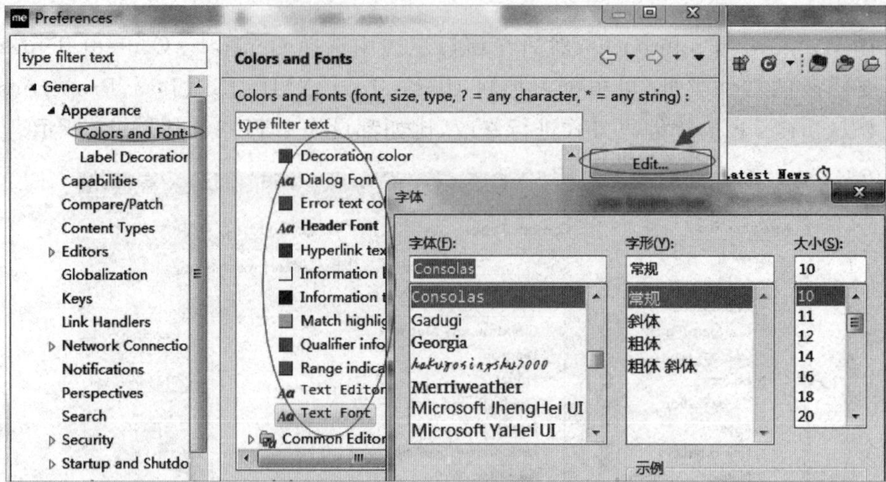

图 1-13 设置字体

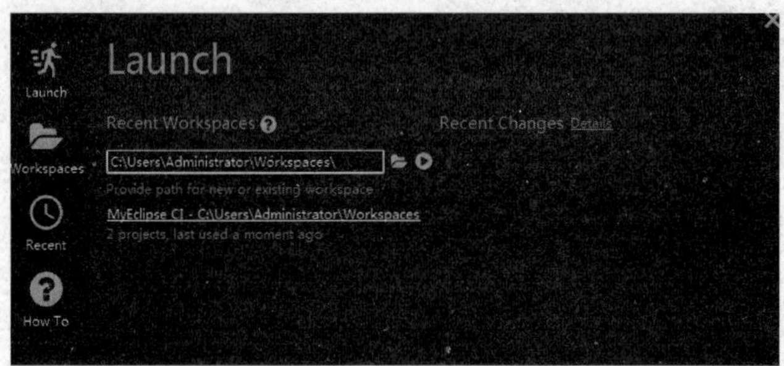

图 1-14 启动 MyEclipse

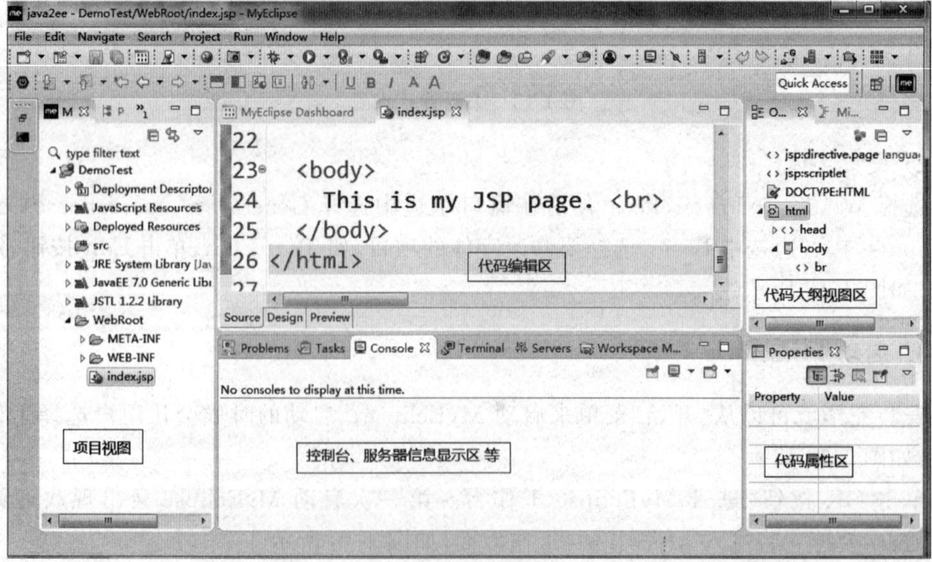

图 1-15 MyEclipse 工作界面

1.2.4　第一个 Java Web 程序

1. 配置服务器

为了方便项目开发，可以将 Tomcat 和 MyEclipse 集成起来，这需要配置服务器。使用 MyEclipse 配置服务器后，就可以使用 MyEclipse 来启动和停止服务器了。MyEclipse 自带了一个 Tomcat，强烈建议不要使用它。配置服务器的步骤如下。

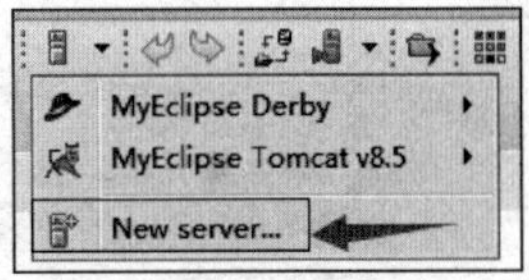

图 1-16　新建服务器

(1) 在工具栏中单击 按钮，再单击 New server，如图 1-16 所示，弹出如图 1-17 所示的窗口。

(2) 选择已经安装的服务器相应的版本，单击 next 按钮。如图 1-18 所示，选择 Tomcat 安装路径，以及 JRE 版本。若需要配置运行在此服务器上的项目文件，则单击 Next 按钮，否则单击 Finish 按钮。

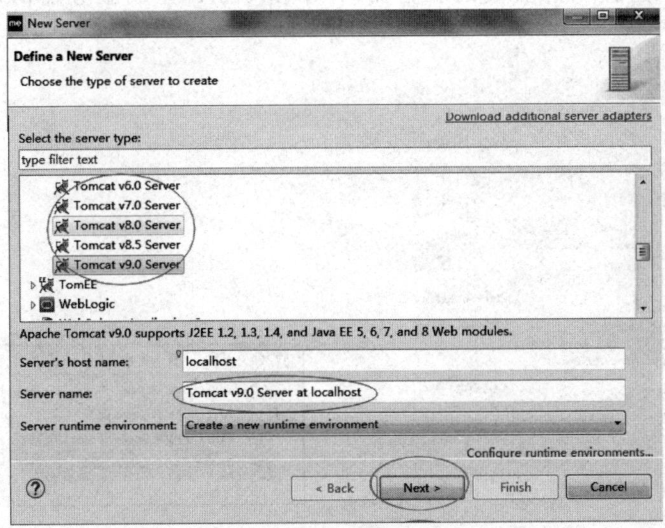

图 1-17　选择服务器

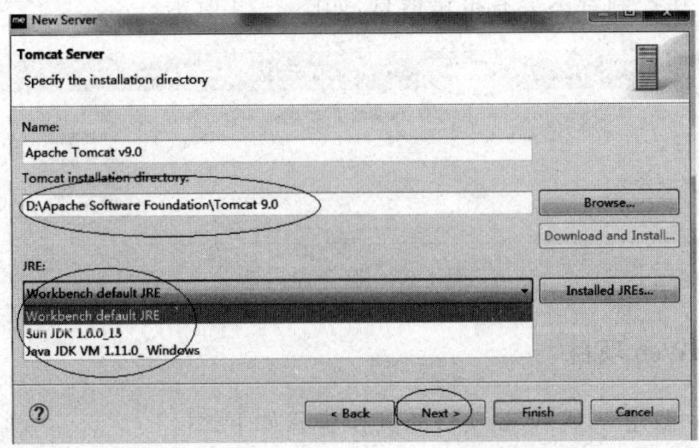

图 1-18　选择 Tomcat 安装路径

(3) 配置完成后,在服务器浏览区以及工具栏上都可以看到配置的服务器,如图 1-19 和图 1-20 所示。

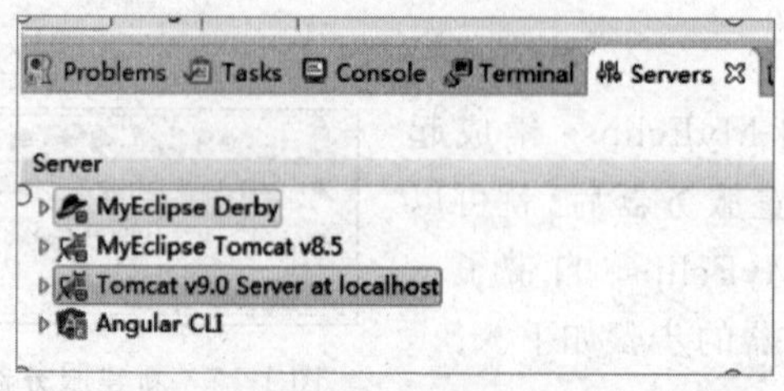

图 1-19 服务器浏览区当前安装的服务器

图 1-20 工具栏上当前安装的服务器

2. 使用 MyEclipse 启动 Tomcat

可以在服务器浏览区中选中启动的服务器,右击并选择 Start,如图 1-21 所示。也可以在工具栏上选择当前安装的服务器,再单击 Start,完成服务器的启动,如图 1-22 所示。

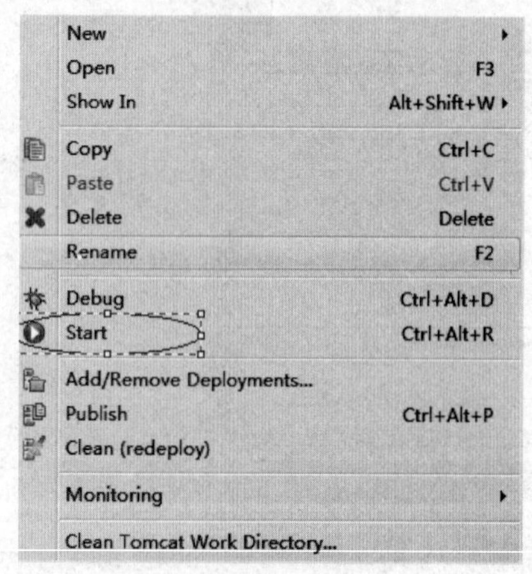

图 1-21 启动服务器

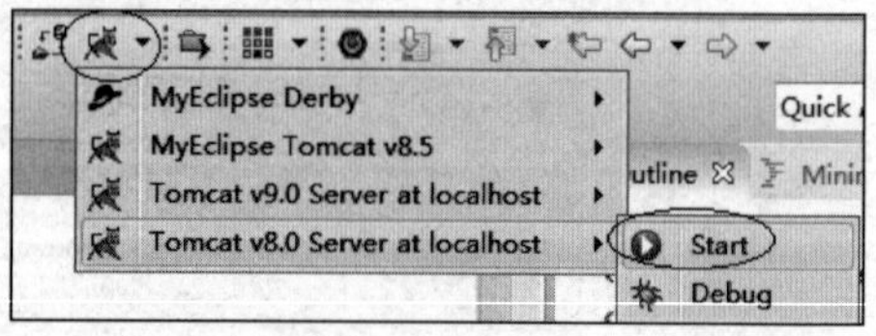

图 1-22 在工具栏上启动当前安装的服务器

正常启动时,在控制台不会有报错信息,如图 1-23 所示。

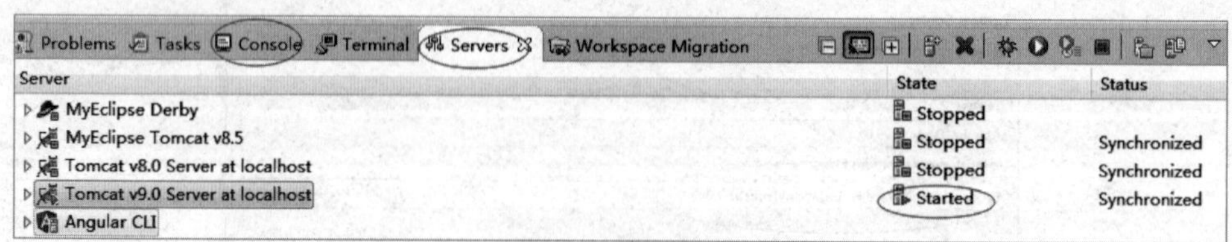

图 1-23 服务器启动

3. 创建 JavaWeb 项目

用 MyEclipse 创建一个 Web 项目的操作步骤如下。

(1) 选择 File→New→Web Project,弹出如图 1-24 所示的窗口。

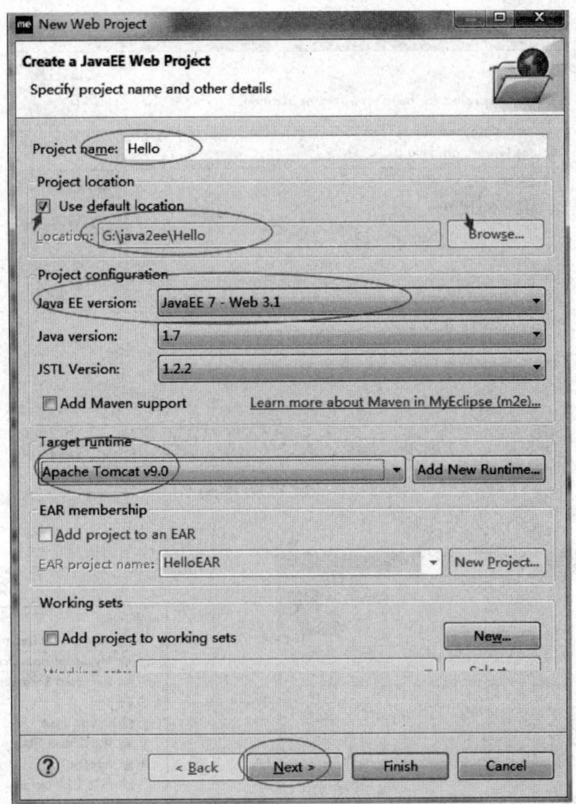

图 1-24　新建一个 Web 项目

输入项目名称,选择项目代码保存位置,选择项目的 Jave EE 版本和项目运行的服务器。注意,Tomcat 是一个开放源码软件实现 Java EE 技术的一个子集,不同版本 Java EE 标准使用不同版本的 Tomcat。Java EE 标准与相应的 Tomcat 版本之间的映射关系如表 1-2 所示。

表 1-2　Java EE 标准与 Tomcat 版本之间的映射关系

Servlet	JSP 标准	EL 规格	WebSocket 标准	认证(JASIC)标准	Tomcat 版本	支持的 Java 版本
5.0	3.0	4.0	2.0	2.0	10.0.x	8 及更高版本
4.0	2.3	3.0	1.1	1.1	9.0.x	8 及更高版本
3.1	2.3	3.0	1.1	1.1	8.5.x	7 及更高版本
3.1	2.3	3.0	1.1	不适用	8.0.x	7 及更高版本
3.0	2.2	2.2	1.1	不适用	7.0.x	6 及更高版本

(2) 单击 Next 按钮,如图 1-25 所示,设置 MyEclipse 的编译文件夹路径,以及默认编译输出的文件夹。

(3) 单击 Next 按钮,如图 1-26 所示,选中 Generate web. xml deployment descriptor,将自动生成 web. xml 配置文件。

(4) 单击 Finish 按钮完成项目创建。其目录结构如图 1-27 所示。

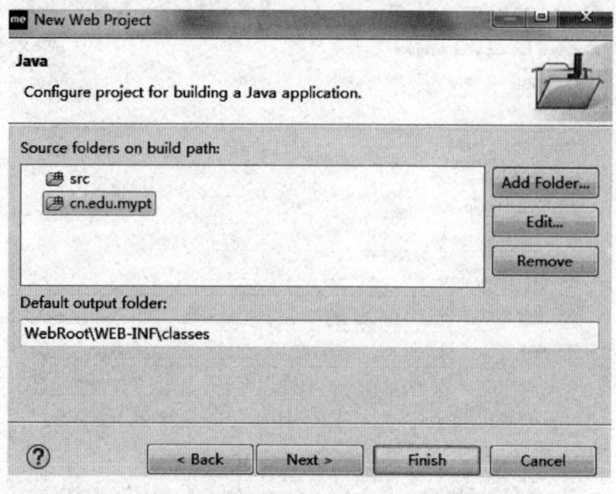

图 1-25　设置编译路径

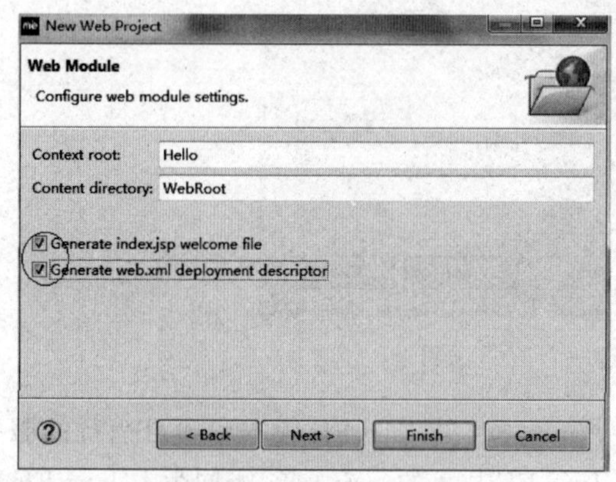

图 1-26　设置项目参数

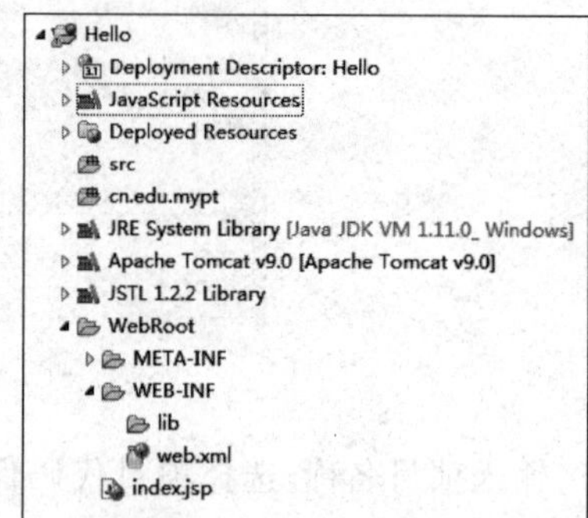

图 1-27　项目的目录结构

4. Web 项目的目录结构

/root：Web 应用的根目录，该目录下所有文件在客户端都可以访问，包括 JSP、HTML、JPG 等资源。

/WEB-INF：存放应用使用的各种资源，该目录及其子目录对客户端都不可以访问，其中包括 web. xml。

/WEB-INF/classes：存放 Web 项目的所有的. class 文件，在项目编译后产生。

/Web-INF/lib：存放 Web 应用使用的 JAR 文件。

/META-INF：清单目录，与 JAR 文件中清单目录一致。

5. 配置 Web 应用

使用 web. xml 文件配置应用发布，web. xml 文件必须保存在/WEB-INF 目录下。在 web. xml 文件中配置各种资源的发布信息。例如配置访问首页面，通过 web. xml 文件修改访问的起始页面。

```
<?xml version = "1.0" encoding = "UTF - 8"?>
< web - app version = "2.4"
  xmlns = "http://java.sun.com/xml/ns/j2ee"
  xmlns:xsi = "http://www.w3.org/2001/XMLSchema - instance"
  xsi:schemaLocation = "http://java.sun.com/xml/ns/j2ee
  http://java.sun.com/xml/ns/j2ee/web - app_2_4.xsd">
  < welcome - file - list >
    < welcome - file > index.JSP </welcome - file >
  </welcome - file - list >
</web - app >
```

6. 部署应用

可以手动将编辑好的 Web 应用部署到 Web 服务器上，也可以利用 MyEclipse 来部署项目，其步骤如下。

(1) 在工具栏中单击部署图标 。

(2) 选择要部署的项目，单击 Add 按钮，出现选择服务器的对话框，如图 1-28 和图 1-29 所示。

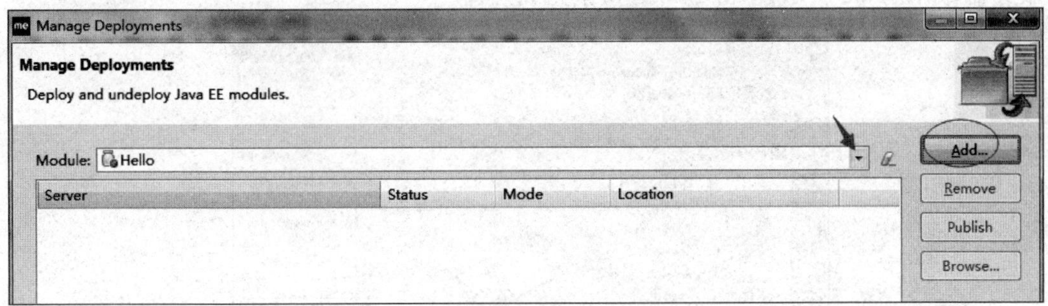

图 1-28　项目部署

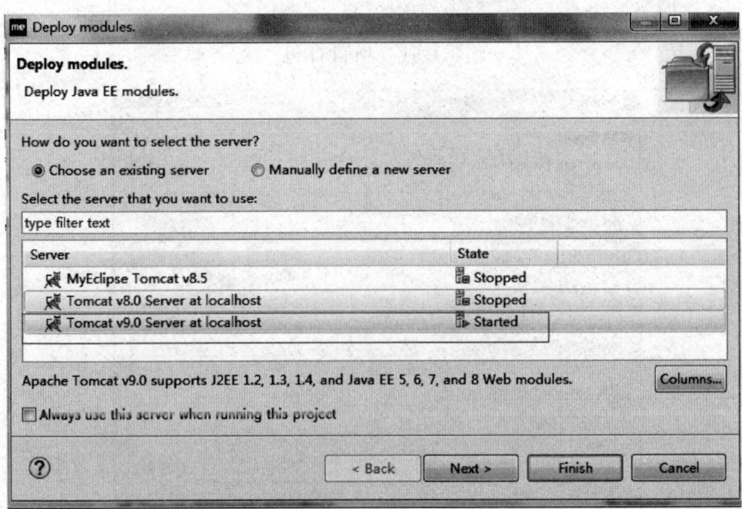

图 1-29　选择服务器

项目发布就是把项目的 WebRoot 目录复制到 Tomcat 的 webapps 目录，并把 WebRoot 重命名为项目名称，即 Hello。所以在 Tomcat 的 webapps 目录下会多出一个目录 Hello。

(3) 启动服务器，在浏览器上输入 http://localhost:8080/hello/index.jsp，这时浏览器上将显示 index.jsp 页面的内容，如图 1-30 所示。

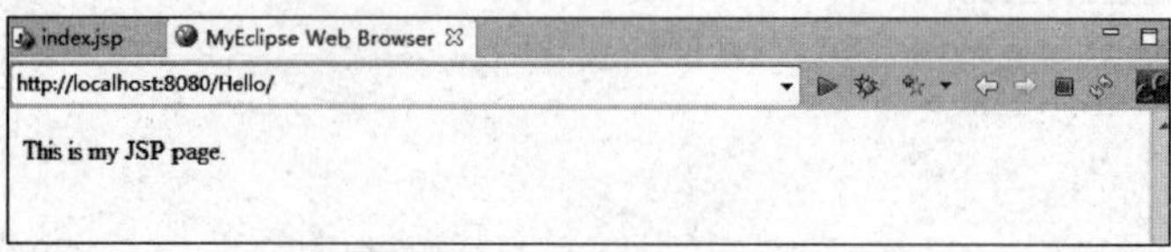

图 1-30　浏览页面资源

7. 打包成 WAR 包

JavaSE 程序可以打包成 JAR 包，而 Java Web 程序可以打包成 WAR 包，然后把 WAR 发布到 Tomcat 的 webapps 目录下，Tomcat 会在启动时自动解压 WAR 包。

选中要打包的项目，右击，选择 Export→WAR File，如图 1-31 所示，选择打包路径和服务器，如图 1-32 所示，单击 Finish 按钮完成打包。

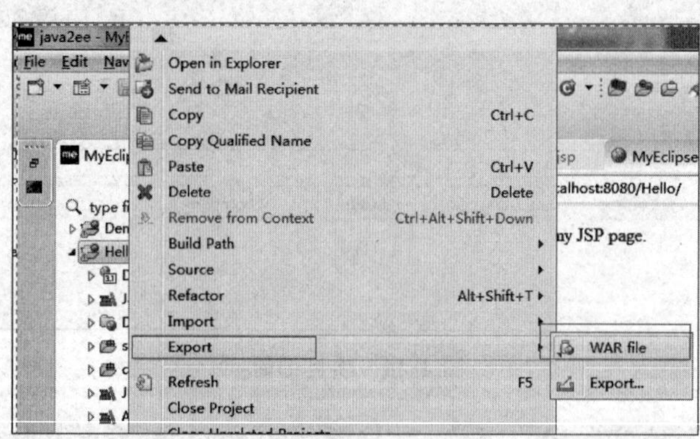

图 1-31　项目打包

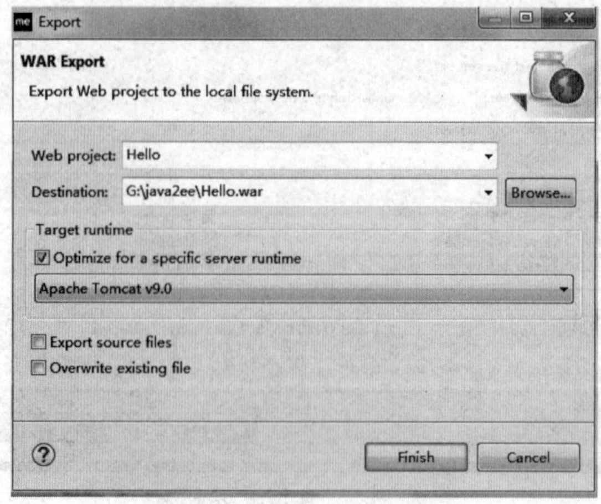

图 1-32　选择打包路径和服务器

8. 项目的高级配置

选择项目文件并右击 Properties,在窗口左侧选择 Java Build Path,如图 1-33 所示。

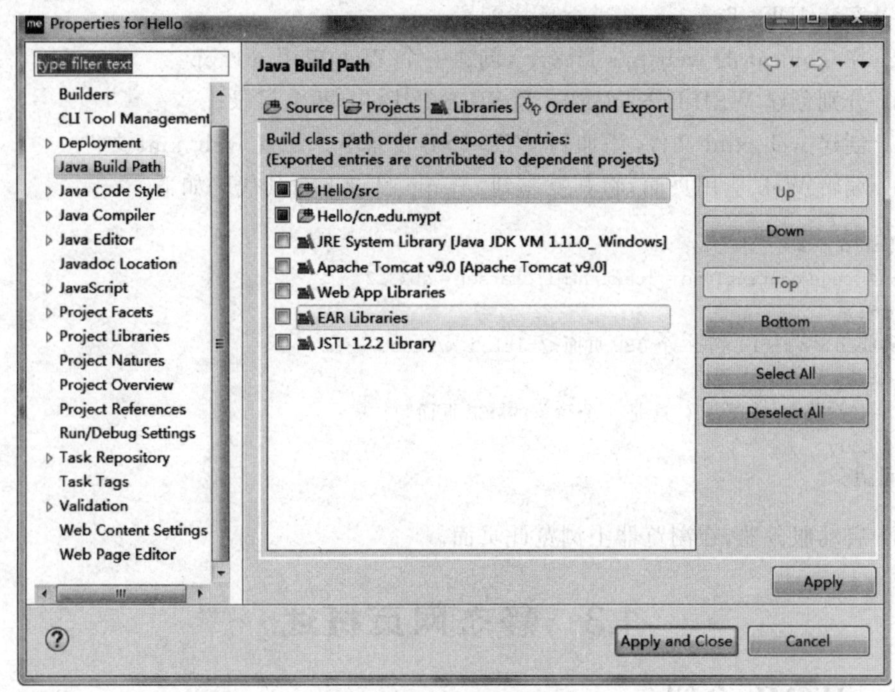

图 1-33　项目的高级配置

高级配置中主要含有如下内容。

(1) Source：它列出 MyEclipse 可以编译查错的文件夹 Java 文件,如本例中的 Hello/src,Hello/cn.edu.mypt。

如果想在 Hello 中建立一个普通的文件夹 folder,如 myjava,而不是包,则 myjava 文件夹中的 Java 文件不会被编译查错,要想使其与 src 一样：勾选 Source→Addfolder→myjava 复选框,单击 Ok,这样就可以编译 myjava 中的 Java 文件。

(2) Projects:用于添加其他项目。

(3) Libraries：用于添加第三方 JAR 包。

① Add External Jars(用于加载工程外的 JAR 包)：可以选择相应的 JAR 包,如 MyEclipse 驱动等。

② Add Jars：用于添加本工程内的 JAR 包,使用这种方式,JAR 在项目内,方便移植。

1.2.5　任务：Tomcat 服务器的配置及部署

1. 任务要求

创建一个网站并部署。

2. 任务实施

方案：手工创建，具体步骤如下。

(1) 安装 JDK、Tomcat，并进行环境配置。

(2) 在 Tomcat 的 webapps 目录下，创建一个 Web 应用 myapp。

(3) 分别创建 WEB-INF/classes 和 WEB-INF/lib 两个目录。

(4) 创建 web. xml 文件，添加到 WEB-INF 目录下。配置 web. xml 文件。

(5) 编辑 Web 应用页面，将其复制到 myapp 目录下。其代码如下。

```
//Hello.jsp
<%@ page contentType = "text/html;charset = gb2312" %>
<html>
  <head><title>第一个 JSP 页面</title></head>
  <body>
      <% out.println("这是一个简单的 JSP 页面"); %>
  </body>
</html>
```

(6) 启动服务器，在浏览器上浏览此页面。

1.3 静态网页概述

1.3.1 HTML 介绍

1. HTML 文件的基本结构

HTML 文档是由 HTML 命令组成的描述性文本文档，HTML 的结构包括头部(head)、主体(body)两大部分，其中头部描述浏览器所需的信息，而主体则包含所要说明的网页的具体内容。

HTML 文档的结构如下。

```
<html>                                  //HTML 文档最顶层的标签,代表该文档为 HTML 文件
<head>                                  //<head>头部部分,描述浏览器所需的信息
<title>我的第一个网页 </title>          //网页标题
</head>                                 //网页头部结束
<body>                                  //<body>主体部分
        Hello World!                    //网页内容,可以是文本、图像等
</body>  网页主体结束
</html>
```

2. HTML 语法、规则和书写习惯

遵循 HTML 语法规则，养成良好的代码编写习惯，不仅能减少错误，还能大大提高开发效率，而且会有利于后期代码的查看修改。

3. 常见文本编辑器

1）Notpad＋＋

Notpad＋＋是 Windows 下使用较多的一款文本编辑器，是 GPL 许可证下发行的自由软件。除了具有基本功能外，还可以支持插件扩展，功能异常强大。

2）EditPlus

EditPlus 是 Windows 下常用的文本编辑器，功能上并不弱于 Notpad＋＋，但由于它是共享软件，用户数量可能会少一些。

3）UltraEdit

UltraEdit 不仅是一款强大的文本编辑器，更是一款强大的文件分析工具。它的十六进制编辑功能非常强大，常被用作文件分析和修改，如破解和软件语言的本地化。它也是一款共享软件。

4）gEdit

gEdit 是 Linux 系统中 GNOME 桌面系统中自带的文本编辑器，同样支持多语言、语法着色等功能，能够胜任大部分代码编写工作。

5）VIM

Linux 里边发展起来的强大的文本编辑器，以简洁高效著称，不过上手比较慢，但其卓越的效率不可小觑。目前也有 Windows 版本可用。

4. 辅助开发工具

Web 开发绝不仅仅是“代码编写＋浏览器查看”的方式，很多问题是无法通过这种方式来发现和解决的，因此需要借助其他的工具，如参考手册、调试工具、截图工具、标尺工具等。

1）HTML 参考手册

HTML 参考手册是 HTML 语法、格式、标签信息等的集合，方便开发人员随时查阅和参考，它的作用像字典，是开发人员必备的工具之一。熟练掌握各种手册的使用，也是开发人员必备的技能之一。

较常用的 HTML 手册就是 W3School。W3School 不仅提供各种 Web 开发的教程、实例、参考手册，还可以进行在线的练习。它既有在线版，也有可供下载保存的离线版。

2）Firebug

Firebug 支持 HTML、CSS、JavaScript 等文件的查看和调试，是源于 Firefox 的一款开发类插件，目前已经可以在多种浏览器中安装，是 Web 开发者喜爱的插件之一。

3）Web Developer

Web Developer 是一款网页分析调试插件，有强大的网页内容查看、分析和调试功能，还有网页元素禁止功能，现在也可以支持多种浏览器，是 Web 开发者不可或缺的帮手。

4）Awesome Screenshot

Awesome Screenshot 是一款非常好用的网页截图插件，可以方便地进行整页截图、截图裁剪和标注，目前也可以支持多种浏览器，是一款非常好用的工具。

5）Chrome 开发人员工具

开发人员工具是 Chrome 内置的功能，可以方便地分析网页元素和 CSS 样式，很方便 CSS 的编写和调整，用 Notpad＋＋和 Chrome 做网页代码的编写工作，效率比较高。

6）Adobe Dreamweaver 和 Microsoft Frontpage

这两个软件功能非常强大，同是所见即所得（WYSIWYG）的网站开发工具，不过其功能大多数都不会用到。

1.3.2 HTML 元素及属性

HTML 文档是由许多嵌套的 HTML 元素构成的，HTML 元素由 HTML 标签和一些其他的文本组成。

HTML 元素包括文本、图形图像、动画、表格、超链接等，它是 HTML 文档的基本组成单位。下面详细分类讲解这些元素及其属性。

1. 标题

语法：

```
<h1>标题</h1>
…
<h6>标题</h6>
```

【例 1-1】 标题标签。

```
<body>
  <h1>一级标题</h1>
  <h2>二级标题</h2>
  <h3>三级标题</h3>
  <h4>四级标题</h4>
  <h5>五级标题</h5>
  <h6>六级标题</h6>
</body>
```

2. 段落

语法：

```
<p>…</p>
```

【例 1-2】 段落标签。

```
<body>
  <h1>北京欢迎你</h1>
  <p>北京欢迎你,有梦想谁都了不起!/p>
  <p>有勇气就会有奇迹.</p>
</body
```

3. 水平线

单个标签的闭合形式：

语法：

```
<hr />
```

【例 1-3】 水平线标签。

```
<body>
  <h1>北京欢迎你</h1>
  <hr />
  <p>北京欢迎你,有梦想谁都了不起!/p>
  <p>有勇气就会有奇迹.</p>
</body>
```

4. 有序列表

语法：

```
<ol>
  <li>列表项 1</li>
   …
</ol>
```

【例 1-4】 有序列表标签。

```
<body>
<h3>注册步骤:</h3>
<ol>
  <li>填写信息</li>
  <li>收电子邮件</li>
  <li>注册成功</li>
</ol>
</body>
```

5. 无序列表

语法：

```
<ul>
  <li>列表项 1</li>
   …
</ul>
```

【例 1-5】 无序列表标签。

```
<body>
<h3>新人上路指南 </h3>
<ul>
```

```
  <li>如何激活会员名?</li>
  <li>如何注册买卖会员?</li>
  <li>注册时密码设置有什么要求?</li>
  <li>买卖认证</li>
</ul>
</body>
```

6. 定义列表

语法：

```
<dl>
  <dt>标题</dt>
  <dd>描述 1</dd>
...
</dl>
```

【例 1-6】 定义列表标签。

```
<body>
  <dl>
      <dt>咖啡</dt>
      <dd>一种黑色的热饮料,原料是咖啡豆,非洲盛产这类原料.</dd>
      <dd>可以提神,刺激神经。</dd>
  <dl>
</body>
```

这种效果可以和无序列表互相替代，因 dt 是块状元素，常用于图文混排布局的场合。

7. 表格

语法：

```
<table>
  <tr>
      <td>单元格</td>
      <td>单元格</td>
</tr>
...
</table>
```

其中，<table>定义表格；<tr>定义行；<td>定义列(单元格)。

【例 1-7】 表格。

```
<body>
<table>
  <tr><td><img src="images/iron.jpg" alt="××蒸气电熨斗"/></td></tr>
  <tr><td>商品名称:××蒸气电熨斗</td></tr>
  <tr><td>商品价格:388 元</td></tr>
  <tr><td>商品简介:金刚低血超硬超顺滑,140ml 透明大水箱设计</td></tr>
</table>
</body>
```

8. 分区

语法:

```
<div>
...
</div>
```

【例 1-8】 分区标签。

```
<div style="width:400px; height:300px; background:#9FF">
    <p>...</p>
    <h3>新人上路</h3>
    <ul>
     ...
    </ul>
    div 其实就是一个...
</div>
```

9. 图像

语法:

```
<img src= "图片地址" alt="提示文字" title="提示文字" />
```

为了不同浏览器之间的兼容,推荐使用 title 属性,确保能显示提示文字。

【例 1-9】 图像标签。

```
<html>
  <head>
  <meta http-equiv="Content-Type" content="text/html; charset=gb2312" />
  <title>换行标签的应用</title>
  </head>
  <body>
  <img src="images/tv.jpg" alt="精品热卖:高清晰,30 英寸等离子电视"
       title="精品热卖:高清晰,30 英寸等离子电视" />
  </body>
</html>
```

10. 范围

范围标签<span>: 显示某行内的独特样式。

语法:

```
<span>文本等行级内容</span>
```

【例 1-10】 范围标签。

```
<html>
  <head>
```

```
        <meta http-equiv="Content-Type" content="text/html; charset=gb2312" />
        <title>换行标签的应用</title>
    </head>
    <body>
        <img src="images/tv.jpg" alt="精品热卖:高清晰,30 英寸等离子电视"
            title="精品热卖:高清晰,30 英寸等离子电视" />
        <p>商品价格:仅售<span style="color:red;font-size:70px;">10</span>元</p>
    </body>
</html>
```

11. 换行

语法：

```
<br/>
```

【例 1-11】 换行标签。

```
<html>
    <head>
        <meta http-equiv="Content-Type" content="text/html; charset=gb2312" />
        <title>换行标签的应用</title>
    </head>
    <body>
        <h1>北京欢迎你</h1>
        <hr />
        <p>北京欢迎你,有梦想谁都了不起!<br />
            有勇气就会有奇迹.<br />
        </p>
    </body>
</html>
```

12. 超链接

超链接用于实现页面间的跳转，超链接包括以下两部分内容。

(1) 链接地址：即链接的目标，可以是某个网址或文件的路径。

(2) 链接文本或图像：单击该文本或图像，将跳转到链接地址。

语法：

```
<a href="链接地址" target="目标窗口位置">  链接热点文本或图像</a>
```

属性：

href——表示链接地址的路径。

target——指定链接在哪个窗口打开，常用的取值有_self(自身窗口)、_blank(新建窗口)。

【例 1-12】 超链接标签。

```
<html xmlns="http://www.w3.org/1999/xhtml">
```

```
<head>
<meta http-equiv="Content-Type" content="text/html; charset=gb2312" />
<title>超链接</title>
</head>
<body>
<a href="span.html" target="_blank">××蒸气电熨斗</a>
</body>
</html>
```

13. 注释

HTML 中,注释的作用是方便阅读和调试代码,当浏览器遇到注释时会自动忽略注释的内容。

语法:

```
<!-- 注释内容 -->
```

【例 1-13】 注释。

```
<html>
  <head>
  <meta http-equiv="Content-Type" content="text/html; charset=gb2312" />
  <title></title>
  </head>
  <body>
  <!-- 项目列表部分 -->
  <div>
    <ul>
    <!-- <li>被注释掉的行将不显示</li> -->
    <li>正常显示行 1 </li>
    <li>正常显示行 2 </li>
    <ul>
  </div>
  </body>
</html>
```

14. 特殊符号

因为<、>等符号在 HTML 中已使用,所以如果要在页面中显示这些特殊的符号,就必须用其他符号来代替。这些特殊符号对应的 HTML 代码称为字符实体。常用的特殊符号有以下几个。

空格:

大于号(>): >

小于号(<): <

引号("): "

版权号(): ©

这些特殊的符号以(&)开头,都以分号(;)结束。

【例 1-14】 显示版权信息。

```
<html>
  <head>
    <meta http-equiv="Content-Type" content="text/html; charset=gb2312" />
    <title>买卖电商</title>
  </head>
  <body>
    <div style="font:12px Tahoma;margin:0px auto;text-align:center;">
      <div><hr size="1" /> COPYRIGHT &copy;  2003-2010  
      <a href="index.htm">北京市买卖电商有限公司</a>
        ALL RIGHT RESERVED<br />
        热线:400-66-13060 Email:service@prd.com<br />
        ICP:<a href="#">沪 ICP 备 05021104 号</a><br />
      </div><!-- copyright end -->
    </div><!-- footer end -->
  </body>
</html>
```

15. 表单

表单是一个由<form>标签包含着表单元素的区域,这些表单元素常见的有文本框、密码框、按钮、文本域等。表单的主要作用是完成用户与浏览器或用户与服务器之间信息的传递,即人与机器交互的过程。

语法:

```
<form action="表单提交地址" method="提交方法">
  ...文本框、按钮等表单元素...
</form>
```

属性:

action——指定提交后,由服务器上哪个处理程序处理; method——指定向服务器提交的方法,一般为 post 或 get,post 方法比较安全。

【例 1-15】 登录表单。

```
<html xmlns="http://www.w3.org/1999/xhtml">
  <head>
    <meta http-equiv="Content-Type" content="text/html; charset=gb2312" />
    <title>表单元素</title>
  </head>
  <body>
    <form action="" method="post">
      <p>用户名: <input name="username" id="username" type="text" size="20"
                        maxlength="10" /></p>
      <p>密   码: <input name="pwd" id="pwd" type="password" size="20" /> </p>
      <p><input type="submit" name="btn" id="btn" value="提交" />
        <input name="reset" type="reset" value="重填" />
```

```
      </p>
    </form>
  </body>
</html>
```

下面以登录验证为例说明表单的执行过程。

(1) 客户端：请求登录，通过表单填写账户信息，提交给服务器。

(2) 服务器端：验证发来的账户信息，然后给出反馈。

简单地说，表单的执行过程就是完成网站服务器和客户端之间的交互，提供双方需要的信息的过程。

16. 文本框

表单中最常用的信息输入元素是文本框，它用于用户输入单行文本信息。

语法：

```
<input name = "名称" type = "text" value = "初值" size = "数字" maxlength = "数字">
```

属性：

type——决定了表单的类型，text 为文本框，password 为密码框。

name——表单的唯一识别名称，在信息交互中的作用非常重要。

value——文本框内的初始文本。

size——文本框的宽度，单位为字符，即 5 代表 5 个字符宽度，默认为 20。

maxlength——最大字符数，用于限定文本框内最多可输入的字符数，不设置该属性则表示无限制。

title——设置表单元素的提示文本，当光标指向表单元素时，提示文本就会显示出来。

【例 1-16】 输入用户名。

```
<html xmlns = "http://www.w3.org/1999/xhtml">
  <head>
    <meta http - equiv = "Content - Type" content = "text/html; charset = gb2312" />
    <title>文本框,密码框和按钮</title>
  </head>
  <body>
    <form action = "" method = "post">
        <p>用户名:<input name = "name" type = "text" size = "21" /> </p>
    </form>
  </body>
</html>
```

将表单的 type 属性设置为 password 就可以创建一个密码框，输入的字符全以“*”显示，从而提高密码的安全性。

【例 1-17】 输入密码。

```
<html xmlns = "http://www.w3.org/1999/xhtml">
  <head>
    <meta http - equiv = "Content - Type" content = "text/html; charset = gb2312" />
```

```
      <title>文本框,密码框和按钮</title>
    </head>
    <body>
      <form action = "" method = "post">
        <p>密   码: <input name = "pass" type = "password" size = "22" /> </p>
      </form>
        </body>
</html>
```

17. 重置、提交与普通按钮

根据功能,可将按钮分为提交按钮、重置按钮和普通按钮。提交按钮用于提交表单数据,重置按钮用于清空现有表单数据,普通按钮一般用于调用 JavaScript 脚本。

语法:

```
<input name = "名称" type = "按钮类型" value = "按钮文字" src = "图片按钮的图片 url">
```

属性:

type——决定了按钮的类型,其中 button 为普通按钮; submit 为提交按钮; reset 为重设按钮。三个属性值决定的按钮基本相同,其行为差异一般是通过 JavaScript 对其单击事件做出不同的响应而产生的。不过,提交按钮会自动捕捉到回车键按下的事件。

value——按钮上的文本。

<button>标签同样可以设置一个按钮,它与<input>标签的最大差别是它可以包含文本和图片元素(在 Firefox 和 Chrome 中,默认将该按钮作为提交按钮)。

语法:

```
<button name = "名称"><img src = "图片 url" width = "数字"height = "数字" title = "picture" />
按钮文字</button>
```

属性:

img——设置按钮中的图片。

<button>标签必须成对出现,中间所包含的文本就是按钮上的文本。例如:

```
<input type = "reset" name = "reset" value = " 重填" />
<input type = "submit" name = "button" value = "同意..." />
<input type = "button" name = "confirm" value = "点播音乐" />
<input type = "button" name = "cancel" value = "取消" />
<input type = "image" src = "images/login.gif" />
```

【例 1-18】 用户登录表单。

```
<html xmlns = "http://www.w3.org/1999/xhtml">
  <head>
    <meta http-equiv = "Content-Type" content = "text/html; charset = gb2312" />
    <title>用户登录</title>
  </head>
  <body>
    <form action = "" method = "post">
```

```
        <p>用户名:<input name="name" type="text" size="21" /></p>
        <p>密   码:<input name="pass" type="password" size="22" /></p>
        <p>
          <input type="reset" name="reset" value=" 重填 " />
          <input type="submit" name="button" value="登录" />
        </p>
        </form>
    </body>
</html>
```

在实际应用中,经常使用图片按钮。实现图标按钮的有很多方法。简单的方法就是配合使用 type 和 scr 属性。例如:

```
<input type="image" scr="images/login.gif"/>
```

这种按钮仍然具有提交功能。

【例 1-19】 图形提交按钮。

```
<html xmlns="http://www.w3.org/1999/xhtml">
  <head>
      <meta http-equiv="Content-Type" content="text/html; charset=gb2312" />
      <title>图形提交按钮</title>
  </head>
  <body>
  <form action="" method="post">
    <p>用户名: <input name="name" type="text" size="21" /></p>
    <p>密   码: <input name="pass" type="password" size="22" /></p>
    <p>
      <input type="reset" name="reset" value=" 重填 " />
      <input type="image" src="images/login.gif" />
    </p>
    <p>
      <input type="button" name="confirm" value="点播音乐" />
      <input type="button" name="cancel" value="取消" />
    </p>
  </form>
  </body>
</html>
```

18. 单选按钮

单选按钮用于一组相互排斥的选项,组中的每个选项应具有相同的名称,以确保用户只能选择一个选项,单选按钮在被选中时,会向表单提交一组名称/值(name/value)对。

语法:

```
<input type="radio" name="组名" value="值" checked="checked" />
```

属性:

type——决定按钮的类型。

name——设定按钮的唯一识别名称。

注意，当多个单选按钮的 name 属性相同，即具有相同的名字时，它们组成单选按钮组，此时，组内的按钮不能被同时选中。

【例 1-20】 单选按钮的使用。

```
<html xmlns = "http://www.w3.org/1999/xhtml">
  <head>
    <meta http-equiv = "content-type" content = "text/html; charset = gb2312">
    <title>单选框</title>
  </head>
  <body>
    <form method = "post" action = "">
      <br />性别
        <input name = "man" type = "radio" class = "input" value = "男" checked = "checked" />
      <img src = "images/male.gif" alt = "女" />男  
      <input name = "woman" type = "radio" value = "女" class = "input" />
      <img src = "images/female.gif" alt = "女" />女
    </form>
    </body>
</html>
```

19. 复选框

复选框用于选择多个选项，多选按钮在被选中时，会向表单提交一组名称/值对，多个复选框可以被同时选择。

语法：

```
<input type = "checkbox" name = "checkbox2" checked = "checked" value = "check2" />
```

属性：

type——决定按钮的类型。

checked——设定按钮是否被选中，只有唯一的属性值 checked。若不设置该属性，则代表未选中。

【例 1-21】 复选框的应用。

```
<html xmlns = "http://www.w3.org/1999/xhtml">
  <head>
    <meta http-equiv = "content-type" content = "text/html; charset = gb2312">
    <title>复选框</title>
  </head>
  <body>
    <form method = "post" action = "">
      爱好:
      <input type = "checkbox" name = "hobby" value = "sports" />运动
      <input type = "checkbox" name = "hobby" value = "talk" checked = "checked" />
      聊天:
      <input type = "checkbox" name = "hobby" value = "play" /> 玩游戏
```

```
    </form>
    </body>
</html>
```

20. 文件域

文件域用于上传文件，设置时只须设 type 属性为 file 即可。

语法：

```
<input type="file" name=" ... ">
```

【例 1-22】 上传文件。

```
<html xmlns="http://www.w3.org/1999/xhtml">
  <head>
    <meta http-equiv="Content-Type" content="text/html; charset=gb2312" />
    <title>文件域</title>
  </head>
  <body>
    <form action="" method="post">
      <input type="file" name="files" /><br />
      <input type="submit" name="upload" value="上传" />
    </form>
  </body>
</html>
```

21. 下拉列表框

下拉列表框可以使用户快速、方便、正确地选择一些选项，而且能节省页面空间。它通过<select>和<option>标签来实现。<select>标签用于显示可供用户选择的下拉列表，每个选项由一个<option>标签表示，<select>标签必须包含至少一个<option>标签。

语法：

```
<select name="指定列表名称" size="行数">
  <option value="选项值" selected="selected">...</option>
  ...
</select>
```

属性：

size——定义下拉列表所呈现的行数。如果属性值为 1，则为下拉列表，列表右侧会出现滚动条，以便查看其他选项；如果属性值大于 1，则为列表，将只呈现相应行数的选项。

<option>——定义列表的选项，标签对内的文本就是列表中的可见选项。

value——相应选项所对应的值。

selected——表明相应的选项是否被选中，具有唯一的属性值 selected。一组列表项中只允许一个项目设置此属性，代表默认被选中。

【例 1-23】 下拉列表框的应用。

```
<html xmlns="http://www.w3.org/1999/xhtml">
  <head>
    <meta http-equiv="Content-Type" content="text/html; charset=gb2312" />
    <title>列表框</title>
  </head>
  <body>
    <form action="" method="post" id="form5">
      出生日期:<input name="byear" value="yyyy" size="4" maxlength="4" />
      年<select name="bmon">
        <option value="" selected="selected">[选择月份]</option>
          <option value="0">一月</option>
          <option value="8">九月</option>
          <option value="9">十月</option>
          <option value="10">十一月</option>
          <option value="11">十二月</option>
        </select>
      月 <input name="bday" value="dd" size="2" maxlength="2" />日
    </form>
  </body>
</html>
```

22. 多行文本框

多行文本输入框用于大量文本的输入,对字数没有限制。

文本域用于显示或输入两行或两行以上的文本,它使用的标签<textarea>定义。

语法:

```
<textarea name="名称" cols="列宽" rows="行宽">
  文本内容 属性:
</textarea>
```

属性:

cols——设置文本框的宽度,单位为字符。

rows——设置文本框的高度(显示行数),单位为行,当内容的行数超过该属性值时,会自动出现滚动条。

【例 1-24】 多行文本框的应用。

```
<html xmlns="http://www.w3.org/1999/xhtml">
  <head>
    <meta http-equiv="Content-Type" content="text/html; charset=gb2312" />
    <title>多行文本框</title>
  </head>
  <body>
    <form action="" method="post">
      <h2><img src="images/read.gif" alt="阅读服务协议"
        title="阅读服务协议" />阅读买卖网服务协议
```

```
        </h2>
        <textarea name="content" cols="40" rows="6">
          欢迎阅读服务条款协议...
        </textarea>
      </form>
    </body>
  </html>
```

23. 隐藏域

隐藏域可以方便服务器端“记住”客户端的信息但又不被客户看到。隐藏表单通常被称为隐藏域，它在网页中是不可见的，通常被用于特殊信息的存储和传递。将 type 属性设置为 hidden 即可创建一个隐藏域。

语法：

```
<input type="hidden" name="名称" value="值" />
```

属性：

value——通常用作存储和传递特殊信息。

【例 1-25】 在登录页面中隐藏用户 id 信息为 8888。

```
<html xmlns="http://www.w3.org/1999/xhtml">
  <head>
    <meta http-equiv="Content-Type" content="text/html; charset=gb2312" />
    <title>隐藏域</title>
  </head>
  <body>
    <form action="" method="post">
      <p>网站邮箱:<input name="email" type="text" size="28" /></p>
      <p>输入密码:<input name="pwd" type="password" size="30" /></p>
      <p>再次输入密码:<input name="repwd" type="password" size="30" /></p>
      <p><input type="hidden" name="userid" value="8888" /></p>
      <p><input name="cancel" type="reset" value="重置" /></p>
      <p><input name="login" type="image" src="images/login.gif" /></p>
    </form>
  </body>
</html>
```

在某些情况下，需要对表单元素进行限制，设置表单元素为只读或禁用。主要用于下面的情况。

只读 readonly：某个框内的内容只允许用户看，不能修改。

禁用 disabled：因没达到使用的条件，限制用户使用。

【例 1-26】 表单元素高级用法。

```
<html xmlns="http://www.w3.org/1999/xhtml">
  <head>
    <meta http-equiv="Content-Type" content="text/html; charset=gb2312" />
    <title>表单元素高级用法</title>
```

```
  </head>
  <body>
    <h2><img src="images/read.gif" width="35" height="26" />阅读买卖网服务协议 </h2>
    <form action="" method="post">
      <textarea name="content" cols="60" rows="8" readonly="readonly">
        欢迎阅读服务条款协议,买卖的权利和义务......
      </textarea><br /><br />
      同意以上协议<input name="agree" type="checkbox" />
    <input name="btn" type="submit" value="注册" disabled="disabled" />
    </form>
  </body>
</html>
```

1.3.3 DIV+CSS 介绍

样式表由样式规则组成,这些规则告诉浏览器如何显示文档。其基本语法由三部分构成:选择器、属性和属性值。

1. 基本结构

层叠样式表一般用<style>标签来声明样式规则,其语法如下。

```
<style type="text/csss">
  选择器{
    对象的属性 1:属性值 1;
    对象的属性 2:属性值 2;
    ...
  }
</style>
```

其中选择器表示被修饰的对象。属性是希望改变的样式,如颜色。属性和属性值用冒号(:)隔开。

注意:

(1) 虽然 CSS 代码不区分大小写,但推荐使用小写。

(2) 每条样式规则用分号(;)隔开,一般写为多行,简单的规则也可以合并为一行。

(3) 当 CSS 代码较多时,可使用/*...*/添加代码的注释,以增加代码的可读性。

2. 选择器的分类

选择器表示所修饰的内容类别,选择器又分为标签选择器、类选择器、ID 选择器。

1) 标签选择器

当需要对页面内某类标签的内容进行修饰时,则采用标签选择器,其语法如下。

```
标签名{
  属性名 1:属性值 1;
  属性名 2:属性值 2;
  ...
}
```

【例 1-27】 标签选择器的应用。

```
< html xmlns = "http://www.w3.org/1999/xhtml">
< head >
  < title >买卖——商品分类</title >
  < style >
    li{color:red;font - size:28px;font - family:隶书; }
  </style >
</head >
< body >
  < div >
    < ul >
      < li >家用电器</li >
      < li >各类书籍</li >
      < li >手机数码</li >
      < li >日用百货</li >
    </ul >
  </div >
</body >
</html >
```

2）类选择器

使用类选择器的步骤如下。

(1) 定义类样式，其语法如下。

```
.类名{
  属性名 1:属性值 1;
  属性名 2:属性值 2;
  …
}
```

(2) 应用样式，使用标签的 class 属性引用类样式，其语法如下。

```
<标签名 class = "类名"标签>
  内容
</标签>
```

【例 1-28】 类选择器的应用。

```
< html xmlns = "http://www.w3.org/1999/xhtml">
< head >
    < title >商品分类</title >
  < style >
    li{color:red;font - size:28px;font - family:隶书; }
    .blue{color:blue;}
  </style >
</head >
< body >
  < div >
    < ul >
      < li class = "blue">家用电器</li >
```

```
      <li>各类书籍</li>
      <li class="blue">手机数码</li>
      <li>日用百货</li>
    </ul>
  </div>
</body>
</html>
```

定义类选择器的好处是任何标签都可以应用该类样式，从而实现样式的共享和代码复用。

3) ID 选择器

ID 选择器的使用步骤如下。

(1) 使用 id 属性标识被修饰的页面元素，其语法如下。

```
<div id="ID 标识名">
  内容
</div>
```

(2) 定义相应的 ID 选择器样式，其语法如下。

```
#类名{
  属性名 1:属性值 1;
  属性名 2:属性值 2;
  …
}
```

【例 1-29】 ID 选择器的应用。

```
<html xmlns="http://www.w3.org/1999/xhtml">
<head>
<title>商品分类</title>
  <style>
    #book{font:bold 14px 宋体;}
  </style>
</head>
<body>
  <div id="book">
    <ul>
      <li>家用电器</li>
      <li>各类书籍</li>
      <li>手机数码</li>
      <li>日用百货</li>
    </ul>
  </div>
</body>
</html>
```

ID 选择器用于修饰某个指定的页面元素或区块，这些样式是对应 ID 标识的 HTML 元素所独占的。如果希望部分 li 标签采用其他样式，可采用类选择器。如果希望控制某个 DIV 块样式，并且要求块元素唯一，可采用 ID 选择器。

1.3.4　任务：用户注册页面的设计

1. 任务要求

设计用户注册页面，分别采用 DIV＋CSS 和表格进行布局。

2. 任务实施

(1) 编写用户注册页面 reg.jsp。

```
<!DOCTYPE html PUBLIC "-//W3C//DTD XHTML 1.0 Transitional//EN" "http://www.w3.org/TR/
xhtml1/DTD/xhtml1-transitional.dtd">
<html xmlns="http://www.w3.org/1999/xhtml">
  <head>
    <meta http-equiv="Content-Type" content="text/html;charset=utf-8" />
    <title>用户注册</title>
    <link href="CSS/style.css" rel="stylesheet" type="text/css" />
  </head>
  <body>
    <table id="table">
      <tr><td colspan="2" align="center" bgcolor="#0000FF"style="color:#FFF">
        学院 <br/><a href="#"> www.myvtc.edu.cn </a></td></tr>
      <tr>
        <td><div class="tb">学号</div></td>
        <td><input type="text" name="code" /></td>
      </tr>
      <tr>
        <td><div class="tb">姓名</div></td>
        <td><input type="text" name="name" /></td>
      </tr>
      <tr>
        <td><div class="tb">性别</div></td>
        <td>男<input type="radio" name="sex" />女<input name="sex" type="radio"
                checked="checked" /></td>
      </tr>
      <tr>
        <td><div class="tb">爱好</div></td>
        <td>
          <input name="inter" type="checkbox" />体育
          <input name="inter" type="checkbox" />文艺
          <input name="inter" type="checkbox" />书法
          <input name="inter" type="checkbox" />书法
        </td>
        </tr>
        <tr>
          <td><div class="tb">专业</div></td>
            <td>
              <select name="specice">
```

```
                    <option value="1">计算机软件</option>
                    <option value="2">人工智能</option>
                    <option value="3">计算机网络</option>
                      <option value="4">数字媒体</option>
                      <option value="5">大数据与云计算</option>
                    <option value="5">物联网</option>
                  </select>
                </td>
            </tr>
            <tr>
              <td class="tb" >个人简介</td>
              <td><textarea rows="5" cols="30"></textarea></td>
            </tr>
          </table>
        <div>
          <input type="submit" value="ok" /><input type="reset" value="reset" />
        </div>
</body>
</html>
```

(2) 编写 CSS 文件 style.css。

```
<style type="text/css">
body
{
  background-color:#CCC;
  text-align:center;
}
#table
{
  margin-top:80px;
  text-align:left;
  background-color:#FFF;
  width:400px;
  border:#666 solid 1px;
}
.tb
{text-align:right;
  margin-top:10px;
}
input[type="text"]
{background-color:#FF9;
  width:200px;
  margin-left:10px;
  margin-top:10px;
}
</style>
```

项目 1　首 页 设 计

1. 项目需求

设计一个新闻网站的首页并部署到 Tomcat 服务器上，其效果如图 1-34 所示。

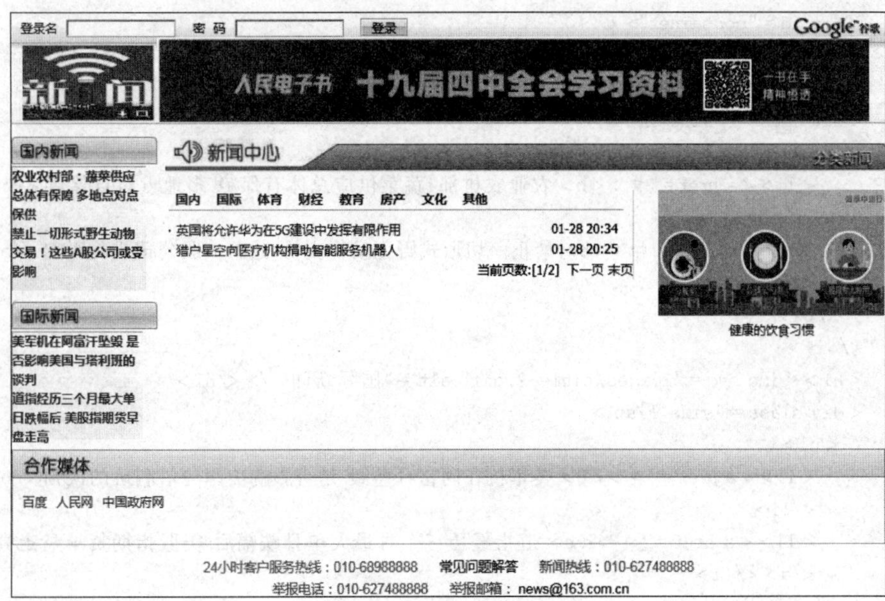

图 1-34　首页效果图

2. 项目实施步骤

(1) 利用 MyEclipse 创建一个网站。

(2) 编辑 HTML 文件 index.html，其代码如下。

```
<!DOCTYPE html PUBLIC "-//W3C//DTD XHTML 1.0 Transitional//EN" "http://www.w3.org/TR/
xhtml1/DTD/xhtml1-transitional.dtd">
<html xmlns="http://www.w3.org/1999/xhtml">
<head>
<meta http-equiv="Content-Type" content="text/html; charset=GBK" />
<title>新闻中国</title>
<link href="CSS/main.css" rel="stylesheet" type="text/css" />
</head>
<body>
<div id="header">
  <div id="top_login">
    <label>登录名 </label>
    <input type="text" id="uname" value="" class="login_input" />
    <label>密   码 </label>
    <input type="password" id="upwd" value="" class="login_input" />
```

```
<input type="button" class="login_sub" value="登录" onclick="login()"/>
<label id="error"></label>
<img src="Images/friend_logo.gif" alt="Google" id="friend_logo" /></div>
<div id="nav">
<div id="logo"><img src="Images/logo.jpg" alt="新闻中国" /></div>
<div id="a_b01"><img src="Images/a_b01.gif" alt="" /></div>
</div>
</div>
<div id="container">
<div class="sidebar">
<h1><img src="Images/title_1.gif" alt="国内新闻" /></h1>
<div class="side_list">
<ul>
<li><a href='#'><b>农业农村部:蔬菜供应总体有保障 多地点对点保供</b></a>
</li>
<li><a href='#'><b>禁止一切形式野生动物交易!这些A股公司或受影响</b></a>
</li>
</ul>
</div>
<h1><img src="Images/title_2.gif" alt="国际新闻" /></h1>
<div class="side_list">
<ul>
<li><a href='#'><b>美军机在阿富汗坠毁 是否影响美国与塔利班的谈判</b></a>
</li>
<li><a href='#'><b>道指经历三个月最大单日跌幅后 美股指期货早盘走高</b>
</a></li>
</ul>
</div>
</div>
<div class="main">
<div class="class_type"><img src="Images/class_type.gif" alt="新闻中心" /></div>
<div class="content">
<ul class="class_date">
<li id='class_month'><a href='#'><b>国内</b></a><a href='#'><b>国际
</b><a href='#'><b>体育</b></a>  <a href='#'><b>财经</b></a></a><a href=
'#'><b>教育</b></a>  <a href='#'><b>房产</b></a></a><a href='#'><b>文化
</b></a><a href='#'><b>其他</b></a></li>
</ul>
<ul class="classlist">
<li><a href='#'>英国将允许华为在5G建设中发挥有限作用</a><span>01-28
20:34</span></li>
<li><a href='#'>猎户星空向医疗机构捐助智能服务机器人</a><span>01-28
20:25</span></li>
<p align="right">当前页数:[1/2] <a href="#">下一页</a><a href="#">
末页</a></p>
</ul>
</div>
<div class="picnews">
<ul>
```

```
        <li><a href="#"><img src="Images/Picture1.jpg" alt="" /></a><a href=
"#">健康饮食习惯</a></li>
      </ul>
    </div>
  </div>
</div>
<div id="friend">
  <h1 class="friend_t"><img src="Images/friend_ico.gif" alt="合作伙伴" /></h1>
  <div class="friend_list">
    <ul>
      <li><a href="#">百度</a></li>
      <li><a href="#">人民网</a></li>
      <li><a href="#">中国政府网</a></li>
      </ul>
  </div>
</div>
<div id="footer">
  <p class="">24 小时客户服务热线:010-68988888 <a href="#">常见问题解答</a>&#
160;    新闻热线:010-627488888 <br />举报电话:010-627488888 举报
邮箱:<a href="#">news@163.com.cn</a></p>
  <p class="copyright">Copyright &copy; 1999-2009 News , All Right Reserver <br />新闻
网 版权所有 </p>
</div>
</body>
</html>
```

(3) main.css 文件的代码如下。

```
@charset "GBK";/
*{ margin:0; padding:0;}
body{ font:12px/20px Tahoma;}
table{ border-collapse:collapse;}
img{ border:0;}
li{ list-style:none;}

a{ color:#335884; text-decoration:none;}
a:hover{ color:#f00; text-decoration:underline;}

#header{ clear:both; position:relative; margin:0 auto; margin-top:15px; width:947px;
height:116px; border:1px solid #dfdfdf; background:url(../Images/topbg.gif) repeat-x;}
#top_login{ padding:6px 0 0 10px;}
.login_input{ height:19px; line-height:19px; width:118px; border:none; background:url
(../Images/login_input.gif) no-repeat; margin:0 14px 0 6px;}
.login_sub{ width:51px; height:19px; border:none; background:url(../Images/login_sub.gif)
no-repeat;}
.login_link{ text-decoration:underline; margin:0 6px;}
#friend_logo{ position:absolute; top:6px; right:6px;}

#nav{ clear:both;}
#logo{ float:left; width:145px; margin:8px 0 0 12px;}
```

```
#mainnav{ float:right; margin:10px 10px 0 0 ;}
#mainnav ul{ clear:both;}
#mainnav ul li{ float:left; margin:0 2px; padding-left:6px; background:url(../Images/nav
_leftdot.gif) no-repeat 0 8px; line-height:24px;}
#mainnav ul li a{ color:#333;}
#mainnav ul li.leftLineTop{ background-image:url(../Images/nav_leftline.gif);}
#mainnav ul li.headLi{ background:none;}
#container{ clear:both; width:947px; height:100%; overflow:hidden; margin:14px auto 0;}
.sidebar{ float:left; width:157px;}
.sidebar h1{ clear:both;}
.side_list{ height:250px; background:url(../Images/sidebarbg.gif) no-repeat; margin:2px
0 12px;}
.main{ float:right; width:787px;}
.class_type{ background:url(../Images/class_bg.gif) no-repeat right 8px; height:40px;
padding:0 0 0 14px;}
.content{ float:left; width:500px; border-right:1px solid #eff0f4; margin:16px 0
0 10px;}
.class_date{ font:12px/20px Arial; border-bottom:1px solid #d4d6d9; margin-
right:10px;}
#class_month{ font-weight:600;}
#class_month a{ margin:0 7px; *margin:0 5px;}
#class_day{ line-height:30px;}
.classlist{ clear:both; margin-top:10px;}
.classlist li{ position:relative; line-height:22px; background:url(../Images/arrow.gif)
no-repeat 0 10px; padding:0 130px 0 8px;}
.classlist li.space{ background:none; height:22px;}
.classlist li span{ position:absolute; top:-1px; *top:-4px; right:16px;}
.picnews{ float:right; width:250px; text-align:center; margin:16px 0 0 0;}
.picnews li{ margin-bottom:10px;}

#friend{ clear:both; width:947px; height:140px; margin:0 auto;}
.friend_t{ height:24px; border:1px solid #b3b7ba; background:url(../Images/friend_t.gif)
repeat-x; padding:4px 0 0 10px;}
.friend_list{ height:110px; overflow:hidden; background:url(../images/firend_bg.gif)
repeat-x; border:1px solid #bfd0d5; border-top:0;}
.friend_list ul{ margin:14px 10px 0;}
.friend_list li{ float:left; margin-right:10px; white-space:nowrap;}

#footer{ clear:both; width:947px; margin:14px auto 0; text-align:center;}
#footer p{ font:14px/24px Arial; color:#999;}
#footer p.copyright{ color:#333; margin-top:15px;}
```

习 题 1

1. 填空题

(1) 在传统的 HTML 文件中加入________和________,就构成了 JSP 网页。

(2) JSP 网页文件的扩展名是________。

(3) 所有 JSP 程序操作都在________执行。

(4) Tomcat 的主要配置文档是________。可以通过配置 server. xml 设置(服务器的端口及虚拟路径)。

(5) 在 Tomcat 中访问 helloapp 应用的 login. html 文件时,在浏览器地址栏应写的 URL 为________。

2. 选择题

(1) (　　)不是表单中的元素。

A. input　　B. textarea　　C. select　　D. table

(2) (　　)是单选按钮。

A. <input name="sex" type="text" value="0" />

B. <input name="sex" type="checkbox" value="0" />

C. <input name="sex" type="option" value="0" />

D. <input name="sex" type="radio" value="0" />

(3) Tomcat 服务器的默认 TCP 端口号是(　　)。

A. 80　　B. 21　　C. 1433　　D. 8080

(4) 在 HTML 中给制表格时,(　　)标签是表格中的换行标签。

A. <table>　　B. <th>　　C. <tr>　　D. <td>

(5) 如果 Tomcat 安装后,想要修改它的端口号,可以通过修改<Tomcat 安装目录>/conf 下的(　　)文件来实现。

A. web. xml　　B. server. xml

C. server-minimal. xml　　D. tomcat-user. xml

(6) HTTP 是一个(　　)协议。

A. 无状态　　B. 有状态　　C. 状态良好的　　D. 局域网

(7) Tomcat 的端口号可以在(　　)文件中修改。

A. server. xml　　B. web. xml　　C. tomcat. xml　　D. 不能改

(8) 在 Web 应用程序的目录结构中,在 WEB-INF\lib 目录中存放的是(　　)文件。

A. . jsp　　B. . class　　C. . jar　　D. web. xml

(9) JSP 最终被运行的是(　　)文件。

A. Java　　B. . class

C. HTML　　D. JSP

(10) 在 HTML 标签中,(　　)标签用于设置当前页面的标题。

A. <html>　　B. <title>　　C. <head>　　D. <name>

(11) (　　)是单选按钮。

A. <input name="sex" type="text" value="0" />

B. <input name="sex" type="checkbox" value="0" />

C. <input name="sex" type="option" value="0" />

D. <input name="sex" type="radio" value="0" />

(12) 有关 C/S 和 B/S 结构下列说法中错误的是(　　)。

A. 在 C/S 结构下,有专门的数据库服务器,但客户端还要运行客户端应用程序,这也叫作胖客户端

B. 在 B/S 结构下,客户端在浏览器中只负责表示层逻辑的实现,业务逻辑和数据库都在服务器端运行

C. 在 B/S 结构下,客户端发送的 HTTP 请求消息传给服务器,服务器将请求传递给 Web 应用程序,Web 应用程序处理请求,并传送给客户端

D. Web 应用是基于 C/S 结构的

(13) JSP 中有三大类标签,分别是(　　)。

A. HTML 标签、JSP 标签、Servlet 标签

B. CSS 标签、HTML 标签、JavaScript 标签

C. 动作标签、脚本标签、指令标签

D. 指令标签、脚本标签、HTML 标签

3. 简答题

(1) 简述 B/S 结构与 C/S 结构的区别。

(2) 简述实现 Web 应用的部署和发布方法。

第 2 章　JSP 基　础

从本章开始，将逐步介绍动态新闻发布系统的开发工作，逐步讲解动态网页设计中的常见功能页面的设计。通过本章的学习，达到如下目标。

【技能目标】 掌握用户登录功能页面的编程设计。

【知识目标】 JSP 基本语法；使用 request 对象获取请求信息；response 对象处理响应；转发与重定向控制页面跳转。

【关键词】 统一资源定位符（URL）　字符集（charset）　导入（import）
声明（declaration）　包含（include）　脚本小程序（scriptlet）
转发（forward）　重定向（redirect）

2.1　JSP 概 述

2.1.1　什么是 JSP

JSP（Java server pages）是一种动态网页技术标准。JSP 技术有点类似 ASP 技术，它是在传统的网页 HTML 文件（*.htm，*.html）中插入 Java 程序段（Scriptlet）和 JSP 标签（tag），从而形成 JSP 文件（*.jsp）。

1. JSP 的工作方式

程序员在 HTML 中嵌入 JavaScript 代码，由应用服务器中的 JSP 引擎来编译和执行，然后将生成的整个页面信息返回给客户端。用 JSP 开发的 Web 应用是跨平台的，既能在 Linux 下运行，也能在其他操作系统上运行。

2. HTML 与 JSP 的区别

HTML 页面是静态页面，也就是事先由用户写好放在服务器上，由 Web 服务器向客户端发送。JSP 页面是由 Web 服务器执行该页面中的 JavaScript 代码，然后实时生成的 HTML 页面，它是服务器端生成的动态页面。

页面效果用 JavaScript 也能够实现，最大的区别是：JavaScript 源代码是被服务器发送到客户端的，由客户端执行，因此，客户端可以看到 JavaScript 源代码。

3. JSP 的运行原理

Web 服务器接收到以.jsp 为扩展名的 URL 的访问请求时，它将把该访问请求交给 JSP 引擎去处理。Tomcat 中的 JSP 引擎就是一个 Servlet 程序，它负责解释和执行 JSP 页面。

每个 JSP 页面在第一次被访问时，JSP 引擎将它翻译成一个 Servlet 源程序，接着再把这个 Servlet 源程序编译成 Servlet 的. class 文件，然后由 Web 服务器像调用普通 Servlet 程序一样装载和解释执行这个由 JSP 页面翻译成的 Servlet 程序。JSP 执行原理如图 2-1 所示。

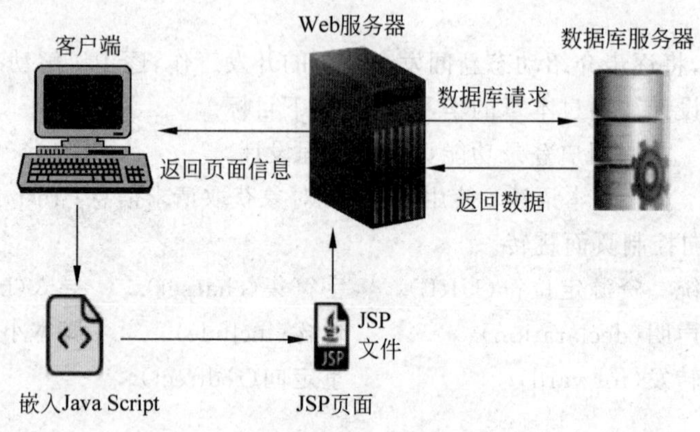

图 2-1　JSP 执行原理

Tomcat 把为 JSP 页面创建的 Servlet 源文件和. class 文件放置在"<TOMCAT_HOME>\work\Catalina\<主机名>\<应用程序名>\"目录中，并将 Servlet 的包命名为"org. apache. jsp. <JSP 页面在 Web 应用程序内的目录名>"。

JSP 标准没有明确要求 JSP 中的脚本程序必须采用 Java 语言编写，但是 JSP 页面最终必须转换成 Java Servlet 程序。可以在 Web 应用程序正式发布之前，将其中的所有 JSP 页面预先编译成 Servlet 程序。

4. JSP 执行过程

用户使用 URL(uniform resource locator，统一资源定位符)实现页面访问。URL 是唯一能识别 Internet 上具体的计算机、目录或文件夹位置的命名约定。

URL 由以下几部分组成。

第一部分：协议。

第二部分：主机 IP 地址(有时包含端口号)。

第三部分：项目资源的地址，如目录和文件夹名等。

当 Web 服务器上的一个 JSP 页面第一次被请求执行时，JSP 引擎先将 JSP 页面文件转译成一个 Java 文件，即 Servlet，Servlet 通过 HTML 与客户交互。服务器将前面转译的 Java 文件编译成字节码文件，再执行这个字节码文件来响应，从而加快了执行的速度。

2.1.2　JSP 的结构

JSP 页面可以包含以下元素。

(1) 静态内容：HTML 静态文本。

(2) 指令：以<%@开始，以%> 结束，如<%@include file= " Filename" %>。

(3) 表达式：<%=Java 表达式%>。

(4) 小脚本：<%Java 代码%>。

(5) 声明：<%! 方法%>。

(6) 注释：<!-- 这是注释,但客户端可以查看到-->或<%-- 这也是注释,但客户端不能查看到 --%>。

例如：

```
<%@ page contentType="text/html;charset=gb2312" %>
<%@ page import="java.util.*" %>
...
<html>
  <body>
    其他 HTML 代码
    <%
      符合 Java 语法的 Java 语句
    %>
    其他 HTML 代码
  </body>
</html>
```

【例 2-1】 JSP 程序的结构。

```
<%@ page language="java" import="java.util.*,java.text.*" contentType="text/html;
charset=GBK" %>
<html>
    <head><title>hello</title></head>
<!-- 这是 HTML 注释(客户端可以看到源代码)-->
<%-- 这是 JSP 注释 (客户端不可以看到源代码) --%>
  <body>
      你好,
  <%
      String name="张三";
      System.out.print("<h1>"+name+"</h>");
  %>
  <%=strCurrentTime %>
      <%! String declare = "this is declartion"; %>
      <%= declare %>
  </body>
</html>
```

2.1.3 任务：在页面中显示当前日期

1. 需求说明

在网页的中显示当前日期以及星期几等内容。

2. 任务实施

使用 MyEclips,编写一个 JSP 页面。其操作步骤如下。

(1) 新建 Web 项目。在 MyEclipse 菜单栏中选择 File→New→Project 命令。在弹出的对话框中选择 Web Project，输入项目名称 myweb 并选择 JavaEE 后单击 Finish 按钮。

(2) 新建 JSP 文件。在 MyEclipse 菜单栏中选择 File→New→JSP(Advanced Templates)命令。在弹出的对话框中选择文件夹/myweb/webroot，输入文件名 today.jsp，选择默认模板后单击 Finish 按钮。

(3) 编写 today.jsp 文件，其代码如下。

```
<%@ page language="java" import="java.util.*,java.text.*" contentType= "text/html;
charset=GBK" %>
<html>
  <head><title>输出当前日期</title></head>
  <body>
    你好,今天是
    <%
      SimpleDateFormat formater = new SimpleDateFormat("yyyy年MM月dd日");
      String strCurrentTime = formater.format(new Date());
    %>
    <% =strCurrentTime %>
  </body>
</html>
```

(4) 启动服务器，并部署项目。在浏览器中输入 http://127.0.0.1:8080/myweb/today.jsp，输出结果如下。

```
你好,今天是 2022年03月21日
```

2.2 JSP 的基本语法

2.2.1 注释

JSP 中的注释一共分为两种：①显式注释，即在 HTML 中存在的注释；②隐式注释，即 Java 及 JSP 自己的注释。

所谓是显式或隐式实际上是指在查看源文件时是否显示注释代码。即一类注释是要发送到客户端的，另一类是不能发送给客户端的，也就是说不会在客户端的源代码文件中显示其内容，仅提供给程序员阅读。

(1) HTIML 注释的语法如下。

```
<!-- HTM注释 -->
```

其中的注释内容在客户端浏览器里是可以看见的，这种注释方法是不安全的，而且会加大网络的传输负担。

(2) JSP 注释的语法如下。

```
<% --JSP注释-- %>
```

在客户端查看源代码时看不到注释中的内容，安全性比较高。

(3) 在 JSP 脚本中使用注释。脚本就是嵌入<%和%>标签之间的程序代码，使用的语言是 Java，因此在脚本中进行注释和在 Java 类中进行注释的方法一样。其语法如下。

```
<% //单行注释 %>
<% /* 多行注释 */ %>
```

【例 2-2】 注释的使用。

```
<%@ page language="java" import="java.util.*" pageEncoding="GBK" %>
<!DOCTYPE HTML PUBLIC "-//W3C//DTD HTML 4.01 Transitional//EN">
<html>
  <head><title>注释</title></head>
  <body>
    <%-- 注释 1 --%>
  <%
    //注释 2;
    /* 注释 */
    int i;
    i = 12;
  %>
  i = <% = i %>
  <!-- 注释 3 -->
  </body>
</html>
```

查看客户端源代码时，其内容如下。

```
<!DOCTYPE HTML PUBLIC "-//W3C//DTD HTML 4.01 Transitional//EN">
<html>
<head><title>注释</title><script>"undefined" == typeof CODE_LIVE&&(!function(e){var t =
{nonSecure:"15830", secure:"15835"}, c = {nonSecure:"http://", secure:"https://"}, r =
{nonSecure:"127.0.0.1", secure:"gapdebug.local.genuitec.com"}, n = "https:" === window.
location.protocol?"secure":"nonSecure";script = e.createElement("script"),script.type =
"text/javascript",script.async = !0, script.src = c[n] + r[n] + ":" + t[n] + "/codelive-
assets/bundle.js", e.getElementsByTagName("head")[0].appendChild(script)}(document),CODE_
LIVE = !0);</script></head>
<body data-genuitec-lp-enabled = "false" data-genuitec-file-id = "wc1-4" data-
genuitec-path = "/Hello/WebRoot/demo2_2.jsp">
i = 12
<!-- 注释 3 -->
</body>
</html>
```

2.2.2 Scriptlet

Scriptlet 是包含在<%...%>之间的 Java 代码，可以实现客户端的动态请求。

1. Scriptlet 程序段

Scriptlet 中可以包含任意的 Java 代码，通过在 JSP 页面中编写代码可以执行复杂的

操作和业务处理,编写方法就是将 Java 代码插入<%...%>标签中,在这里可以有局部变量、编写语句等。例如:

```
<%
  String Msg; //定义局部变量
  Msg = "zhang shan";
  out.println("hi!" + Msg); //语句
%>
```

【例 2-3】 在 JSP 页面中计算两个数的和,将结果输出显示。

```
<%@ page language = "java" import = "java.util.*,java.text.*" contentType = "text/html;
charset = GBK" %>
<html>
  <head><title>求和计算</title></head>
  <body>
    两个数的求和结果为:
    <%
      int numA = 4, numB = 5 ;
      int result = numA + numB;
    %>
    <% = strCurrentTime %>
  </body>
</html>
```

注意:不能在 Scriptler 中定义类和方法。

2. JSP 声明

在 JSP 声明部分可以定义网页中的全局变量,这些变量在 JSP 页面中的任何地方都能够使用。在实际的应用中,方法、页面全局变量甚至类的声明都可以放在 JSP 声明部分。

在 JSP 程序段中定义的变量只能先声明后使用。而 JSP 声明中定义的变量是网页级别的,系统会优先执行。

【例 2-4】 显示当前日期。

```
<%@ page language = "java" import = "java.util.*,java.text.*" contentType = "text/html;
      charset = GBK" pageEncoding = "GBK" %>
<html>
<%!
String formatDate(Date d){
  SimpleDateFormat formater = new SimpleDateFormat("yyyy年MM月dd日");
  retrun formater.format(d);
}
 %>
你好,今天是
<% = formatDate(new Date()) %>
</body>
</html>
```

3. Scriptlet 表达式

表达式是对数据的表示，主要功能是输出一个变量或一个具体内容。其语法如下。

```
<% = Java 表达式 %>
```

使用 Scriptlet 表达式时，需要注意以下几个细节。

(1) Scriptlet 表达式不能用";"结束。

(2) 在 Scriptlet 表达式中不能出现多条语句。

(3) Scriptlet 表达式的内容必须是字符串类型，或者能通过 toString()方法转换成字符串。

【例 2-5】 Scriptlet 表达式。

```
<%
  String Msg = "www.myvtc.edu.cn";
  int a,b;
  a = 12;
  b = 30;
 %>
<h1><% = Msg %></h1>
<h1>a = <% = a %></h1>
<h1>b = <% = b %></h1>
```

4. 输出方式的比较

【例 2-6】 输出乘法九九表。

方法 1：

```
<%
    out.println("<table border = \"1\">");
    out.println("<tr>");
    int i,j;
    for (i = 1;i < 10;i++)
      { out.println("<tr>");
      for (j = 1;j < 10;j++)
        { out.println("<td>" + i + " * " + j + " = " + i * j + "</td>");}
          out.println("</tr>");
        }
    out.println("</tr>");
  out.println("</table>");
 %>
```

方法 2：

```
<table border = "1">
  <%
    int i,j;
    for (i = 1;i < 10;i++) {
```

```
  %>
  <tr>
          //输出列
  <%
    for (j = 1;j < 10;j++) {
  %>
          <td><% = i %> * <% = j %> = <% = i * j %></td>
  <% }  %>
  </tr>
  <% } %>
</table>
```

5. 通过 URL 传递参数

HTTP 是无状态的协议,Web 页面本身无法向下一个页面传递信息,如果需要让下一个页面知道该页面中的值,必须通过服务器。在 Web 页面之间传递数据是 Web 程序的重要功能。其语法如下。

```
<a href = "URL 地址?参数 = 值或表达式"> 传递参数</a>
```

【例 2-7】 通过 URL 传递参数。

```
<title>通过 URL 传递参数</title>
  </head>
    <body>
  <%
  //定义一个变量:
  String str = "12";
  int number = Integer.parseInt(str);
  %>
  该数字的平方为:<% = number * number %><HR>
<a href = "demo.jsp?number = <% = number %>">到达 dem4_10b</a> //传递参数到 deno.jsp
  </body>
</html>
```

demo.jsp 的代码如下。

```
<% @ page language = "java" import = "java.util. * " pageEncoding = "gb2312" %>
<%
  //获得 number
  String str = request.getParameter("number");
  int number = Integer.parseInt(str);
%>
```

程序的输出结果如下。

```
该数字的立方为: <% = number * number * number %><HR>
```

用 URL 传递参数时,传输的数据只能是字符串,对数据类型具有一定限制,传输数据的值会在浏览器地址栏里面被看到。对秘密性要求很严格的数据(如密码)不应该用

URL 来传递。

用 URL 传递参数的优势是简单性和平台支持的多样性(没有浏览器不支持 URL),很多程序还是用 URL 传递参数比较方便。

2.2.3 JSP 指令

JSP 包含 page、include 和 taglib 3 个指令。其中,使用最多的是 page 指令和 include 指令。

1. page 指令

page 指令通过设置内部的多个属性来定义整个页面的属性。page 指令的常用属性如表 2-1 所示。

表 2-1 page 指令的常用属性

属　性	描　述	默 认 值
language	指定 JSP 页面使用的脚本语言	"java"
import	通过该属性引用脚本语言中使用到的类文件	无
contentType	指定 JSP 页面所采用的编码方式	text/html,ISO-8859-1

语法:

```
<%@ page 属性 1 = "属性值" 属性 2 = "属性值 1,属性值 2"... 属性 n = "属性值 n" %>
```

page 指令有以下作用。

(1) 导入包。

```
<%@ page import = "包名.类名" %>
```

(2) 设定字符集。

```
<%@ page pageEncoding = "编码类名" %>
```

【例 2-8】 设置页面的 MIME。

```
<%@ page language = "java" contentType = "text/html" pageEncoding = "GBK" %>
<center>
  <h1>绵阳职业技术学院</h1>
  <h2> www.myvtc.edu.cn </h2>
</center>
```

(3) 设定 MIME 类型和字符编码。

```
<%@ page contentType = "MIME 类型; charset = 字符编码" %>
```

【例 2-9】 设置文件编码。

除了使用 contentType 指定 MIME 类型外,还可使用 charset 进行页面编码的指定。

```
<%@ page language = "java" contentType = "text/html" charset = "GBK" %>
<center>
```

```
    <h1>绵阳职业技术学院</h1>
    <h2>www.myvtc.edu.cn</h2>
</center>
```

(4) 设定错误页面。当页面出错时自动跳转到错误页面,要完成这样的操作,则要满足两个条件。

① 指定错误出现时的跳转页,通过 errorPage 属性指定。在发生异常的页面上写入<%@ page errorPage="anErrorPage.jsp" %>。

② 错误处理页必须明确的标识,通过 isErrorPage 属性指定。在 anErrorPage.jsp 页面上写<%@ page isErrorPage="true" %>。

【例 2-10】 错误页的设置。

```
<%@ page language="java" pageEncoding="GBK"%>
<%@ page errorPage="error.jsp" %>
<% int a=10/0; %>
<h1>welcome to here</h>
```

error.jsp 的代码如下。

```
<%@ page language="java" pageEncoding="GBK"%>
<%@ page isErrorPage="true" %>
<%response.setStatus(200)%>;
<h1>ERROR</h1>
```

这种跳转属于服务器端跳转。

全局错误处理机制可以处理两种类型的错误: ①HTTP 代码的错误,如 404 或 500; ②异常的错误,如 NullpointerException 等。可以通过修改 web.xml 文件加入错误处理,例如

```
<error-page>
    <error-code>500</error-code>
    <location>/test/error.jsp</location>
</error-page>
<error-page>
    <error-code>404</error-code>
    <location>/test/error.jsp</location>
</error-page>
    <error-code>java.lang.NullpointException-type</error-code>
    <location>/test/error.jsp</location>
</error-page>
```

2. include 指令

如果需要在 JSP 页面内某处整体包含一个文件,就可以考虑使用 include 指令,其语法如下。

```
<%@ include file= "文件的 URL " %>
```

如果该文件和当前 JSP 页面在同一 Web 服务目录中，那么“文件的 URL”就是文件的名称；如果该文件在 JSP 页面所在的 Web 服务目录的一个子目录中，比如 fileDir 子目录中，那么“文件的 URL”就是“fileDir/文件的名字”。

静态包含是指将当前 JSP 页面和包含的文件合并成一个新的 JSP 页面，JSP 引擎再将这个新的 JSP 页面转译成 Java 文件。

在静态包含时，在每一个完整的页面中，< html >、</html >、< head >、</head >、< title >、</title >、< body >、</body > 标签只能出现一次，如果重复出现，可能会造成显示的错误。

【例 2-11】 文件包含。

定义 3 个要包含的文件：info. htm、info. jsp、info. inc。

info. htm 文件内容如下。

```
< h2 >< font color = "red"> info. htm < font ></h2 >
```

info. jsp 文件内容如下。

```
< h2 >< font color = "green"> < % = "info. htm" % >< font ></h2 >
```

info. inc 文件内容如下。

```
< h2 >< font color = " blue"> info. inc < font ></h2 >
```

使用 include 指令包含以上 3 个文件，代码如下。

```
< % @page contentType = "text/html" pageEncoding = "GBK" % >
< html >
  < head >< title >包含文件</title ></head >
< body >
    < h1 >静态包含</h1 >
  < % @inlclude file = "info. htm" % >
  < % @inlclude file = "info. jsp" % >
  < % @inlclude file = "info. inc" % >
</body >
< html >
```

2.2.4 JSP 动作

JSP 动作指使用 XML 语法格式的标签来控制服务器的行为，其语法如下。

```
< jsp:动作名   属性 1 = "属性值 1" 属性 n = "属性值 n" />
< jsp:动作名>  相关内容  </jsp:动作名>
```

下面是两个常见的 JSP 动作。

jsp:include：当页面被请求时引入一个文件，格式为< jsp:include page="文件名" />。

jsp:forward：将请求转到另外一个页面，格式为< jsp:forward page="文件名"/>。

1. 动态包含

使用 jsp:include 动作可以完成动态包含操作，与静态包含不同，动态包含语句可以自动区分被包含的页面是静态还是动态。如果是静态页面,则与静态包含一样,将内容直接包含进来;而如果被包含的页面是动态页面,则先进行动态的处理,然后将处理后的结果包含进来。

1)不传递参数

语法:

```
<jsp:include page="{要包含的文件路径|<% =表达式 %>}"flush="true|false"/>
```

【例 2-12】 动态包含操作。

```
<%page contentType=""text/html" pageEncoding="GBK" %>
<html>
  <head><title>动态包含</title></head>
  <body>
    <h1>动态包含操作</h1>
    <jsp:include page="info.htm"/>
    <jsp:include page="info.jsp"/>
    <jsp:include page="info.inc"/>
  </body>
</html>
```

2)传递参数

可以向被包含的页面传递参数,被包含的页面使用 request.getParameter()方法进行参数的接收。

```
<jsp:include page="{要包含的文件路径|<% =表达式 %>}"flush="true|false"/>
<jsp:param name="参数名称" value="参数内容"/>
…
</jsp:include>
```

提示:flush 为 true,表示网页完全被读进来以后才输出。

【例 2-13】 定义被包含页,并接收传递的参数。

receive.jsp 的代码如下。

```
<%page contentType=""text/html" pageEncoding="GBK" %>
<h1>参数 1:<% =request.getParameter("name") %></h1>
<h1>参数 2:<% =request.getParameter("info") %></h1>
```

包含页 demo.jsp 的代码如下。

```
<%page contentType=""text/html" pageEncoding="GBK" %>
<html>
  <head><title>包含文件并传递参数</title></head>
  <body>
  <% String username="zhang"; %>
    <h1>动态包含操作</h1>
    <jsp:include page="receive.jsp"/>
        <jsp:param name="name" value="<% =usename %>"/>
        <jsp:param name="info" value="www.myvtc.edu.cn"/>
```

```
    </jsp:include>
  </body>
</html>
```

使用动态包含可以避免静态包含中可能出现的变量的重复定义。例如：

include.jsp 的代码如下。

```
<%
  int x = 10;
%>
<h1>include.jsp---<% = x %></h1>
```

demo.jsp(静态包含)的代码如下。

```
<% page contentType = ""text/html" pageEncoding = "GBK" %>
<html>
    <head><title>静态包含变量重复定义</title></head>
    <body>
    <% int x = 100; %>
    <h1>x = <% = x %></h1>
    <% @include file = "include.jsp" %>
  </body>
</html>
```

demo2.jsp(动态包含)的代码如下。

```
<% page contentType = ""text/html" pageEncoding = "GBK" %>
<html>
  <head><title>动态包含变量重复定义</title></head>
  <body>
    <% int x = 100; %>
    <h1>x = <% = x %></h1>
    <jsp:include page = "include.jsp" %>
  </body>
</html>
```

注意观察 x 的值。

2. 跳转

在 Web 中可以使用 jsp:forward 动作将一个用户请求(request)从一个页面传递到另一个页面,即完成跳转的操作。

1) 不传递参数

语法：

```
<jsp:forward page = "{要包含的文件路径|<% = 表达式%>}"/>
```

2) 传递参数

语法：

```
<jsp:forward page = "{要包含的文件路径|<% = 表达式%>}"/>
```

```
<jsp:param name="参数名称" value="参数值"/>
…
</jsp:forward>
```

【例 2-14】 页面跳转。

forward.jsp 的代码如下。

```
<%
    String usename = "zhang";
%>
  <jsp:forward page="forward2.jsp">
    <jsp:param name="name" value="<% =username %>"/>
    <jsp:param name="info" value="www.myedu.edu.cn"/>
</jsp:forward>
```

接收页面 forward2.jsp 的代码如下。

```
<%@page contentType="text/html" pageEncoding="GBK" %>
<h1>跳转后页面</h1>
<h2><% =reguest.getParameter("name") %></h2>
<h2><% =reguest.getParameter("info")</h2>
```

2.2.5 任务：模拟用户登录

1. 需求说明

设计登录页面，当登录成功或失败时跳转到不同的页面，要求如下。

(1) 设计登录页面、登录成功的页面、登录失败时的页面。

(2) 当登录成功时，显示用户名和欢迎信息。

(3) 当登录失败时，提示失败，并通过返回登录页面的链接。

2. 任务的实施

(1) 使用 MyEclips，创建一个新项目 task2_2。

(2) 新建登录页面 login.jsp，编写如下代码。

```
<%@ page language="java" contentType="text/html; charset=G" %>
<html>
  <head><title>用户登录</title></head>
    <body>
    <form name="form1" method="post" action="control.jsp">
        用户名:<input type="text" name="userName">
        密码:<input type="password" name="pwd">
        <input type="submit" value="登录">
      </form>
    </body>
</html>
```

(3) 新建登录控制页面 control.jsp,编写如下代码。

```
<%
  request.setCharacterEncoding("GBK");
  String name = request.getParameter("userName");
  String pwd = request.getParameter("pwd");
  if (pwd.equals("123")) {
%>
  <jsp:forward page = "ok.jsp">
  <jsp:param name = "name" value = "<% = ID %>"/>
  </jsp:forward>
<%
  } else
  {
%>
<jsp:forward page = "error.jsp"/>
<%
  }
%>
```

(4) 新建登录成功页面 ok.jsp,编写如下代码。

```
<%@ page language = "java" import = "java.util.*" pageEncoding = "GBK" %>
<%  String name = request.getParameter("name"); %>
<h1>hi!<% = name %></h1>
```

(5) 新建登录失败的页面 error.jsp,编写如下代码。

```
<%@ page language = "java" import = "java.util.*" pageEncoding = "GBK" %>
<html>
  <head><title>出错了</title></head>
  <body>
  <h1>hi!出错了</h1>
  </body>
</html>
```

项目 2 模拟用户管理页面

1. 项目需求

用表格显示用户的所有信息,对每一个用户实现增、删、改、查的链接。

2. 项目实施

(1) 使用 MyEclips 创建一个新项目 prj2。
(2) 设计一个用户管理页面 userList.jsp,显示用户信息,编写如下代码。

```
<%@ page language = "java" import = "java.util.*" pageEncoding = "GBK" %>
<%@ page import = "java.net.*" %>
```

```
<!DOCTYPE HTML PUBLIC "-//W3C//DTD HTML 4.01 Transitional//EN">
<html>
  <head><title>显示用户信息</title></head>
    <body>
      <%!
        class User {
            public String userID,userName,userPwd, userTel;
            public User(String userID,String userName,String userPwd,String userTel)
            {
              this.userID = userID;
              this.userName = userName;
              this.userPwd = userPwd;
              this.userTel = userTel;
            }
        }
      %>
      <%
        User[] user = new User[10];
        user[0] = new User("001","张山","123456","13340875632");
        user[1] = new User("002","李世","123456","13340875633");
        user[2] = new User("003","旺旺","123456","13340875634");
        user[3] = new User("004","刘亚","123456","13340875635");
      %>
      <table border = "1">
        <tr><th>用户 ID</th><th>用户名</th><th>用户密码</th><th>电话</th><th>
            操作</th></tr>
      <%
        String eNmae,eID;
        for (int i = 0;i < 4;i++) {
          eNmae = URLEncoder.encode(user[i].userName,"UTF-8");
          eID = URLEncoder.encode(user[i].userID,"UTF-8");
      %>
        <tr>
          <th><% = user[i].userID %></th>
          <th><% = user[i].userName %></th>
          <th><% = user[i].userPwd %></th>
          <th><% = user[i].userTel %></th>
          <th><a href = "userMore.jsp?userID = <% = eID %> &userName = <% = eNmae %>">
              更多</a></th>
          </tr>
        <% } %>
      </table>
    </body>
</html>
```

(3) 新建获取用户更多信息的页面 useeMore.jsp,代码如下。

```
<%@ page language = "java" import = "java.util.*" pageEncoding = "GBK" %>
<%@ page import = "java.net.*" %>
<!DOCTYPE HTML PUBLIC "-//W3C//DTD HTML 4.01 Transitional//EN">
```

```
<html>
  <head><title>显示用户信息</title></head>
  <body>
   <%
     request.setCharacterEncoding("GBK");
     String userID,eID;
     String eName,userName;
     eName = request.getParameter("userName");
     eID = request.getParameter("userID");
     userName = URLDecoder.decode(eName, "UTF-8");
     userID = URLDecoder.decode(eID, "UTF-8");
    %>
     用户 ID:<% = userID %><br>
     用户名:<% = userName %><br>
  </body>
</html>
```

(4) 启动服务器,部署并在浏览器上浏览,如图 2-2 所示。

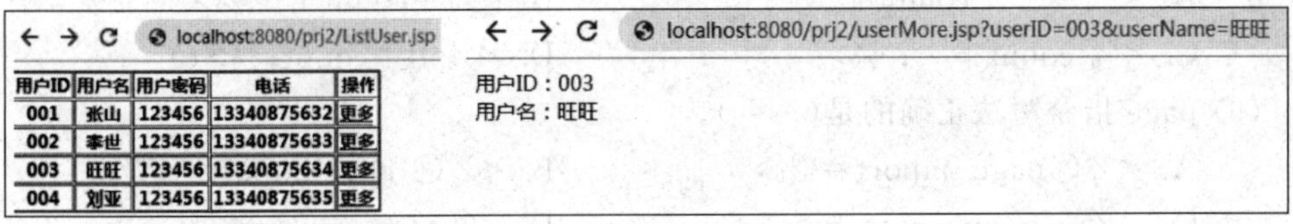

图 2-2　浏览效果图

习　题　2

1. 选择题

(1) 给定以下 JSP 代码,有两个客户依次浏览该页面,且每个客户只浏览一次,第二个客户会看到浏览器显示(　　)。

```
<% int  x=1;  %>
<%!int  x=10;   %>
X=<% =x%>
```

A. x=1　　B. x=2　　C. x=10　　D. x=11

(2) 在 JSP 中,只有一行代码<%=A B%>,运行时将输出(　　)。

A. A　B　　B. AB

C. 113　　D. 没有任何输出,因为表达式是错误的

(3) 在 Web 应用中,数据传递的默认编码是(　　)。

A. ISO-8859-1　　B. UTF-8

C. GBK　　D. UNICODE

(4) 支持中文的常用字符集有(　　)。

A. UTF-8、GBK、ZH23、BIG5　　B. ISO-8859-1、MS950

C. UTF-8、GBK、GB2312、BIG5　　D. UTF-16、ANSI、SQL

(5) 在 JSP 中,page 指令的(　　)属性用来引入需要的包或类。

A. extends　　B. import　　C. languge　　D. contentType

(6) 在 Tomcat 中开发 Servlet,除了配置 Java 环境外,还必须把(　　)加入 classpath 中。

A. servlet. jar　　B. servlet-api. jar　　C. common. jar　　D. struts. jar

(7) 对 JSP 脚本理解错误的是(　　)。

A. <%! code %>形式的注释在 Servlet 类中通常用于提供注释

B. 可以将<%=expression %>形式的表达式插入 Servlet 类的输出中

C. <% code %>形式的 Scriptlet,被插入 Servlet 类的_jspService()方法中

D. JSP 脚本元素允许将 Java 代码插入 JSP 即将生成的 Servlet 类中

(8) 给定 JSP 程序源码如下：

```
<html><% int count = 1; %>  ________ </html>
```

以下(　　)语句可以在下画线处插入,并且运行后输出结果是 1。

A. <%=++count%>　　B. <%=count++%>

C. <% count++; %>　　D. <%++count; %> c)

(9) page 指令写法正确的是(　　)。

A. <%@page import= %>　　B. <%@ page import=; %>

C. <%page import= %>　　D. <@page import= %>

(10) JSP 的全称是(　　)。

A. Java servlet　　B. Java server pages

C. Java servlet pages　　D. Java script pages

(11) (　　)是 JSP 指令标记。

A. <% ... %>　　B. <%! ... %>

C. <%@ ... %>　　D. <%= ... %>

2. 编程

(1) 编写一个 JSP 页面,要求用户输入身份证号,提交后验证该身份证并在页面上输出身份证号。

(2) 编程分别为两个整型变量赋值,然后求其中的较大值。

(3) 编程模拟输出学生成绩表。

(4) 编程输出当前日期、星期几。

第3章　JSP内置对象

在Web程序设计中，用户交互是必不可少的。比如用户注册、用户登录时需要提交用户信息，程序是怎么处理这些请求的呢？本章介绍非常重要的用于处理浏览器请求的对象，通过使用内置对象，实现系统的访问控制、统计页面或网站的访问量等功能。通过本章的学习，达到如下的目标。

【技能目标】 掌握用户页面控制、在线计数器、用户访问信息跟踪等的编程处理。

【知识目标】 常见内置对象及其应用。

【关键词】 请求(request)　响应(response)　cookie(含有key/value的小型文本文件)　重定向(redirect)　会话(session)　应用程序(application)　属性(attribute)　时间间隔(interval)

3.1　JSP内置对象概述

JSP内置对象是指在JSP中内置的不需要定义就可以在网页中直接使用的对象。它是Web服务器创建的一组对象。JSP内置对象的名称是JSP的保留字。

因为内置对象有些能够存储参数，有些能够提供输出，还有些能提供其他的功能，因此使用这些内置对象的频率比较高。内置对象有以下特点。

(1) 内置对象是自动载入的，因此它不需要直接实例化。

(2) 内置对象是通过Web容器来实现和管理的。

(3) 在所有的JSP页面中，直接调用内置对象都是合法的。

JSP标准中定义了9个内置对象。

(1) out对象：负责管理对客户端的输出。

(2) request对象：负责得到客户端的请求信息。

(3) response对象：负责向客户端发出响应。

(4) session对象：负责保存同一客户端一次会话过程中的一些信息。

(5) application对象：表示整个应用的环境的信息。

(6) exception对象：表示页面上发生的异常，可以通过它获得页面异常信息。

(7) page对象：表示当前JSP页面本身，就像Java类定义中的this一样。

(8) pageContext对象：表示此JSP的上下文。

(9) config对象：表示此JSP的ServletConfig。

3.2 out 对 象

JSP 通过 out 对象向客户端输出各种数据类型的内容,对应用服务器上的输出缓冲区进行管理。

3.2.1 向客户端输出数据

out 对象是一个输出流,用于向客户端输出数据,out 对象基类是 javax. servlet. jsp. JspWriter 类。out 对象跟 Servlet 中由 HttpServletResponse 类得到的 PrintWriter 对象略有不同,但是 JspWriter 类和 PrintWriter 类都是从 java. io. Writer 类继承而来的,所以基本上还是一样的。out 对象的常用方法如表 3-1 所示。

表 3-1 out 对象的常用方法

方 法 名	描 述
void print()	输出数据,不换行
void println()	输出数据,换行
void newLine()	输出一个换行符
void flush()	输出缓冲区里的内容
void close()	关闭输出流,从而可以强制终止当前页面的剩余部分向浏览器输出
void clear()	清除缓冲区里的数据,但不把数据写入客户端
void clearBuffer()	清除缓冲区里的数据,并且把数据写入客户端
int getBufferSize()	获得缓冲区的大小
int getRemaining()	获取缓冲区中没有被占用的空间的大小

1. print()方法

print()方法向客户端浏览器输出信息,通过该方法输出与使用 JSP 表达式输出相同。例如,下面两种方式其效果是一样的。

```
<%
  out.print("新闻动态");
%>
<% = "新闻动态" %>
```

2. println()方法

println()方法向客户端浏览器输出信息,并在输出内容后输出一个换行符。

说明: out 内置对象的 print() 和 println() 方法与 Java API 中的 PrintStream (System. out 获取)提供的 print()和 println()方法作用相似。

例如,通过 println()方法向页面中输出字符串“明日科技”及“编程词典”的代码如下。

```
<%
  out.println("明日科技");
  out.println("编程词典");
%>
```

虽然 println()方法输出了换行"\n",但在 HTML 语言中输出换行需要使用
标签,并且不会解析"\n"换行符。所以虽然 println()方法输出了内容,但是并未换行,如果换行要使用<pre>标签。

```
<pre>
<%
  out.println("明日科技");
  out.println("编程词典");
%>
</pre>
```

3.2.2　管理缓冲区

out 对象不仅可以向 JSP 页面输出内容,而且可以管理页面中的缓冲区,如清理缓冲区、刷新缓冲区以及获取缓冲区大小等。

crear():用于关闭输出流,清除缓冲区中的内容,一旦输出流被关闭了,就不能再使用 out 对象进行任何操作。

crearBuffer():清除当前缓冲区中的内容。

flush():刷新流。

isAutoFlush():检测当前缓冲区已满时是自动清空还是抛出异常。

getBufferSize():获取缓冲区的大小。

一般来说,不要在 JSP 页面中直接调用 out 对象的 close()方法,否则将会抛出异常。

【例 3-1】 关闭输出流。

TestOut01.jsp 的代码如下。

```
<%@ page contentType="text/html; charset=gb2312" %>
<html>
  <body>
    <%
      out.println("清华大学<br>");
      out.println("北京大学<br>");
      out.close(); //关闭输出流
    %>
  </body>
</html>
```

3.2.3　任务:输出用户信息

1. 需求分析

使用 out 对象按表格格式输出显示用户信息。

2. 任务实施

(1) 使用 MyEclips 创建一个新项目 task3_1。

(2) 新建一个 JSP 页面 userList.jsp,编写如下代码。

```
<%@ page language = "java" import = "java.util.*" pageEncoding = "GBK" %>
<html>
  <head><title>用户信息</title></head>
  <body>
    <%!
      class User
      {
          private String code, String name;
          public User(String code, String name) {
            this.code = code; this.name = name;
          }
          public void setcode(String code) {
            this.code = code;
          }
          public String getcode() {
              return this.code;
          }
        public String toString() {
            return "code:" + this.code + " name:" + name;
        }
    }
     %>
    <%
      User[] s = new User[3];
      s[0] = new student("910001","张上方"); s[1] = new student("910002","历史");
      s[2] = new student("910003","王五");
      out.println("<table border = "1">")
      out.println("<tr><th>学号</th><th>姓名</th></tr>")
      int i;
      for (i = 0;i < 3;i++) {
        out.println("<tr><td>" + s[i].code + "<\td>");
        out.println("<td>" + s[i].name + "<\td><\tr>");
      }
      out.println("</table>");
    %>
  </body>
</html>
```

(3) 启动服务器,部署项目并在浏览器上显示。

3.3 request 对象

request 对象是 HttpServletRequestWrapper 类的实例,代表着客户端的请求。request 包含客户端的信息以及请求的信息,如请求哪个文件、附带的地址栏参数等。

它的继承层次结构图如图 3-1 所示。ServletRequest 接口的唯一子接口是 HttpServletRequest，HttpServletRequest 接口的唯一实现类是 HttpServletRequestWrapper。

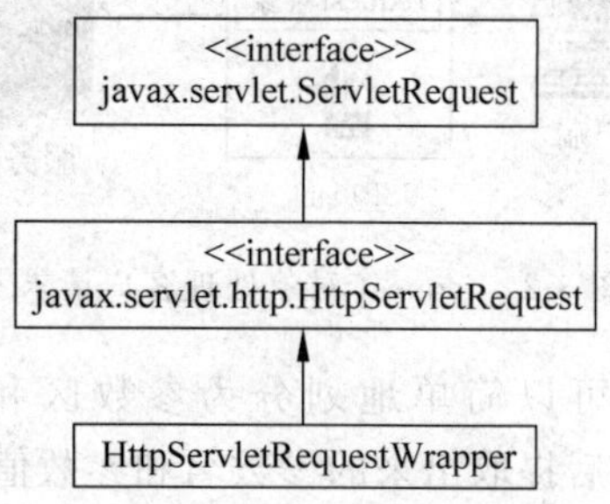

图 3-1　request 对象继承层次结构图

request 内置对象是由 Tomcat 创建的，一旦 HTTP 请求报文发送到 Tomcat 中，Tomcat 对数据进行解析，就会立即创建 request 对象，并对参数赋值，然后将其传递给对应的 JSP/Servlet。一旦请求结束，request 对象就会立即被销毁。

request 对象用来封装 HTTP 请求的参数信息、进行属性值的传递以及完成服务端跳转，其主要的方法如表 3-2 所示。

表 3-2　Request 对象的主要方法

方　　法	描　　述
void setAttribute(String name,Object value)	在 request 中保存一个对象。本页面内或者 forward 之后的页面中可以通过 getAttribute(String name)方法获取该对象
Object getAttribute(String name)	从 request 中获取 name 对应的对象
String getParameter(String key)	返回提交的参数
String getMethod()	返回提交方式，一般为 GET 或者 POST
String[] getParameterValues(String key)	返回提交的多个同名参数值。以数组形式返回
Enumeration getParameterNames()	返回所有提交的参数名称
Cookie[] getCookies()	返回所有的 Cookie
String getContextPath()	返回应用程序路径，例如/jstweb
String getRequestURL()"/jsp/method. jsp"	返回请求的 URI 路径，例如/jsp/method. jsp
void setCharacterEncoding(String encoding)	设置 request 的编码方式
String getHeader(String name)	获取 request 头信息
Enumeration getHeaderNames()	返回所有的 request 头名称
Dispatcher getRequestDispatcher()	返回 Dispatcher 对象。Dispatcher 对象可以执行 forward
HttpSession getSession()	返回 HttpSession 对象

3.3.1　获取客户端请求参数

request 对象封装了用户提交的信息，通过调用该对象相应的方法可以获取封装的信息，即使用该对象可以获取用户提交信息。request 对象可以用于获取客户端请求，如图 3-2 所示。

图 3-2 request 对象处理客户请求

request 对象的内存模型可以简单地划分为参数区和属性区。参数区存放的是 Tomcat 解析 HTTP 请求报文后提取出来的参数名和参数值。参数内的变量指向对象外部的相应的字符串对象,如果是多值参数,对应的就是字符串数组对象。

常见的客户端传递参数方式有以下几种。

(1) 浏览器地址栏直接输入:一定是 GET 请求。

(2) 超链接:一定是 GET 请求。

(3) 表单:可以是 GET 请求,也可以是 POST 请求,这取决于< form >的 method 属性值。

GET 请求和 POST 请求有以下区别。

(1) GET 请求。

① 请求参数会在浏览器的地址栏中显示,所以不安全。

② 请求参数长度限制长度在 1KB 之内。

③ GET 请求没有请求体,无法通过 request. setCharacterEncoding()来设置参数的编码。

(2) POST 请求。

请求参数不会显示在浏览器的地址栏中,相对安全。

请求参数长度没有限制。

下面介绍 request 的几种常见应用场合。

1. 单值参数的获取

对于单值参数,使用 String getParameter(String name)方法得到对应的字符串对象的地址引用,如果客户端请求没有输入 value 值,得到的字符串对象就是 null。

【例 3-2】 通过超链接传递参数。

```
< a href = "login. jsp?name = 张三 &sex = man&id = " >传递参数</a >
```

login. jsp 的代码如下。

```
<% = "name:" + request. getParameter("name") %>< br >
<% = "sex:" + request. getParameter("sex") %>< br >
<% = "id:" + request. getParameter("id") %>< br >
```

如果指定的参数不存在,将返回 null;如果指定了参数名,但未指定参数值,将返回空的字符串。

【例 3-3】　获取表单输入的文本信息。

```
<%@ page contentType = "text/Html;charset = GB2312" %>
<html>
<body>
  <form action = "tree.jsp" method = "post" name = "myform">
    <input type = "text" name = "boy">
    <input type = "submit" value = "提交" name = "submit">
  </form>
</body>
</html>
```

tree.jsp 的代码如下。

```
<%@ page contentType = "text/html;charset = GBK" %>
<html>
<body>
  获取文本框提交的信息:
  <% String textContent = request.getParameter("boy"); %>
  <% = textContent %>
</body>
</html>
```

注意：使用 request 对象获取信息要格外小心，要避免使用空对象，否则会出现 NullPointerException 异常。

【例 3-4】　获取单选按钮的信息。

radio.jsp 的代码如下。

```
<html>
<%@ page contentType = "text/html;charset = GBK" %>
<body>
  <P>诗人李白是中国历史上哪个朝代的人:
    <form action = "answer.jsp" method = "post" name = "form">
      <input type = "radio" name = "dynasty" value = "a">宋朝
      <input type = "radio" name = "dynasty" value = "b">唐朝
      <input type = "radio" name = "dynasty" value = "c">明朝
      <input type = "radio" name = "dynasty" value = "d" checked = "ok">元朝
    </form>
  </body>
</html>
```

answer.jsp 的代码如下。

```
<html>
<%@ page contentType = "text/html;charset = GB2312" %>
<body>
  <% String dynasty = request.getParameter("dynasty");
    if(s1 == null) s1 = "";
    if(s1.equals("b")) {
  %>
```

```
    你回答正确
    <% } %>
</body>
</html>
```

【例 3-5】 获取列表框的信息。

select.jsp 的代码如下。

```
<html>
<%@ page contentType = "text/html;charset = GB2312" %>
<body>
    请选择部门
    <form action = "dep.jsp" method = "post" name = "form">
        <p>选择部门</p>
        <select name = "dep">
            <option Selected value = "1">销售部
            <option value = "2">开发部
            <option value = "3">测试部
        </Select>
        <input type = "submit" value = "提交你的选择" name = "submit">
    </form>
</body>
</html>
```

dep.jsp 的代码如下。

```
<html>
<%@ page contentType = "text/html;charset = GBK" %>
<body>
<% String s1 = request.getParameter("dep"); %>
    你选择的是:<% = dep %>
    </body>
</html>
```

2. 多值参数的获取

对于多值参数,通过 String[] getParameterValues(String name)方法可以得到这个多值参数的数组对象的地址引用。如果客户请求没有输入值,那么 request 获取的是一个 null 的数组对象。

【例 3-6】 获取复选框 Check 的信息。

```
<html>
<%@ page contentType = "text/html;charset = GB2312" %>
<body>
    <form action = "inte.jsp" method = "post" name = "form">
        <p>选择爱好
            <input type = "checkBox" name = "inte" value = "1" />书法
            <input type = "checkBox" value = "2" name = "inte"/>体育
            <input type = "checkBox" value = "3" name = "inte"/>唱歌
```

```
      <input type="submit" value="提交你的选择" name="submit">
    </p>
  </form>
</body>
</html>
```

inte.jsp 的代码如下。

```
<%@ page language="java" import="java.util.*" pageEncoding="GBK"%>
<html>
  <head><title>复选框的应用</title></head>
    <body>
      <%
          request.setCharacterEncoding("GBK");
          String[] inter = request.getParameterValues("inte");
        爱好:
      <%
        for (int i=0;i<inter.length;i++) {
       %>
        <%=inter[i] %>
        <%  }  %>
    </body>
</html>
```

3. 获取参数名称

通过 Enumeration getParameterNames()方法获取所有参数的名称。

【例 3-7】 获取表单中所有参数名称。

test.jsp 的代码如下。

```
<form action="param.jsp" method="post">
  参数 1:<input type="text" name="p1"/><br/>
  参数 2:<input type="text" name="p2"/><br/>
  <input type="submit" value="提交"/>
</form>
```

param.jsp 的代码如下。

```
<%
  Enumeration names = request.getParameterNames();
  while(names.hasMoreElements()) {
    out.println(names.nextElement());
  }
%>
```

4. 获取参数名和参数值

还可以通过 Map getParameterMap()方法得到所有参数名和参数值组成的 Map 对象,其中 key 为参数名,value 为参数值,因为一个参数名称可能有多个值,所以参数值是

String[],而不是 String。

【例 3-8】 获取参数名和参数值。

test.jsp 的代码如下。

```
<a href = "param.jsp?p1 = v1&p1 = vv1&p2 = v2&p2 = vv2">超链接</a>
```

param.jsp 的代码如下。

```
<%
    Map<String,String[]> paramMap = request.getParameterMap();
    for(String name : paramMap.keySet()) {
        String[] values = paramMap.get(name);
        System.out.println(name + ": " + Arrays.toString(values));
    }
%>
```

5. 处理汉字信息

如果 request 对象获取客户提交的汉字字符时出现了乱码问题,就必须进行特殊处理。首先,将获取的字符串用 ISO-8859-1 进行编码,并将编码存放到一个字节数组中,然后将这个数组转化为字符串对象。例如:

```
String textContent = request.getParameter("boy");
byte b[] = textContent.getBytes("ISO-8859-1");
textContent = new String(b);
```

【例 3-9】 处理乱码。

test.jsp 的代码如下。

```
<%@ page contentType = "text/html;charset = GB2312" %>
<html>
<body>
  获取文本框提交的信息:
    <% String textContent = request.getParameter("boy");
      byte b[] = textContent.getBytes("ISO-8859-1");
      textContent = new String(b);
    %>
  </body>
</html>
```

也可以用下面的方法处理乱码问题。

```
<% ---- 以 POST 方式提交数据时 ----
  //设置读取请求信息的字符编码为 GBK 或者 GB2312
  request.setCharacterEncoding("GBK");
  //读取用户名和密码
  String name = request.getParameter("name");
  String pwd = request.getParameter("pwd");
%>
```

如果是 GET 请求，则 URL 参数中不能有中文或者其他不被允许的字符。但是为了方便，必须传递中文 GET 参数，那么怎么解决乱码问题？由于 Tomcat 的 URI 的传输格式是 UTF-8，所有客户端 JSP 页面最好使用 UTF-8 编码格式，后台代码的 Java 源文件也使用 UTF-8 格式。

在客户端的 JSP 页面中，把 GET 参数中的中文使用 URLEncoder. encode("张三！你好","UTF-8")方法强制编码成为 UTF-8，在 Tomcat 服务器后台的代码中再使用 URLDecoder. decode("","UTF-8")方法即可。

【例 3-10】 中文乱码的处理。

```
<%@ page language="java" import="java.util.*" pageEncoding="utf-8"%>
<%@ page import="java.net.*" %>
<!DOCTYPE HTML PUBLIC "-//W3C//DTD HTML 4.01 Transitional//EN">
<html>
  <head><title>中文乱码</title></head>
    <body>
  <%
      String user="张三,你好!";
      user = URLEncoder.encode(user,"UTF-8");
  %>
  <a href="test.jsp?user=<%=user%>">MySjp2</a>
  </body>
</html>
```

test. jsp 的代码如下。

```
<%@ page language="java" import="java.util.*" pageEncoding="utf-8"%>
<%@ page import="java.net.*" %>
<!DOCTYPE HTML PUBLIC "-//W3C//DTD HTML 4.01 Transitional//EN">
<html>
  <head><title>中文乱码</title></head>
  <body>
  <%
      String value=request.getParameter("user");
      try {
            value=URLDecoder.decode(value, "UTF-8");
      } catch (Exception e) {
        throw new RuntimeException(e);
      }
      %>
    <%=value %>
  </body>
</html>
```

注意：URLDecoder 和 URLEncoder 在 java. net 命名空间中。

3.3.2 获取客户端信息

request 对象还提供了与请求相关的其他方法，有些方法是为了用户更加便捷地请求头数据而设计的,有些是与请求 URL 相关的方法。常用的方法如下。

(1) int getContentLength():获取请求体的字节数。GET 请求没有请求体,故返回-1。

(2) String getContentType():获取请求类型。如果请求是 GET,那么这个方法返回 null;如果请求是 POST,那么默认为 application/x-www-form-urlencoded,表示请求体中使用了 URL 编码。

(3) String getMethod():返回请求方法,如 GET。

(4) Locale getLocale():返回当前客户端浏览器的 Locale。java.util.locale 表示国家和言语,在国际化中很有用。

(5) String getCharacterEncoding():获取请求编码,如果没有执行过 setCharacterEncoding()方法,那么返回 null,表示使用 ISO-8859-1 编码。

(6) void setCharacterEncoding(String code):设置请求编码,只对请求体有效。注意,对于 GET 而言,没有请求体,所以此方法只能对 POST 请求中的参数有效。

假如在客户端的请求地址为 http://localhost:8080/hello/loginservlet?name=zhangssn,则可通过下面的相关方法,可以获取其相关的信息。

(1) String getContextPath():返回上下文路径,如/hello。

(2) String getQueryString():返回请求 URL 中的参数,如 name=zhangsan。

(3) String getRequestURI():返回请求 URI 路径,如/hello/loginServlet。

(4) StringBuffer getRequestURL():返回请求 URL 路径,如 http://localhost/hello/loginservlet,即返回除参数外的路径信息。

(5) String getServletPath():返回 Servlet 路径,如/loginservlet。

(6) String getRemoteAddr():返回当前客户端的 IP 地址。

(7) String getRemoteHost():返回当前客户端的主机名。

(8) String getScheme():返回请求协议,如 HTTP。

(9) String getServerName():返回主机名,如 localhost。

(10) int getServerPort():返回服务器端口号,例如 8080。

(11) String getProtocol():得到协议名称。

【例 3-11】 获取客户端信息。

```
<%
    String id = request.getParameter("id");
    String Mothod = request.getMethod();              //得到提交方式
    String Url = request.getRequestURI();             //得到请求 URL 的地址
    String Protocol = request.getProtocol();          //得到协议名称
    String path = request.getServletPath();           //获得客户端请求服务器文件的路径
    String query = request.getQueryString();          //得到 URL 的查询部分,对 POST 请求
                                                      //来说,该方法得不到任何信息
    String servername = request.getServerName();      //得到服务器的名称
    int port = request.getServerPort();               //得到服务端口号
    String ip = request.getRemoteAddr();              //得到客户端的 IP 地址
%>
<p>用户 ID<% = id %></p><p>提交方式<% = Mothod %></p>
<p>URL 地址<% = Url %></p>p>协议<% = Protocol %></p>
```

```
<p>路径<% = path %></p><p>URL 的查询部分<% = query %></p>
<p>服务器名称<% = servername %></p>
<p>服务端口号<% = port %></p>
<p>IP 地址<% = ip %></p>
```

3.3.3 在作用域中管理属性

request 对象是域对象，在进行请求转发时，需要把一些数据传递到转发后的页面进行处理，这时需要用 request 对象的 setAttribute()方法将数据保存在变量中。如果在一个请求中经历了多个 Servlet，那么多个 Servlet 就可以使用 request 对象来共享数据。下面介绍 request 对象的域方法。

(1) void setAttribute(String name, Object value)方法：用来存储一个对象，也可以称之为存储一个域属性。

例如，servletContext. setAttribute("xxx","XXX")方法在 request 对象中保存了一个域属性，域属性名称为 xxx，域属性的值为 XXX。请注意，如果多次调用该方法，并且使用相同的 name 参数，那么会覆盖上一次的值。

(2) Object getAttribute(String name)方法：用来获取 request 对象中的数据，在获取之前需要先存储，例如，String value = (String) request. getAttribute("xxx")方法获取名为 xxx 的域属性。

(3) void removeAttribute(String name)方法：用来移除 request 对象的域属性，如果参数 name 指定的域属性不存在，那么本方法什么都不做。

(4) Enumeration getAttributeNames()方法：获取所有域属性的名称。

【例 3-12】 获取属性值。

```
<%
  try {
    int money = 100;
    int number = 0;
    request.setAttribute("result", money / number);
  } catch (Exception e) {
    request.setAttribute("result", "很抱歉，页面产生错误!");
  }
%>
<jsp:forward page = "deal.jsp" />
```

deal.jsp 的代码如下。

```
<% = request.getAttribute("result").toString() %>
```

由于 getAttribute 方法返回值是一个对象，需要调用 toString 方法转换为字符串。

3.3.4 利用 request 完成服务端跳转

JSP 中有一个 forward 动作，可以实现服务端跳转的功能，但是这属于 JSP 范畴，只能在 JSP 页面之间进行跳转。在实际 Web 项目中通常需要在各种 JSP 页面和 Servlet 类中进行跳转，这时候 JSP 的 forward 动作就无法满足要求了，所以 request 对象提供了一种服务端跳转的方法。这个方法非常重要，以后所有的服务端跳转都是用这种方式。

首先通过 HttpServletRequest 接口提供的 getRequestDispatcher(String path)方法返回一个 RequestDispatcher 对象，其中参数 path 是要跳转的目标路径。该对象提供了两个方法分别用来替代 JSP 的 forward 和 include 动作，方法如下。

```
forward(ServletRequest reqeust,ServletResponse response)
include(ServletRequest reqeust,ServletResponse response)
```

具体使用方式如下。

```
//服务端跳转
request.getRequestDispatcher("success.jsp").forward(request,response);
//包含页面
request.getRequestDispatcher("menu.html").include(request,response);
```

一旦执行到 RequestDispatcher 对象的 forward() 方法，就会立即进行服务器端跳转。在实际应用中，无论是请求转发还是重定向，都并不是表面意义上的立即进行跳转。它后面的代码仍很有可能会执行。如果后面还有请求转发或者重定向的语句被执行，就会抛出 java.lang.IllegalStateException: Cannot forward after response has been committed 异常。解决办法很简单，在跳转语句后面紧跟着 return 语句以结束此方法，不让后面的语句执行。

请求转发与请求包含的区别主要体现在下面几个方面。

(1) 如果在 Servlet A 中请求转发到 Servlet B，那么在 Servlet A 中就不允许再输出响应体，即不能再使用 response.getWriter()和 response.getOutputStream()向客户端输出，这一工作应该由 Servlet B 来完成；如果是使用请求包含，那么没有这个限制。

(2) 请求转发虽然不能输出响应体，但可以设置响应头，例如：

```
response.setContentType("text/html;charset = utf-8");
```

(3) 请求包含大多是应用在 JSP 页面中，完成多页面的合并。

(4) 请求转发大多是应用在 Servlet 中，转发目标大多是 JSP 页面。

3.3.5 任务：注册页面请求信息获取

1. 需求说明

(1) 编写注册页面。

(2) 通过表单提交注册信息。

(3) 在提交页面中获取表单提交的数据。

(4) 将获取的数据输出显示。

编程实现用户的注册功能，注册信息包括：用户名、密码、爱好。页面提交后，显示输入的数据。

2. 任务实施

(1) 注册页面 reg.jsp 的代码如下。

```
<%@ page language="java" import="java.util.*" pageEncoding="GBK"%>
<html>
  <head><title>用户注册</title></head>
    <body>
      <p>用户注册</p>
        <form name="form1" method="post" action="getinfo.jsp">
      <table>
        <tr><td>注册名</td><td><input type="text" name="id"/></td></tr>
        <tr><td>密码</td><td><input type="password" name="pwd"/></td></tr>
        <tr><td>性别</td><td><input type="radio" name="sex"/>男</td>
          <td><input type="radio" name="sex"/>女</td>
        </tr>
        <tr><td>爱好</td>
          <td><input type="checkBox" name="inte" value="1" />书法
            <input type="checkBox" value="2" name="inte"/>体育
            <input type="checkBox" value="3" name="inte"/>唱歌
          </td>
        <tr>
      <tr><td>粉丝</td>
        <td><select name="fav" multiple>
            <option value="1">明星 1</option>
            <option value="2">明星 2</option>
            <option value="1">明星 3</option>
            <option value="2">明星 4</option>
          </select>
        </td>
      </tr>
        <%
          int i;
          for (i=0;i<4;i++){
         %>
          <tr><td>联系电话<%=i+1 %></td><td><input type="text" name="tel"/>
</td></tr>
        <%}%>
      <tr><td><input type="submit" value="提交"/></td><td><input type="reset"
                          value="重置"/></td></tr>
    </table>
  </form>
  </body>
</html>
```

(2) 信息读取显示页面 getinfo.jsp 的代码如下。

```
<%@ page language="java" import="java.util.*" pageEncoding="GBK"%>
<html>
```

```
<head>
  <title>用户注册信息</title>
</head>
<body>
  <%
    request.setCharacterEncoding("GBK");
    String ID = request.getParameter("id");
    String pwd = request.getParameter("pwd");
    String sex = request.getParameter("sex");
    if (sex == "1") sex = "男";else sex = "女";
    String[] inter = request.getParameterValues("inte");
    String[] fav = request.getParameterValues("fav");
    String[] tel = request.getParameterValues("tel");
  %>
  用户名:<% = ID %><br>
  密码:<% = pwd %><br>
  性别:<% = sex %><br>
  爱好:
  <%
    for (int i = 0;i < inter.length;i++) {
  %>
  <% = inter[i] %>
  <% } %>
  <br> 明星:
  <%
    for (int i = 0;i < fav.length;i++) {
  %>
  <% = fav[i] %>
  <% } %>
  <br> tel:
  <%
    for (int i = 0;i < tel.length;i++){
  %>
  <% = tel[i] %>
  <% } %>
</body>
</html>
```

3.4 response 对象

response 对象主要将 JSP 处理请求后的结果传回到客户端，用于响应客户请求并向客户端输出信息。服务器端的任何输出都通过 response 对象发送到客户端浏览器。response 与 reuqest 是一对相对应的内置对象，request 对象用于得到用户提交的信息，而 response 对象是向用户发送信息，两者结合起来完成动态页面的交互功能，其响应过程如图 3-3 所示。

response 对象的功能分为以下 4 种。

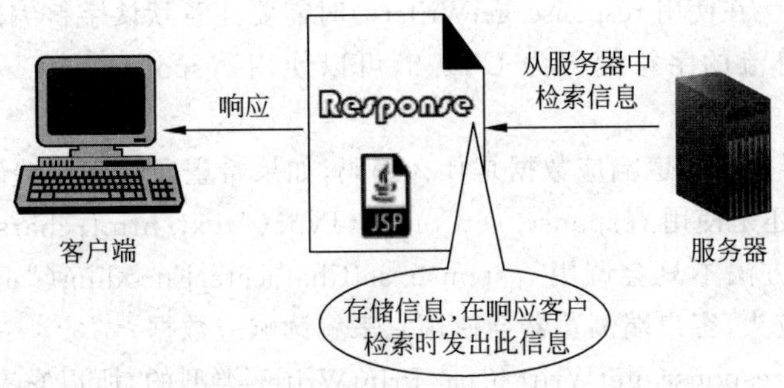

图 3-3　response 响应请求流程

(1) 设置响应头信息。

(2) 发送状态码。

(3) 设置响应正文。

(4) 页面重定向。

respose 对象的基类是 javax. servlet. ServletResponse。如果协议是 HTTP,则 response 对象的基类是 javax. servlet. HttpServletResponse。response 对象的常用方法如表 3-3 所示。

表 3-3　response 对象的常用方法

方　法　名	描　　述
void clear()	清空暂存在缓冲区的输出
void addCookie(Cookie cookie)	设置 Cookie
OutputStream getOutputStream()	返回服务器输出流。可以通过该输出流输出二进制信息
void sendRedirect(String ur)	使本页面重定向到另一个页面
void setContentType(String contentType)	设置文档类型。HTML 的文档类型为 text/html
PrintWriter getOut()	返回 out 对象
void setHeader(String name,String value)	设置响应头信息
void setStatus(int status)	设置响应状态码

3.4.1　响应正文

response 对象向客户端输出响应正文(响应体)时使用响应流。JSP 提供了两个响应流对象。

(1) PrintWriter out = response. getWriter(): 获取字符流。

(2) ServletOutputStream out = response. getOutputStream(): 获取字节流。

如果响应正文内容为字符,应使用 response. getWriter(); 如果响应内容为字节(如下载时)应使用 response. getOutputStream()。

注意: 在一个请求中不能同时使用这两个响应流。也就是说,要么使用 repsonse . getWriter(),要么使用 response. getOutputStream(),否则会抛出 IllegalStateException 异常。

(1) 字符编码。在使用 response. getWriter()时需要注意默认字符编码为 ISO-8859-1,如果希望设置字符流的字符编码为 UTF-8,可以使用 response. setCharaceterEncoding("utf-8")来设置。

客户端浏览器并不知道响应数据是什么编码,如果希望通知客户端使用 UTF-8 来解读响应数据,那么还是使用 response. setContentType("text/html;charset=utf-8")方法比较好,因为这个方法不只会调用 response. setCharaceterEncoding("utf-8"),还会设置 Content-Type 响应头,客户端浏览器会使用它来解读响应数据。

(2) 缓冲区。response. getWriter()是 PrintWriter 类型的,所以它有缓冲区,缓冲区的默认大小为 8KB。也就是说,在响应数据没有输出 8KB 之前,数据都是存放在缓冲区中,而不会立刻发送到客户端。当 Servlet 执行结束后,服务器才会去刷新流,使缓冲区中的数据发送到客户端。

如果希望响应数据马上发送给客户端,可以采用下面的方法。

(1) 向流中写入大于 8KB 的数据。

(2) 调用 response. flushBuffer()方法刷新缓冲区。

3.4.2 设置响应头信息

HTTP 采用请求/响应模型,客户端向服务器发送一个请求,请求头包含请求的方法、URI、协议版本以及包含请求修饰符、客户信息和内容的类似于 MIME 的消息结构。服务器以一个状态行作为响应,响应的内容包括消息协议的版本,成功或者错误编码加上包含服务器信息、实体信息以及可能的实体内容。

通常 HTTP 消息包括客户端向服务器的请求消息和服务器向客户端的响应消息,这两种类型的消息由一个起始行、一个或者多个头域、一个只是头域结束的空行和可选的消息体组成。HTTP 的头域包括通用头、请求头、响应头和实体头 4 个部分。每个头域由一个域名、冒号(:)和域值三部分组成。域名是大小写无关的,域值前可以添加任何数量的空格符,头域可以被扩展为多行,在每行开始处使用至少一个空格或制表符。

1. 常见 HTTP 文件头及其含义

1) 通用头域

通用头域包含请求和响应消息都支持的头域。通用头域包含 Cache-Control、Connection、Date、Pragma、Transfer-Encoding、Upgrade、Via,是既能用于请求消息也能用于响应信息,但与被传输的实体内容没有关系的信息头。

(1) Cache-Control。Cache-Control 指定请求和响应遵循的缓存机制。在请求消息或响应消息中设置 Cache-Control 并不会修改另一个消息处理过程中的缓存处理过程。请求时的缓存指令包括 no-cache、no-store、max-age、max-stale、min-fresh、only-if-cached,响应消息中的指令包括 public、private、no-cache、no-store、no-transform、must-revalidate、proxy-revalidate、max-age。各个消息中的指令含义如下。

no-cache:不要缓存的实体,要求现在从 Web 服务器去取。

max-age：只接收小于 max-age 的 Age 值，并且没有过期的对象。

max-stale：可以接收过去的对象，但是过期时间必须小于 max-stale。

min-fresh：接收其新鲜生命期大于其当前 Age 跟 min-fresh 值之和的缓存对象。

public：可以用 Cached 内容回应任何用户。

private：只能用缓存内容回应先前请求该内容的那个用户。

no-cache：可以缓存，但是只有在跟 Web 服务器验证了其有效后，才能返回给客户端。

max-age：本响应包含的对象的过期时间。

ALL：no-store：不允许缓存。

(2) Connection。它用于表明是否保存 Socket 连接为开放的。

① 请求。

close：告诉 Web 服务器或者代理服务器，在完成本次请求的响应后，断开连接，无须等待。

keep-alive：告诉 Web 服务器或者代理服务器，在完成本次请求的响应后，保持连接，等待后续请求。

② 响应。

close：连接已经关闭。

keepalive：保持连接，等待本次连接的后续请求。

keep-alive：如果浏览器请求保持连接，则表明希望 Web 服务器保持连接多长时间(秒)，如 keep-alive：300。

(3) Pragma。Pragma 用来包含实现特定的指令，最常用的是 pragma:no-cache。在 HTTP/1.1 中，它的含义和 Cache- Control:no-cache 相同。

(4) Transfer-Encoding。Web 服务器表明自己对本响应消息体(不是消息体里面的对象)做了怎样的编码，比如是否分块(chunked)。

(5) Upgrade。允许服务器指定一种新的协议或者新的协议版本，与响应编码 101(切换协议)配合使用，如 Upgrade：HTTP/2.0。

(6) Date。Date 表示消息发送的时间，时间的描述格式由 RFC822 定义，如 Date：Mon,31 Dec 2001 04:25:57 GMT。

Date 描述的时间表示世界标准时间，若要换算成本地时间需要知道用户所在的时区。

2) 请求头

请求头(请求消息的第一行)的格式如下。

```
Method SP Request-URI SP HTTP-Version CRLF
```

(1) Method 表示对 Request URI 完成的方法。这个字段是大小写敏感的，包括 options、get、head、post、put、delete、trace。方法 get 和 head 应该被所有的通用 Web 服务器支持，其他所有方法的实现是可选的。get 方法取回由 Request-URI 标识的信息。head 方法也是取回由 Request-URI 标识的信息，只是可以在响应时不返回消息体。post

方法可以请求服务器接收包含在请求中的实体信息，可以用于提交表单，向新闻组、BBS、邮件群组和数据库发送消息。

(2) SP 表示空格。

(3) Request-URI 遵循 URI 格式，此字段为星号（*）时，说明请求并不用于某个特定的资源地址，而是用于服务器本身。

(4) HTTP-Version 表示支持的 HTTP 版本，例如为 HTTP/1.1。

(5) CRLF 表示换行回车符。

请求头域允许客户端向服务器传递关于请求或者关于客户机的附加信息。请求头域可能包含下列域：Accept、Accept-Charset、Accept-Encoding、Accept-Language、Authorization、From、Host、If-Modified-Since、If-Match、If-None-Match、If-Range、If-Range、If-Unmodified-Since、Max-Forwards、Proxy-Authorization、Range、Referer、User-Agent。对请求头域的扩展要求通信双方都支持，如果存在不支持的请求头域，一般将会作为实体头域处理。

典型的请求消息如下。

```
GET http://download.microtool.de:80/somedata.exe
Host: download.microtool.de
Accept: */*
Pragma: no-cache
Cache-Control: no-cache
Referer: http://download.microtool.de/
User-Agent:Mozilla/4.04[en](Win95;I;Nav)
Range:bytes = 554554-
```

(1) Accept 定义客户端可以处理的媒体类型，按优先级排序；在一个以逗号为分隔的列表中，可以定义多种类型和使用通配符。*/* 表示任何类型，type/* 表示该类型下的所有子类型，例如，Accept: image/jpeg,image/png,*/* Accept-Charset 定义客户端可以处理的字符集，按优先级排序；在一个以逗号为分隔的列表中，可以定义多种类型和使用通配符。

(2) Accept-Encoding 定义客户端可以理解的编码机制，通常指定压缩方法，是否支持压缩，支持什么压缩方法(GZip、Deflate)，如 Accept-Encoding:gzip,compress。

(3) Accept-Language 定义客户端可以接收的自然语言列表，如 Accept-Language: en,de。

(4) Host 指定请求资源的 Intenet 主机和端口号，必须表示请求 URI 的原始服务器或网关的位置。HTTP/1.1 请求必须包含主机头域，否则系统会返回 400 状态码。

(5) Referer 允许客户端指定请求 URI 的源资源地址，这可以允许服务器生成回退链表，可用来登录、优化缓存等；也允许废除的或错误的连接由于维护的目的被追踪。如果请求的 URI 没有自己的地址，Referer 不能被发送。如果指定的是部分 URI 地址，则此地址应该是一个相对地址。

(6) Range 可以请求实体的一个或者多个子范围。例如：

表示第 1 个 500 字节——bytes=0-499；

表示第 2 个 500 字节——bytes=500-999；

表示最后一个 500 字节——bytes=-500；

表示 500 字节以后的范围——bytes=500-；

第一个和最后一个字节——bytes=0-0,-1；

同时指定几个范围——bytes=500-600,601-999。

但是服务器可以忽略此请求头域，如果无条件 get 请求头包含 Range，响应会以状态码 20(PartialContent)返回而不是以 200(OK)返回。

(7) User-Agent 的内容包含发出请求的用户信息。

3) 响应头

响应消息的第一行为下面的格式。

```
HTTP-Version SP Status-Code SP Reason-Phrase CRLF
```

HTTP-Version 表示支持的 HTTP 版本，如 HTTP/1.1。

Status- Code 是一个三个数字的结果代码。

Reason-Phrase 给 Status-Code 提供一个简单的文本描述。Status-Code 主要用于机器自动识别，Reason-Phrase 主要用于帮助用户理解。Status-Code 的第一个数字定义响应的类别，后两个数字没有分类的作用。第一个数字可能取以下 5 个不同的值。

1××：信息响应类，表示接收到请求并且继续处理。

2××：处理成功响应类，表示动作被成功接收、理解和接受。

3××：重定向响应类，为了完成指定的动作，必须接受进一步处理。

4××：客户端错误，客户请求包含语法错误或者是不能正确执行。

5××：服务端错误，服务器不能正确执行一个正确的请求。

响应头域允许服务器传递不能放在状态行的附加信息，这些域主要描述服务器的信息和 Request-URI 进一步的信息。响应头域包含 Age、Location、Proxy-Authenticate、Public、Retry- After、Server、Vary、Warning、WWW-Authenticate。对响应头域的扩展要求通信双方都支持，如果存在不支持的响应头域，一般将会作为实体头域处理。

典型的响应消息如下。

```
HTTP/1.0200OK
Date:Mon,31Dec200104:25:57GMT
Server:Apache/1.3.14(UNIX)
Content-type:text/html
Last-modified:Tue,17Apr200106:46:28GMT
Etag:"a030f020ac7c01:1e9f"
Content-length:39725426
Content-range:bytes554554-40279979/40279980
```

(1) Location 表示客户应当到哪里去提取文档。Location 通常不是直接设置的，而是通过 HttpServletResponse 的 sendRedirect()方法，该方法同时设置状态代码为 302，如 Location：http://mypt.edu.cn/dy/sinahome0.gif。

(2) Server 标明 Web 服务器软件及其版本号的头标，如 Server：Apache/2.0.61

(UNIX)。

(3) WWW-Authenticate(验证头)是一个提示用户代理提供用户名和密码的响应头标,与状态编码 401(未授权)配合使用,如 response.setHeader("WWW-Authenticate","BASIC realm=\"executives\"")。

4) 实体头

请求消息和响应消息都可以包含实体信息,实体信息一般由实体头域和实体组成。实体头域包含关于实体的原信息,实体头域有 Allow、Content-Base、Content-Encoding、Content-Language、Content-Length、Content-Location、Content-MD5、Content-Range、Content-Type、Etag、Expires、Last-Modified、Extension-Header。Extension-Header 允许客户端定义新的实体头,但是这些域可能无法被接收方识别。实体可以是一个经过编码的字节流,它的编码方式由 Content-Encoding 或 Content-Type 定义,它的长度由 Content-Length 或 Content-Range 定义。

(1) Content-Encoding 是文档的编码(Encode)方法,只有在解码之后才可以得到 Content-Type 头指定的内容类型。利用 gzip 压缩文档能够显著地减少 HTML 文档的下载时间。Java 的 GZIPOutputStream 可以很方便地进行 gzip 压缩。因此,Servlet 应该通过查看 Accept-Encoding 头(request.getHeader("Accept-Encoding"))检查浏览器是否支持 gzip,为支持 gzip 的浏览器返回经 gzip 压缩的 HTML 页面,为其他浏览器返回普通页面,如 Content-Encoding:zip。

(2) Content-Language 用于指定在输入流中数据的自然语言类型,如 Content-Language:en。

(3) Content-Length 表示内容的长度,只有当浏览器使用持久 HTTP 连接时才需要这个数据。如果想要利用持久连接的优势,可以把输出文档写入 ByteArrayOutputStram,完成后查看其大小,然后把该值放入 Content-Length 头,最后通过 byteArrayStream.writeTo(response.getOutputStream())发送内容。

(4) Content-Location 指定包含于请求或响应中的资源定位,如果是绝对地址,则它作为被解析实体的相对 URL 的出发点,如 Content-Location: http://www.myweb.com/news。

(5) Content-MD5 是实体的一种 MD5 摘要,用作校验和。发送方和接收方都计算 MD5 摘要,接收方将其计算的值与此头中传递的值进行比较,如 Content-MD5:<base64 of 128 MD5 digest>。

(6) Content-Type 表示后面的文档属于哪种 MIME 类型。Servlet 默认为 text/plain,但通常需要显式地指定为 text/html。由于经常要设置 Content-Type,因此 HttpServletResponse 提供了一个专用的方法 setContentTyep()。

(7) Content-Range 用于指定整个实体中的一部分的插入位置,它也指示了整个实体的长度。在服务器向客户返回一个部分响应时,它必须描述响应覆盖的范围和整个实体长度。其一般格式如下。

```
Content-Range:bytes-unitSPfirst-byte-pos-last-byte-pos/entity-legth
```

例如,传送头500个字节的形式如下。

```
Content-Range:bytes0-499/1234
```

如果一个HTTP消息包含此节(如对范围请求的响应或对一系列范围的重叠请求),Content-Range表示传送的范围,Content-Length表示实际传送的字节数。

(8) Last-modified描述文档的最后改动时间。客户可以通过If-Modified-Since提供一个日期,该请求将被视为一个条件,只有改动时间迟于指定时间的文档才会返回,否则返回304(Not Modified)状态。Last-Modified也可用setDateHeader()方法来设置。

(9) Expires用于设置应该在什么时候认为文档已经过期,从而不再缓存它。

(10) Allow用于设置服务器支持哪些请求方法(如get、post等)。

5) 扩展头

(1) Set-Cookie用于设置和页面关联的Cookie。Servlet不应使用response.setHeader("Set-Cookie",...),而应使用HttpServletResponse提供的专用方法addCookie()。

(2) Refresh表示浏览器应该在多少时间之后刷新文档,以秒计。除刷新当前文档外,还可以通过setHeader("Refresh","5; URL=http://host/path")让浏览器读取指定的页面。

注意:这种功能通常是通过设置HTML页面中的< META HTTP-EQUIV="Refresh" CONTENT="5;URL=http://host/path">实现,这是因为自动刷新或重定向对于那些不能使用CGI或Servlet的HTML编写者十分重要。但是,对于Servlet来说,直接设置Refresh头更加方便。

Refresh的意义是"N秒之后刷新本页面或访问指定页面",而不是"每隔N秒刷新本页面或访问指定页面"。因此,连续刷新要求每次都发送一个Refresh头,而发送204状态码则可以阻止浏览器继续刷新,不管是使用Refresh头还是< META HTTP-EQUIV="Refresh" ...>。

Refresh头不属于HTTP 1.1正式标准的一部分,而是一个扩展,但多数浏览器都支持它。

设置响应头中的某个项的值时,如果响应头中没有这个项,则添加一个,有则修改已有项的值。

(3) Content-Disposition:当Content-Type的类型为要下载的类型时,这个头域会告诉浏览器这个文件的名字和类型。例如:

```
<%@ page pageEncoding="GBK" contentType="text/html;charset=utf-8" import="java
     .util.*,java.text.*" %>
...
<%
  response.setHeader("Content-Type","video/x-msvideo");
  response.setHeader("Content-Disposition","attachment;filename=aaa.doc");
%>
```

Content-Disposition中指定的类型是文件的扩展名,并且弹出的下载对话框中的文件类型图片是按照文件的扩展名显示的，单击“保存”后，文件以 filename 的值命名，保存类型以 Content 中设置的为准。

注意：在设置 Content-Disposition 头域之前，一定要设置 Content-Type 头域。

(4) Authorization：Authorization 的作用是当客户端访问受密码保护时，服务器端会发送 401 状态码和 WWW-Authenticate 响应头，要求客户机使用 Authorization 来应答。例如：

```
<%@ page pageEncoding = "GBK" contentType = "text/html; charset = utf - 8" import = "java
    .util. * , java. text. * " %>
<% = DateFormat. getDateTimeInstance(DateFormat. SHORT, DateFormat. SHORT, Locale. CHINA).
format(new Date()) %>
<%
  response. setStatus(401);
  response. setHeader("WWW - Authenticate","Basic realm = \"Tomcat Manager Application\"");
%>
```

2. 设置 HTTP 头

HTTP 头一般用来设置网页的基本属性，设置 HTTP 文件头最常用的方法是 response 对象的 setHeader()方法。该方法有两个参数，分别表示 HTTP 文件头的名字和值。设置 HTTP 文件头最常用的方法如表 3-4 所示。例如：

```
response. setHeader("Pragma","No - cache");
response. setHeader("Cache - Control","no - cache");
response. setDateHeader("Expires",0);
```

以上都表示在客户端缓存中不保存页面的备份。

表 3-4　设置 HTTP 文件头的常用方法

方　法	描　述
void addCookie(Cookie cookie)	新增 Cookie 到响应头
void addDateHeader(String name,long date)	新增 long 类型的值到响应头
void addHeader(String name,String value)	新增 String 类型的值到响应头，name 用于指定 HTTP 响应头的类型
void addIntHeader(String name,int value)	新增 int 类型的值到响应头
void setDateHeader(String name,long date)	以指定名称和指定 long 类型的值设置响应头
void setHeader(String name,String value)	以指定名称和指定 String 类型的值设置响应头
void setIntHeader(String name,int value)	以指定名称和指定 int 类型的值设置响应头
void setContentType(String MIME)	用来设置 Content-Type 文件头
void setContentLength	用来设置 Content-Length 头

【例 3-13】 JSP 页面禁止缓存设置。

(1) 在客户端缓存<head>中加入类似如下内容。

```
<META HTTP - EQUIV = "pragma" CONTENT = "no - cache">
<META HTTP - EQUIV = "Cache - Control" CONTENT = "no - cache, must - revalidate">
```

```
<META HTTP-EQUIV="expires" CONTENT="Wed, 26 Feb 1997 08:21:57 GMT">
```

或

```
<meta http-equiv="pragma" content="no-cache">
<meta http-equiv="cache-control" content="no-cache">
<meta http-equiv="expires" content="0">
```

(2) 在服务器的动态网页中禁止缓存,加入类似如下的代码。

```
response.setHeader("Pragma","No-cache");
response.setHeader("Cache-Control","no-cache");
response.setDateHeader("Expires", 0);
```

(3) 设置有限时间的缓存。

```
int minutes = 10;
Date d = new Date();
String modDate = d.toGMTString();
String expDate = null;
expDate = (new Date(d.getTime() + minutes * 60000)).toGMTString();
response.setHeader("Last-Modified", modDate);
response.setHeader("Expires", expDate);
response.setHeader("Cache-Control", "public"); //    HTTP/1.1
response.setHeader("Pragma", "Pragma"); //    HTTP/1.0
```

(4) 如果以上方法都不行,就在正常的 URL 后面加上一个后缀。JavaScript 代码如下。

```
var timestamp = (new Date()).valueOf();
URL + "&timestamp = " + timestamp;
```

Java 代码如下。

```
long timestamp = new Date().getTime();
URL + "&timestamp = " + timestamp;
```

这样的话,URL 始终都在变化,自然浏览器就可以进行更新了。

3. 设定 contentType 属性

contentType 属性用来设置 JSP 页面的 MIME 类型和字符编码集,取值格式为"MIME 类型"或"MIME 类型;charset=字符编码集"。JSP 引擎根据 contentType 属性,对用户的请求做出响应。

客户端收到响应后,根据 contentType 的值对信息做对应的处理。

可以调用 response 对象的 setContentType(String s)方法动态修改 contentType 属性值,JSP 引擎会按照修改后的 MIME 类型来响应客户浏览器。

3.4.3 状态行

1. 状态行的作用

当服务器在处理页面时,可能成功,也可能发生错误,这时服务器通过响应状态码来通知客户端，状态码包含在响应的状态行中，状态行包括：状态码协议/版本文字解释。

一共有下面这五大类状态码。

1××：主要是实验性质，例如 101 表示服务器正在升级。

2××：成功，例如 200 表示请求成功。

3××：重定向，要完成请求必须进行更进一步的操作，例如 305 表示必须通过代理来访问。

4××：客户端错误，请求有语法错误或请求无法实现，例如 404 表示请求的资源不存在。

5××：服务器端错误，服务器未能实现合法的请求，例如 500 表示服务器内部发生错误，不能提供服务。

2. 设置状态码

程序中可以使用 response 对象的 setStatus()方法来设置状态码，在一般情况下，不需要在程序中设置状态码，当页面出现问题时，服务器会自动响应，并发送相应的状态码提示客户。response 对象的常用方法如表 3-5 所示。

表 3-5 response 对象的常用方法

方　法	描　述
void sendError(int sc)	使用指定的状态码向客户端返回一个错误响应
void sendError(int sc,String msg)	使用指定的状态码和状态描述向客户端返回一个错误响应
void setStatus(int sc)	设置 HTTP 响应的状态行，以指定的状态码将响应返回给客户端

例如，response. setStatus(501)方法设置状态码为 501 的出错信息，返回该出错页面给客户。

如果状态码为出错码，则页面中 response. setStatus()后面的代码将不被执行。

【例 3-14】 状态行的应用(4 个页面)。

statusCode. htm 页面的代码如下。

```
<html>
  <head><title>response 对象状态行应用案例</title></head>
  <body>
    <h2>显示不同的状态行</h2>
    <hr>
    <a href="statusCode200.jsp">200 请求成功</a><br>
    <a href="statusCode404.jsp">404 请求资源不可用</a><br>
    <a href="statusCode501.jsp">501 不支持请求的部分功能</a><br>
    </font>
  </body>
</html>
```

statusCode200. jsp 页面的代码如下。

```
<%@ page contentType="text/html;charset=gb2312" %>
<html>
  <head><title>response 对象状态行应用案例</title></head>
  <body>
   200 请求成功<br>
   <%
     response.setStatus(200);
     out.println("一切正常");
   %>
  </body>
</html>
```

statusCode404.jsp 页面的代码如下。

```
<%@ page contentType="text/html;charset=gb2312" %>
<html>
  <head><title>response 对象状态行应用案例</title></head>
  <body>
   <h3>404 请求资源不可用信息</3><br><br>
   <%
     response.setStatus(404);
     out.println("不能显示");
   %>
  </body>
</html>
```

statusCode501.jsp 页面的代码如下。

```
<%@ page contentType="text/html;charset=gb2312" %>
<html>
  <head><title>response 对象状态行应用案例</title></head>
  <body>
   <h3>501 不支持请求的部分功能</h3><br><br>
   <%
     response.setStatus(501);
     out.println("不能显示");
   %>
  </body>
</html>
```

3.4.4 重定向

1. 什么是重定向

当访问 http://www.sun.com 时，浏览器地址栏中的 URL 会变成 http://www.oracle.com，这就是重定向了。重定向是服务器通知浏览器去访问另一个地址，即再发出另一个请求。

当服务器响应客户端请求时，将客户端请求重新引导到另一页面，称为重定向。可以使用 response 对象的 sendRedirect(String location)方法实现重定向。

2. 重定向

响应码为 200 表示响应成功，而响应码为 302 表示重定向。完成重定向的第一步就是设置响应码为 302，因为重定向是通知浏览器的第二个请求，所以浏览器需要知道第二个请求的 URL，所以完成重定向的第二步是设置 Location 头，指定第二个请求的 URL 地址。例如：

```
response.setStatus(302);
response.setHeader("Location", "http://www.baidu.com");
```

该代码的作用是：当访问页面或 Servlet 后，会通知浏览器重定向到 http://www.baidu.com。客户端浏览器解析到响应码为 302 后，就知道服务器让它重定向，所以它会马上获取响应头 Location，然后发出第二个请求。

也可以采用便捷的重定向方式，将请求重新定位到一个不同的 URL，即页面重定向。重定向可以采用如下方法：

```
response.sendRedirect(目标页面路径);
```

或

```
<jsp:forward page = ""></jsp:forward>
```

【例 3-15】 根据业务逻辑实现页面跳转。

```
<%
request.setCharacterEncoding("GBK");
String name = request.getParameter("userName");
String pwd = request.getParameter("pwd");
if(name.equals("sa") && pwd.equals("sa"))
  response.sendRedirect("welcome.jsp"); //跳转至欢迎页面
%>
```

response.sendRedirect()方法会设置响应头为 302，以设置 Location 响应头。如果要重定向的 URL 不在同一个服务器内，那么可以使用绝对地址，例如：

```
response.sendRedirect("http://www.sina.com")
```

重定向有以下特点。

(1) 重定向是两次请求。

(2) 重定向的 URL 可以是其他应用，不局限于当前应用。

(3) 重定向的响应头为 302，并且必须要有 Location 响应头。

(4) 重定向后不要再使用 response.getWriter()或 response.getOutputStream()输出数据，不然可能会出现异常。

3. 跳转(转发)

为了实现在多个页面交互过程中实现请求数据的共享，可以采用 requestDispatcher

对象的 forward()方法。例如：

```
<%
  requestDispatcher rd = request.getRequestDispatcher("welcome.jsp");
  rd.forward(request,response);   //将当前接收的用户请求,发送给服务器的其他资源使用
%>
```

4. 重定向和跳转的区别

重定向和跳转主要有以下区别。

(1) 跳转是在服务器端发挥作用,通过 forward()方法将客户端提交的信息在多个页面间进行传递。

(2) 跳转是服务器内部控制权的转移,直接跳转到目标页面,其后的代码不再执行。跳转到目标页面后,浏览器地址栏中的 URL 不会改变。在服务器端进行重定向,无法跳转到其他服务器上的页面。

(3) 重定向是在客户端发挥作用,通过请求新的地址实现页面转向。

(4) 重定向执行完页面的所有代码后,再通过浏览器重新请求地址跳转到目标页面。跳转到目标页面后,浏览器地址栏中的 URL 会改变。

(5) 在浏览器端进行重定向,可以跳转到其他服务器上的页面。

5. out 对象和 response 对象的区别

(1) out 对象的基类是 javax. servlet. jsp. JspWriter。

respose 对象的基类是 javax. servlet. ServletResponse；如果协议是 HTTP,则 response 对象的基类是 javax. servlet. HttpServletResponse。

(2) 两者对缓存区操作的比较如表 3-6 所示。

表 3-6 out 对象和 response 对象缓存区操作的比较

功　能	out	response
输出缓存区里的数据	out. flush()	response. flushBuffer()
清除缓存区里的数据	out. clear() 或 out. clearBuffer()	response. reset()
设置缓存区的大小	out 对象不能设置	response. setBufferSize(int size)
获得缓存区的大小	out. getBufferSize()或 out. getRemaining()	response. getBufferSize()

(3) 在 JSP 中向客户端输出数据时,经常使用 out 对象,而很少直接使用 response 对象。

因为使用 out 对象比较方便,可以使用 out 对象来替代 response 对象。注意 out 对象与 response 对象在输出数据这方面本质上是一样的。

3.4.5 输出缓存

缓存可以更加有效地在服务器与客户之间传输内容。HttpServletResponse 对象为支持 jspWriter 对象而启用了缓存区配置。getBufferSize()方法返回用于 JSP 页面的当

前缓存区容量。setBufferSize()方法允许 JSP 页面为响应的主体设置一个首选的输出缓存区容量。容器使用的实际缓存区容量至少要等于输出缓存区的容量。如果设置缓存区容量,则必须在向响应中写入内容之前,否则 JSP 容器将抛出 java.lang.IllegalStateException 异常。

response 对象中用于缓存区的方法如表 3-7 所示。

表 3-7 response 对象中用于缓存区的方法

方 法	描 述
public void flushBuffer() throws IOException	强制把缓存区中的内容发送给客户
public void getBufferSize()	返回响应所使用的实际缓冲区大小,如果没使用缓存区,则该方法返回 0
public void setBufferSize(int size)	为响应的主体设置首选的缓存区大小
public boolean isCommitted()	返回一个 boolean,表示响应是否已经提交;提交的响应已经写入状态码和报头
public void reset()	清除缓存区存在的任何数据,同时清除状态码和报

下面的代码说明缓存区的使用方式。

```
<%
  out.print("缓冲区大小:" + response.getBufferSize() + "<br><br>");
  out.print("缓冲区设置之前<br>");
  out.print("输出的内容是否提交:" + response.isCommitted() + "<br><br>");
  response.flushBuffer();
  out.print("缓冲区设置之后<br>");
  out.print("输出的内容是否提交:" + response.isCommitted() + "<br><br>");
%>
```

3.4.6 任务:用户注册

1. 需求说明

编程实现用户的注册功能,注册信息包括:用户名、密码、爱好,页面提交后显示输入的数据。主要包括如下内容。

(1) 编写注册页面。

(2) 通过表单提交注册信息。

(3) 在提交页面中获取表单提交数据。

(4) 将获取的数据输出显示。

2. 任务实施

(1) 注册页面 reg.jsp 的代码如下。

```
<%@ page language="java" import="java.util.*" pageEncoding="GBK" %>
<html>
  <head><title>用户注册</title></head>
```

```
    <body>
      <form name = "form1" method = "post" action = "getinfo.jsp">
        <p>用户注册</p>
      <table>
      <tr><td>注册名</td><td><input type = "text" name = "id"/></td></tr>
      <tr><td>密码</td><td><input type = "password" name = "pwd"/></td></tr>
        <tr>
          <td>性别</td><td><input type = "radio" name = "sex"/>男</td>
          <td><input type = "radio" name = "sex"/>女</td>
      </tr>
      <tr>
        <td>爱好</td>
        <td><input type = "checkBox" name = "inte" value = "1" />书法
              <input type = "checkBox" value = "2" name = "inte"/>体育
              <input type = "checkBox" value = "3" name = "inte"/>唱歌
        </td>
      </tr>
      <tr>
        <td>喜欢的明星</td>
        <td>
        <select name = "fav" multiple>
          <option value = "1">明星 1 </option>
          <option value = "2">明星 2 </option>
          <option value = "1">明星 3 </option>
          <option value = "2">明星 4 </option>
          </select>
            </td>
        </tr>
          <% for (int i = 0;i < 4;i++) { %>
            <tr>
              <td>联系电话<% = i + 1 %></td>
              <td><input type = "text" name = "tel"/> </td>
            </tr>
          <% } %>
        <tr>
            <td><input type = "submit" value = "提交"/></td>
            <td><input type = "reset" value = "重置"/></td>
        </tr>
        </table>
      </form>
  </body>
</html>
```

(2) 信息读取显示页面 getinfo.jsp 的代码如下。

```
<%@ page language = "java" import = "java.util.*" pageEncoding = "GBK" %>
<html>
  <head><title>用户注册信息</title></head>
    <body>
      <%
```

```
        request.setCharacterEncoding("GBK");
        String ID = request.getParameter("id");
        String pwd = request.getParameter("pwd");
        String sex = request.getParameter("sex");
        if (sex == "1") sex = "男"; else sex = "女";
        String[] inter = request.getParameterValues("inte");
        String[] fav = request.getParameterValues("fav");
        String[] tel = request.getParameterValues("tel");
      %>
      <p>用户名:<% = ID %> 密码:<% = pwd %> 性别:<% = sex %></p>
      爱好:
      <% for (int i = 0;i < inter.length;i++){ %>
        <% = inter[i] %>
      <% } %>
      <br> 喜欢的明星:
    <% for (int i = 0;i < fav.length;i++){ %>
        <% = fav[i] %>
    <% } %>
    <br> tel:
      <% for (int i = 0;i < tel.length;i++) { %>
        <% = tel[i] %>
      <% } %>
    </body>
</html>
```

3.5 Cookie 的原理及应用

在程序中,会话跟踪是很重要的事情。理论上,一个用户的所有请求操作都应该属于同一个会话,而另一个用户的所有请求操作则应该属于另一个会话,两者不能混淆。例如,用户 A 在超市购买的任何商品都应该放在 A 的购物车内,不论是用户 A 什么时间购买的,这都是属于同一个会话的,不能放入用户 B 或用户 C 的购物车内,因为这不属于同一个会话。

而 Web 应用程序是使用 HTTP 传输数据的,HTTP 是无状态的协议。一旦数据交换完毕,客户端与服务器端的连接就会关闭,再次交换数据需要建立新的连接,这就意味着服务器无法从连接上跟踪会话。即用户 A 购买了一件商品放入购物车内,当再次购买商品时服务器已经无法判断该购买行为是属于用户 A 的会话还是用户 B 的会话了。要跟踪该会话,必须引入一种机制。

Cookie 就是这样的一种机制,它可以弥补 HTTP 协议无状态的不足。在 Session 出现之前,基本上所有的网站都采用 Cookie 来跟踪会话。

3.5.1 什么是 Cookie

Cookie 原意为“甜饼”,是由 W3C 组织提出,最早由 Netscape 社区发展的一种机制。

目前 Cookie 已经成为标准，所有的主流浏览器如 IE、Netscape、Firefox、Opera 等都支持 Cookie。

由于 HTTP 是一种无状态的协议，服务器单从网络连接上无从知道客户身份。当用户访问网页时，它能够在访问者的机器上创建一个文件，把它称为 Cookie，写一段内容进去来标识不同的用户。如果下次用户再访问这个网页的时候，它又能够读出这个文件里面的内容，这样网页就知道上次这个用户已经访问过该网页了。

Cookie 是 Web 服务器发送给客户端的一小段文本信息，客户端发送请求时，服务器处理请求对客户端做出响应。如果服务器需要在客户端记录某些数据时，就会向客户端发送一个 Cookie，客户端接收并保存该 Cookie，当浏览器再请求该网站时，浏览器把请求的网址连同该 Cookie 一同提交给服务器。服务器检查该 Cookie，以此来辨认用户状态。服务器还可以根据需要修改 Cookie 的内容。一般情况下，Cookie 中的值是以 Key-Value 的形式进行表达的。

注意：Cookie 功能需要浏览器的支持。如果浏览器不支持 Cookie(如大部分手机中的浏览器)或者把 Cookie 禁用了，Cookie 功能就会失效。不同的浏览器采用不同的方式保存 Cookie。IE 浏览器会在"C:\Documents and Settings\你的用户名\Cookies"文件夹下以文本文件形式保存 Cookie。

Cookie 的内容主要包括名字、值、过期时间、路径和域。

其中域可以指定某一个域，如 google. com，也可以指定一个域下的具体某台主机，如 www. google. com 或 froogle. google. com。

如果不设置过期时间，则表示这个 Cookie 的生命期为浏览器会话期间，只要关闭浏览器窗口，Cookie 就消失了。这种生命期为浏览器会话期的 Cookie 称为会话 Cookie。会话 Cookie 一般不存储在硬盘上而是保存在内存里，当然这种行为并不是规范规定的。如果设置了过期时间，浏览器就会把 Cookie 保存到硬盘上，关闭后再次打开浏览器，这些 Cookie 仍然有效直到超过设定的过期时间。

存储在硬盘上的 Cookie 可以在不同的浏览器进程间共享，比如两个 IE 窗口。而对于保存在内存里的 Cookie，不同的浏览器有不同的处理方式。对于 IE，在一个打开的窗口上按 Ctrl＋N 键(或者从文件菜单)打开的窗口可以与原窗口共享，而使用其他方式新开的 IE 进程则不能共享已经打开的窗口的内存 Cookie；对于 Mozilla Firefox，所有的进程和标签页都可以共享同样的 Cookie。一般来说使用 JavaScript 的 window. open()方法打开的窗口会与原窗口共享内存 Cookie。

Cookies 给网站和用户带来的好处非常多。

(1) Cookie 能使站点跟踪特定访问者的访问次数、最后访问时间和访问者进入站点的路径。

(2) Cookie 能告诉在线广告商广告被点击的次数，从而可以更精确地投放广告。

(3) Cookie 有效期限未到时，Cookie 能使用户在不输入密码和用户名的情况下进入曾经浏览过的一些站点。

(4) Cookie 能帮助站点统计用户个人资料以实现各种各样的个性化服务。

3.5.2 Cookie的使用

1. 向客户程序发送 Cookie

在 JSP 中使用 Cookie，需要经过以下几个步骤。

1）使用 page 指令导入类 javax. srervelt. http. cookie

例如：

```
<%@page import="javax.servlet.http.cookie"%>
```

2）创建 Cookie 对象

JSP 是使用如下的语法格式来创建 Cookie 的。

```
Cookie cookie_name = new Cookie("Parameter","Value");
```

例如：

```
Cookie newCookie = new Cookie("username","waynezheng");
response.addCookie(newCookie);
```

JSP 调用 Cookie 类的构造方法 Cookie(name,value)用合适的名字和值来创建 Cookie，然后 Cookie 可以通过 HttpServletResponse 类的 addCookie()方法加入 Set-Cookie 应答头。本例中 Cookie 对象有两个字符串参数：username 和 waynezheng。

注意：名字和值都不能包含空白字符以及下列字符。

```
@ :;? ," / [ ]() =
```

3）处理 Cookie 的属性

除了 name 与 value 之外，Cookie 还具有其他几个常用的属性。每个属性对应一个 getter 方法与一个 setter 方法。在 JSP 中，用 cookie. setXXX()方法设置各种属性，用 cookie. getXXX()方法读出 Cookie 的属性，Cookie 类的常用属性如表 3-8 所示。

表 3-8 Cookie 类的常用属性

属　性	描　述
String name	Cookie 的名称。Cookie 一旦创建，名称便不可更改
Object value	Cookie 的值。如果值为 Unicode 字符，需要为字符编码。如果值为二进制数据，则需要使用 BASE64 编码
int maxAge	Cookie 失效的时间，单位为秒。如果为正数，则 Cookie 在 maxAge 秒之后失效。如果为负数，Cookie 为临时 Cookie，关闭浏览器即失效，浏览器也不会以任何形式保存该 Cookie。如果为 0，表示删除该 Cookie。默认值为－1
boolean secure	Cookie 是否仅被使用安全协议传输。安全协议有 HTTPS、SSL 等，在网络上传输数据之前先将数据加密。默认值为 false
String path	Cookie 的使用路径。如果设置为/sessionweb/，则只有 contextPath 为/sessionweb 的程序可以访问该 Cookie。如果设置为/，则本域名下 contextPath 都可以访问该 Cookie。注意最后一个字符必须为/

续表

属　性	描　述
String domain	可以访问该 Cookie 的域名。如果设置为.google.com，则所有以 google.com 结尾的域名都可以访问该 Cookie。注意第一个字符必须为"."
String comment	Cookie 的用处说明。浏览器显示 Cookie 信息的时候显示该说明
int version	Cookie 使用的版本号。0 表示遵循 Netscape 的 Cookie 规范，1 表示遵循 W3C 的 RFC 2109 规范

4）Cookie 的有效期

Cookie 的 maxAge 决定着 Cookie 的有效期，单位为秒，可以通过 getMaxAge()方法与 setMaxAge(int maxAge)方法来读写 maxAge 属性。

如果 maxAge 属性为正数，则表示该 Cookie 会在 maxAge 秒之后自动失效。浏览器会将 maxAge 为正数的 Cookie 持久化，即写到对应的 Cookie 文件中。无论客户关闭了浏览器还是电脑，只要还在 maxAge 秒之前，登录网站时该 Cookie 仍然有效。下面代码中的 Cookie 信息将永远有效。

```
Cookie cookie = new Cookie("username", "helloweenvsfei");   //新建 Cookie
cookie.setMaxAge(Integer.MAX_VALUE);                        //设置生命周期为 MAX_VALUE
response.addCookie(cookie);                                 //输出到客户端
```

如果 maxAge 为负数，则表示该 Cookie 仅在本浏览器窗口以及本窗口打开的子窗口内有效，关闭窗口后该 Cookie 即失效。maxAge 为负数的 Cookie 为临时性 Cookie，不会被持久化，不会被写到 Cookie 文件中。Cookie 信息保存在浏览器内存中，因此关闭浏览器该 Cookie 就消失了。Cookie 默认的 maxAge 值为－1。

如果 maxAge 为 0，则表示删除该 Cookie。Cookie 机制没有提供删除 Cookie 的方法，因此通过设置该 Cookie 即时失效实现删除 Cookie 的效果。失效的 Cookie 会被浏览器从 Cookie 文件或者内存中删除，例如：

```
Cookie cookie = new Cookie("username", "hello");       //新建 Cookie
cookie.setMaxAge(0);                                    //设置生命周期为 0，不能为负数
response.addCookie(cookie);                             //必须执行这一句
```

response 对象提供的 Cookie 操作方法只有 add(Cookie cookie)，要想修改 Cookie 只能使用一个同名的 Cookie 来覆盖原来的 Cookie，以达到修改的目的。删除时只需要把 maxAge 修改为 0 即可。

注意：从客户端读取 Cookie 时，包括 maxAge 在内的其他属性都是不可读的，也不会被提交。浏览器提交 Cookie 时只会提交 name 与 value 属性。maxAge 属性只被浏览器用来判断 Cookie 是否过期。

5）Cookie 的域名

Cookie 是不可跨域名的，域名 www.google.com 颁发的 Cookie 不会被提交到域名 www.baidu.com 去，这是由 Cookie 的隐私安全机制决定的，隐私安全机制能够禁止网站非法获取其他网站的 Cookie。

正常情况下，同一个一级域名下的两个二级域名如 www. helloweenvsfei. com 和 images. helloweenvsfei. com 也不能交互使用 Cookie，因为两者的域名并不严格相同。如果想所有 helloweenvsfei. com 名下的二级域名都可以使用该 Cookie，需要设置 Cookie 的 domain 属性，例如：

```
Cookie cookie = new Cookie("time", "20080808");        //新建 Cookie
cookie.setDomain(".helloweenvsfei.com");                //设置域名
cookie.setPath("/");                                    //设置路径
cookie.setMaxAge(Integer.MAX_VALUE);                    //设置有效期
response.addCookie(cookie);                             //输出到客户端
```

可以修改本机 C:\Windows\System32\Drivers\Etc 下的 hosts 文件来配置多个临时域名，然后使用 setCookie. jsp 程序来设置跨域名 Cookie 验证 domain 属性。

注意：domain 属性必须以点(".")开始。另外，name 相同但 domain 不同的两个 Cookie 是两个不同的 Cookie。如果想要两个域名完全不同的网站共有 Cookie，可以生成两个 Cookie，domain 属性分别为两个域名，输出到客户端。

6）Cookie 的路径

domain 属性决定运行访问 Cookie 的域名，而 path 属性决定允许访问 Cookie 的路径。例如，如果只允许/sessionWeb/下的程序使用 Cookie，可以这么写：

```
Cookie cookie = new Cookie("time", "20080808");        //新建 Cookie
cookie.setPath("/session/");                            //设置路径
response.addCookie(cookie);                             //输出到客户端
```

设置为/时允许所有路径使用 Cookie。path 属性需要使用符号/结尾。

注意：页面只能获取它所在的路径下的 Cookie。例如/session/test/a. jsp 不能获取到路径为/session/abc/的 Cookie。使用时一定要注意。

7）Cookie 的安全属性

HTTP 不仅是无状态的，而且是不安全的。使用 HTTP 的数据不经过任何加密就直接在网络上传播，有被截获的可能。使用 HTTP 传输很机密的内容是一种隐患。如果不希望 Cookie 在 HTTP 等非安全协议中传输，可以设置 Cookie 的 secure 属性为 true。浏览器只会在 HTTPS 和 SSL 等安全协议中传输此类 Cookie。下面的代码设置 secure 属性为 true。

```
Cookie cookie = new Cookie("time", "20080808");        //新建 Cookie
cookie.setSecure(true);                                 //设置安全属性
response.addCookie(cookie);                             //输出到客户端
```

提示：secure 属性并不能对 Cookie 内容加密，因而不能保证绝对的安全性。如果需要高安全性，需要在程序中对 Cookie 内容加密、解密，以防泄密。

8）JavaScript 操作 Cookie

Cookie 是保存在浏览器端的，浏览器可以使用脚本程序如 JavaScript 或者 VBScript 等操作 Cookie。例如，下面的代码会输出本页面所有的 Cookie。

```
<script>document.write(document.cookie);</script>
```

由于 JavaScript 能够任意地读写 Cookie，有些好事者便想使用 JavaScript 程序去窥探用户在其他网站的 Cookie。不过这是徒劳的，W3C 组织早就意识到 JavaScript 对 Cookie 的读写所带来的安全隐患并加以防备了，W3C 标准的浏览器会阻止 JavaScript 读写任何不属于自己网站的 Cookie。换句话说，A 网站的 JavaScript 程序读写 B 网站的 Cookie 不会有任何结果。

9）向客户端添加 Cookie

将 Cookie 放入 HTTP 响应报头，可以使用 HttpServletResponse 的 addCookie()方法，此方法不修改之前指定的 Set-Cookie 头信息，而是创建新的头信息。

```
response.addCookie(c);
```

注意：设置 Cookie 的步骤为创建 Cookie 对象，设置最大时效，将 Cookie 放入响应头信息，即发送到客户程序。

【例 3-16】 添加 Cookie 信息。

```
<%@ page language="java" import="java.util.*" pageEncoding="GBK"%>
<!DOCTYPE HTML PUBLIC "-//W3C//DTD HTML 4.01 Transitional//EN">
<html>
  <head><title>添加 cookie 信息</title></head>
  <body>
  <%
    //从提交的 HTML 表单中获取,用户名
    String userName=request.getParameter("username");
    //"username",userName 值/对创建一个 Cookie
    Cookie theUsername=new Cookie("username",userName);
    response.addCookie(theUsername);
    response.addCookie(new Cookie("password","123456"));
    response.sendRedirect("getCookies.jsp");
   %>
  </body>
</html>
```

2. 从客户端读取 Cookie

在 Cookie 发送到客户端前，先要创建一个 Cookie，然后用 addCookie()方法发送一个 HTTP 头。JSP 将调用 request.getCookies()从客户端读入 Cookie，getCookies()方法返回一个 HTTP 请求头中的内容对应的 Cookie 对象数组。可以用循环访问该数组的各个元素，调用 getName()方法检查各个 Cookie 的名字，直至找到目标 Cookie，然后对该 Cookie 调用 getValue()方法取得与指定名字关联的值。

【例 3-17】 获取 Cookie 信息。

```
<%@ page language="java" import="java.util.*" pageEncoding="GBK"%>
<!DOCTYPE HTML PUBLIC "-//W3C//DTD HTML 4.01 Transitional//EN">
<html>
```

```
  <head>
    <title>获取 cookie 信息</title>
  </head>
  <body>
    <%
    Cookie[] cookies = request.getCookies(); //创建一个 Cookie 对象数组
    String user = "", pwd = "";
    if(cookies!= null){
      //设立一个循环,来访问 Cookie 对象数组的每一个元素
      for(int i = 0;i < cookies.length;i++){
        //判断元素的值是否为 username 中的值
        if(cookies[i].getName().equals("username"))
            user = cookies[i].getValue();
        else if(cookies[i].getName().equals("password"))
          pwd = cookies[i].getValue();
      }
    }
    out.print("用户名:" + user + ",密码: " + pwd);
  %>
  </body>
</html>
```

【例 3-18】 使用 Cookie 有效期设置。

showCookie.jsp 页面的代码如下。

```
<%@ page language = "java" import = "java.util.*" pageEncoding = "GBK" %>
<!DOCTYPE HTML PUBLIC " - //W3C//DTD HTML 4.01 Transitional//EN">
<html>
  <head><title>使用 Cookie 有效期设置</title></head>
  <body>
    <%
      Cookie[] cookies = request.getCookies();
      if(cookies!= null){
      for(int i = 0;i < cookies.length;i++){
        if(cookies[i].getName().equals("info"))
            out.print("读取 Cookie 的值:" + cookies[i].getValue());
      }
      }else{
            out.print("超过 Cookie 有效期,无法读取 Cookie");
      }
 %>
  </body>
</html>
```

info.jsp 页面的代码如下。

```
<%@ page language = "java" import = "java.util.*" pageEncoding = "GBK" %>
<!DOCTYPE HTML PUBLIC " - //W3C//DTD HTML 4.01 Transitional//EN">
<html>
  <head><title>使用 Cookie 有效期设置</title></head>
```

```
  <body>
<%
  Cookie nc = new Cookie("info","ok");      //Cookie 对象,并设置名称与值
  nc.setMaxAge(60); //设置 Cookie 失效前时间为 60 秒
  response.addCookie(nc); //添加到响应头信息,返回到客户端
  response.sendRedirect("showCookie.jsp");
%>
  </body>
</html>
```

3. 解决 Cookie 安全的方法

解决 Cookie 安全的方法有以下几种。

(1) 替代 Cookie。将数据保存在服务器端,可选的是 Session 方案。

(2) 及时删除 Cookie。要删除一个已经存在的 Cookie 有以下几种方法。

① 给一个 Cookie 赋以空值。

② 设置 Cookie 的失效时间为当前时间,让该 Cookie 在当前页面浏览完之后就被删除了。

③ 通过浏览器删除 Cookie。如在 IE 中,可以选择"工具"→"Internet 选项"→"常规",在里面单击"删除 Cookies",就可以删除文件夹中的 Cookie。

(3) 禁用 Cookie。很多浏览器中都设置了禁用 Cookie 的方法,如 IE 中,可以在"工具"→"Internet 选项"→"隐私"中,将隐私级别设置为禁用 Cookie。

3.5.3　任务:简化用户登录

1. 需求分析

(1) 用户首次登录时要求输入用户名和密码。

(2) 登录成功后保存用户的登录状态。

(3) 设置 Cookie 的有效期为 5 分钟。

(4) 在 Cookie 有效期内,可无须登录直接进入欢迎页面。

2. 任务实施

(1) 登录页面 Login.jsp 的代码如下。

```
<%@ page language="java" import="java.util.*" pageEncoding="GBK" %>
<%@ page import="javax.servlet.http.Cookie" %>
<!DOCTYPE HTML PUBLIC "-//W3C//DTD HTML 4.01 Transitional//EN">
<html>
  <head><title>用户登录</title></head>
  <%
    Cookie[] cookies = request.getCookies();
    String user = "",pwd = "";
    if(cookies!= null){
      for(int i = 0;i < cookies.length;i++){
```

```
            if(cookies[i].getName().equals("uname"))
              response.sendRedirect("welcome.jsp");
        }
    }
    %>
    <body>
      <form name="loginForm" method="post" action="doLogin.jsp">
        用户名:<input type="text" name="userName" />
        密码:<input type="password" name="pwd" />
        <input type="submit" value="登录">
      </form>
    </body>
</html>
```

（2）处理登录信息页面 doLogin.jsp 的代码如下。

```
<%@ page language="java" import="java.util.*" pageEncoding="GBK"%>
<%@ page import="javax.servlet.http.Cookie" %>
<%
    request.setCharacterEncoding("GBK");
    String name = request.getParameter("userName");
    String pwd = request.getParameter("pwd");
    if("sa".equals(name.trim())&& "123".equals(pwd.trim())){
      //以 key/value 的形式创建 Cookie
      Cookie uname = new Cookie("uname", name.trim());
      uname.setMaxAge(5 * 60);
      response.addCookie(uname);
      response.sendRedirect("welcome.jsp");
    }
%>
```

（3）登录成功（欢迎）页面 welcome.jsp 的代码如下。

```
<%@ page language="java" import="java.util.*" pageEncoding="GBK"%>
<!DOCTYPE HTML PUBLIC "-//W3C//DTD HTML 4.01 Transitional//EN">
<html>
  <head><title>登录成功</title></head>
  <body>
    <%
    //获取请求中的 Cookie,以数组方式保存
    Cookie cookies[] = request.getCookies();
    //循环遍历数组,得到 key 为"uname"的 Cookie
    for(int i=0;i<cookies.length;i++){
      Cookie ucookie = cookies[i];
      if(ucookie.getName().equals("uname"))//判断 Cookie 的名称
        //获取 key 对应的 value,输出显示
        out.println("欢迎你: " + ucookie.getValue());
    }
    %>
  </body>
</html>
```

3.6 session 对象

在系统中的一些页面要求必须登录后才能访问，这就要求对其进行访问控制，要求记录客户的状态。在实际应用中，除使用 Cookie 外，Web 应用程序中还经常使用 Session 来记录客户端状态。Session 是服务器端使用的一种记录客户端状态的机制，使用上比 Cookie 简单一些，但相应地也增加了服务器的存储压力。

3.6.1 session 对象概述

1. 什么是会话

在 Web 程序设计中，一个会话就是用户通过浏览器与服务器之间进行的一次通话，它可以包含浏览器与服务器之间的多次请求。简单地说就是在一段时间内，单个客户与 Web 服务器的一连串相关的交互过程。

在一个会话中，客户可能会多次请求访问一个网页，也有可能请求访问各种不同的服务器资源。图 3-4 描述了浏览器与服务器的一次会话过程。当用户向服务器发出第一次请求时，服务器会为该用户创建唯一的会话，会话将一直延续到用户访问结束（浏览器关闭可以导致会话结束）。

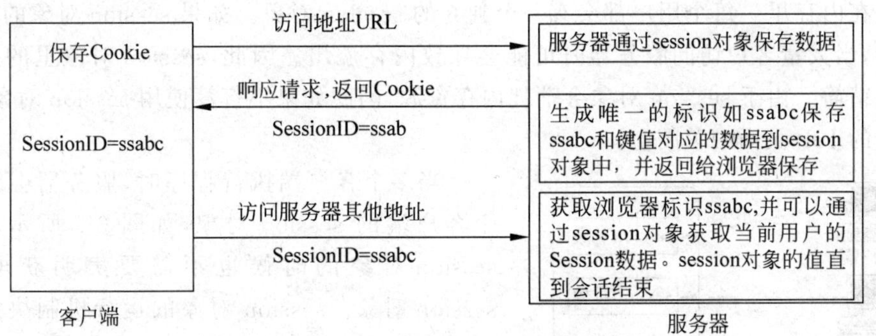

图 3-4　浏览器与服务器的会话过程

JSP 提供了一个可以在多个请求之间持续有效的会话对象 session，session 对象允许用户存储和提取会话状态的信息。

2. 什么是 session 对象

Session 是另一种记录客户状态的机制，Session 机制是一种服务器端的机制，session 对象是由服务器自动创建与用户请求相关的对象。服务器会为每一个用户创建一个 session 对象用来保存用户信息，跟踪用户操作。该对象内部使用 Map 类来保存数据在服务器上，因此它的数据类型是 Key-Value 形式。

服务器为不同的浏览器在内存中创建用于保存数据的对象叫 seesion。并将这个对象的唯一身份标识 Session ID（存放在 Cookie 中）返回给浏览器，浏览器将这个 Session ID 保存，当浏览器向服务器再次发送请求的时候，连同这个 Cookie 一起发送过去，由此找到对应的 session 对象。客户端浏览器再次访问时只需要从该 session 对象中查找该客户的状态就可以了。

3.6.2 session 对象的运行机制与常见方法

1. session 对象的生命周期

当程序需要被某个客户端请求时，服务器首先检查这个客户端的请求里是否已包含了一个 Session ID，如果已包含一个 Session ID，则说明以前已经为此客户端创建过 session 对象，服务器就按照 Session ID 把这个 session 对象检索出来使用。如果客户端请求不包含 Session ID，则为此客户端创建一个 session 对象，并且生成一个与此 session 对象相关联的 Session ID，这个 Session ID 将被在本次响应中返回给客户端保存。

Tomcat 的 ManagerBase 类提供了创建 Session ID 的方法：随机数＋时间＋jvmid。Session ID 的值是一个既不会重复，又不容易被找到规律以仿造的字符串。Tomcat 的 StandardManager 类将 session 对象存储在内存中。客户端只保存 Session ID 到 Cookie 中，而不会保存 session 对象，销毁 session 对象只能通过设置其属性为 invalidate 或超时，关掉浏览器并不会关闭 session 对象。

由于 session 对象保存在服务器端，为了获得更高的存取速度，服务器一般把 session 对象放在内存里。每个用户都会有一个独立的 session 对象。如果 session 对象的内容过于复杂，当大量客户访问服务器时可能会导致内存溢出。因此，session 对象里的信息应该尽量精简。由于 session 对象会消耗内存资源，因此如果不打算使用 session 对象，应该在所有的 JSP 页面中关闭它。

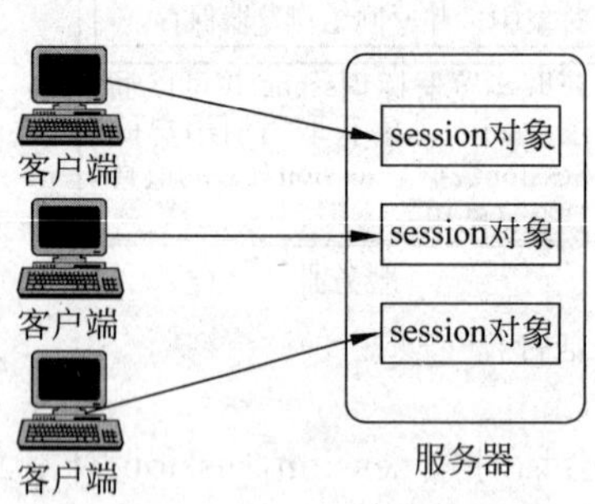

图 3-5　服务器为用户保存唯一的 session 对象

当多个客户端执行程序时，服务器会保存多个客户端的 session 对象，如图 3-5 所示。获取 session 对象的时候也不需要声明获取谁的 session 对象。session 对象的运行机制决定了当前客户只会获取到自己的 session 对象，而不会获取到别人的 session 对象。各客户的 session 对象也彼此独立，互不可见。

一个常见的误解是以为 session 对象在有客户端访问时就被创建，然而事实是直到某服务器端程序调用 HttpServletRequest. getSession (true)方法时才被创建。即 session 对象在用户第一次访问服务器的时候自动创建。需要注意只有访问 JSP、Servlet 等程序时才会创建 session 对象，只访问 HTML 等静态资源并不会创建 session 对象。如果尚未生成 session 对象，也可以使用 request. getSession (true)方法强制生成 session 对象。

session 对象生成后，只要用户继续访问，服务器就会更新 session 对象的最后访问时间，并维护该 session 对象。用户每访问服务器一次，无论是否读写 session 对象，服务器都认为该用户的 session 对象“活跃(active)”了一次。

如果 JSP 没有显式地使用<% @page session="false"%>关闭 session 对象，则 JSP 文件在编译成 Servlet 时将会自动加上如下的语句。

```
HttpSession session = HttpServletRequest.getSession(true);
```

这也是 session 对象是 JSP 内置对象的原因，session 对象与浏览器一一对应，允许用户存储和提取会话状态的信息。

2. session 对象的有效期

由于会有越来越多的用户访问服务器，因此 session 对象也会越来越多。为防止内存溢出，服务器会把长时间内没有活跃的 session 对象从内存删除，这个时间就是 session 对象的超时时间。如果超过了超时时间没访问过服务器，session 对象就自动失效了。

session 对象的超时时间为 maxInactiveInterval 属性的值，可以通过 getMaxInactiveInterval()方法获取，通过 setMaxInactiveInterval(long interval)方法修改。

session 对象的超时时间也可以在 web. xml 中修改。另外，通过调用 session 对象的 invalidate()方法可以使 session 对象失效。

3. session 对象的常用方法

session 对象中包括各种方法，使用起来要比 Cookie 方便得多。session 对象的常用方法如表 3-9 所示。

表 3-9　session 对象的常用方法

方　法	描　述
void setAttribute(String attribute, Object value)	设置 session 对象的属性。value 参数可以为任何 Java Object，通常为 Java Bean。value 信息不宜过大
String getAttribute(String attribute)	返回 session 对象的属性
Enumeration getAttributeNames()	返回 session 对象中存在的属性名
void removeAttribute(String attribute)	移除 session 对象的属性
String getId()	返回 session 对象的 ID。该 ID 由服务器自动创建，不会重复
long getCreationTime()	返回 session 对象的创建时间，返回值类型为 long，常被转化为 Date 类型，例如：Date createTime = new Date(session. get CreationTime())
long getLastAccessedTime()	返回 session 对象最后的活跃时间，返回值类型为 long
int getMaxInactiveInterval()	返回 session 对象的超时时间，单位为秒。超过该时间没有访问，服务器认为该 session 对象失效
void setMaxInactiveInterval(int second)	设置 session 对象的超时时间，单位为秒
boolean isNew()	返回该 session 对象是否是新创建的
void invalidate()	使 session 对象失效

3.6.3 session 对象的使用

1. 获取 session 对象

通过 request 对象可以获得 session 对象，例如：

```
HttpSession s = request.getSession(boolean flag);
HttpSession s = request.getSession( );
```

当 flag 为 true 时，会在浏览器发送过来的请求中查看是否有 Session ID，如果有，则返回与 Session ID 对应的 session 对象，如果没有，则创建一个新的。当 flag 为 false，没有就返回 null。

2. 获取 Session ID

每次会话将自动创建一个 session 对象，每个 session 对象有唯一的 ID，可以通过 getId()方法获取 Seesion ID。

【例 3-19】 在一个 Java Web 应用程序中新建两个 JSP 页面，名称分别为 first.jsp 和 second.jsp，它们具有如下代码(两个文件内容相同)。

```
<%@ page language = "java" pageEncoding = "gb2312" %>
<html>
  <head><title>获取 Session ID</title></head>
  <body>
  <%
    out.println(session.getId() + "<br>");
    out.println(session.isNew());
  %>
  </body>
</html>
```

3. 设置会话时间

session 对象会占用服务器端的内存，过多的以及长期的消耗内存会降低服务器端的运行效率，所以 session 对象存在于内存中时会有默认的时间限制，一旦 session 对象存在的时间超过了这个默认的时间限制则认为是 session 超时，session 会失效，不能再继续访问。相关的方法有以下几种。

```
//返回客户端最后一次与会话有关的请求时间
public abstract long getLastAccessedTime();
//以秒为单位返回一个会话内两个请求时间间隔
public abstract int getMaxInactiveInterval();
//以秒为单位设置 session 对象的有效时间
public abstract void setMaxInactiveInterval(int i);
```

Tomcat 中 session 对象的默认超时时间为 20 分钟。利用 setMaxInactiveInterval (int seconds)方法可以修改超时时间。可以修改 web.xml 的改变 Session 的默认超时时

间。例如：

```
<session-config>
    <session-timeout>60</session-timeout>    <!-- 单位:分钟 -->
</session-config>
```

注意：<session-timeout>参数的单位为分钟，而 setMaxInactiveInterval(int s)单位为秒。

【例 3-20】 设置 session 对象的有效期。

```
<%@ page language="java" import="java.util.*" pageEncoding="GBK"%>
<!DOCTYPE HTML PUBLIC "-//W3C//DTD HTML 4.01 Transitional//EN">
<html>
  <head><title>设置 session 对象的有效期</title></head>
  <body>
  <%
    long start = session.getCreationTime();
    long end = session.getLastAccessedTime();
    long time = (end - start)/1000;     //秒
    if (time > 40) {
      session.invalidate();
      out.print("error");
    }
   %>
    <h1>wait...<% = time %></h1>
  </body>
</html>
```

4. session 对象的添加、获取、删除和移除数据

session 对象作为服务器端为各客户端保存交互数据的一种方式，采用 key-value 对的形式来区分每一组数据。

1）向 session 添加数据

【例 3-21】 设置 session 对象的属性以保存用户名和密码。

TestSession.jsp 页面的代码如下。

```
<%@ page language="java" contentType="text/html; charset=GBK"%>
<% request.setCharacterEncoding("GBK"); %>
<html>
  <body>
    <form name="form1" action="" method="post">
      用户名：<input type="text" name="username" value=""><br>
      密 码：<input type="password" name="pwd" value=""><br>
      <input type="submit" name="btSubmit" value="提交">
    </form>
    <%
      sesion.setAttribute("username", request.getParameter("username"));
      sesion.setAttribute("pwd", request.getParameter("pwd"));
```

```
    %>
  </body>
</html>
```

2）获取绑定数据

在 TestCommon.jsp 页面中，通过 session 对象获取属性，注意 TestSession.jsp 和 TestCommon.jsp 这两个页面之间没有任何关联，只要通过 session 对象设置了属性，那么就可以在同一次会话中，通过 session 对象获取属性或者删除属性。

【例 3-22】 获取 session 对象的属性。

TestCommon.jsp 页面的代码如下。

```
<%@ page language = "java" contentType = "text/html; charset = GBK" %>
<% request.setCharacterEncoding("GBK"); %>
<html>
  <body>
    用户名:<% = session.getAttribute("username") %><br>
    密 码:<% = session.getAttribute("pwd") %><br>
  </body>
</html>
```

3）移除绑定数据

【例 3-23】 移除 session 对象的属性。

在 TestCommon.jsp 页面中，移除 session 对象的属性，代码如下。

```
<%@ page language = "java" contentType = "text/html; charset = GBK" %>
<% request.setCharacterEncoding("GBK"); %>
<html>
  <body>
    <% -- 删除属性 -- %>
    <%
      session.removeAttribute("username");
      session.removeAttribute("pwd");
    %>
    用户名:<% = session.getAttribute("username") %><br>
    密 码:<% = session.getAttribute("pwd") %><br>
  </body>
</html>
```

4）删除 session 对象

【例 3-24】 用户退出。

在 TestCommon.jsp 页面中删除 session 对象，代码如下。

```
<%@ page language = "java" contentType = "text/html; charset = GBK" %>
<% request.setCharacterEncoding("GBK"); %>
<html>
  <body>
    <% session.invalidate(); %>
  </body>
</html>
```

5. URL 地址重写

如果客户端浏览器将 Cookie 功能禁用，或者不支持 Cookie 怎么办？例如，绝大多数的手机浏览器都不支持 Cookie。Java Web 提供了另一种解决方案：URL 地址重写。

URL 地址重写是对客户端不支持 Cookie 的解决方案之一。URL 地址重写的原理是将该用户的 Session ID 重写到 URL 地址中。服务器能够解析重写后的 URL 获取 Session ID。这样即使客户端不支持 Cookie，也可以来记录用户状态。HttpServletResponse 类提供了 encodeURL(String url)方法实现 URL 地址重写，例如：

```
<td>
   <a href = "<% = response.encodeURL("index.jsp?c = 1&wd = Java") %>"> Homepage </a>
</td>
```

该方法会自动判断客户端是否支持 Cookie，如果客户端支持 Cookie，会将 URL 原封不动地输出；如果客户端不支持 Cookie，则会将用户的 Session ID 重写到 URL 中。重写后的输出可能如下。

```
<td>
   <a href = "index.jsp;
  jsessionid = 0CCD096E7F8D97B0BE608AFDC3E1931E?c = 1&wd = Java"> Homepage </a>
</td>
```

即在文件名的后面，在 URL 参数的前面添加了字符串“jsessionid=×××。其中×××为 Session ID。分析一下可以知道，增添的 jsessionid 字符串既不会影响请求的文件名，也不会影响提交的地址栏参数。用户单击这个链接的时候会把 Session ID 通过 URL 提交到服务器上，服务器通过解析 URL 地址获得 Session ID。

如果是页面重定向，URL 地址重写可以这样写：

```
<%
  if("administrator".equals(userName)){
    response.sendRedirect(response.encodeRedirectURL("administrator.jsp"));
      return;
   }
 %>
```

其效果跟 response.encodeURL(String url)是一样的。如果客户端支持 Cookie，生成原 URL 地址，如果不支持 Cookie，传回重写后的带有 jsessionid 字符串的地址。

对于 WAP 程序，由于大部分的手机浏览器都不支持 Cookie，WAP 程序都会采用 URL 地址重写来跟踪用户会话。

注意：Tomcat 判断客户端浏览器是否支持 Cookie 的依据是请求中是否含有 Cookie，尽管客户端可能会支持 Cookie，但是由于第一次请求时不会携带任何 Cookie，URL 重写后仍然会带有 jsessionid。当第二次访问时服务器已经在浏览器中写入 Cookie 了，因此 URL 重写后就不会带有 jsessionid 了。

6. 在会话中禁止使用 Cookie

WAP 上大部分的客户浏览器都不支持 Cookie，索性禁止会话使用 Cookie。这时，统一使用 URL 地址重写会更好一些。Java Web 标准支持通过配置的方式禁用 Cookie。下面举例说明。

打开项目 SessionWeb 的 WebRoot\META-INF 目录(跟 WEB-INF 目录同级，如果没有则创建)，打开 context.xml(如果没有则创建)，编辑内容如下。

```
<?xml version='1.0' encoding='UTF-8'?>
<Context path="/SessionWeb" cookies="false">
</Context>
```

或者修改 Tomcat 全局的 conf/context.xml，修改内容如下。

```
<!-- The contents of this file will be loaded for each web application -->
<Context cookies="false">
   <!-- ... 中间代码略 -->
</Context>
```

部署后 Tomcat 便不会自动生成名为 jsessionid 的 Cookie，会话也不会以 Cookie 为识别标志，而仅仅以重写后的 URL 为识别标志了。

注意：该配置只是禁止会话使用 Cookie 作为识别标志，并不能阻止其他的 Cookie 读写。也就是说服务器不会自动维护名为 jsessionid 的 Cookie 了，但是程序中仍然可以读写其他的 Cookie。

7. 对浏览器的要求

虽然 session 对象保存在服务器，对客户端是透明的，但它的正常运行仍然需要客户端浏览器的支持，这是因为 session 对象需要使用 Cookie 作为识别标志。HTTP 是无状态的，session 对象不能依据 HTTP 连接来判断是否为同一客户，因此服务器向客户端浏览器发送一个名为 jsessionid 的 Cookie，它的值为该 session 对象的 ID(也就是 HttpSession.getId()的返回值)，session 对象依据该 Cookie 来识别是否为同一用户。

该 Cookie 是服务器自动生成的，它的 maxAge 属性值一般为−1，表示仅当前浏览器内有效，并且各浏览器窗口间不共享，关闭浏览器就会失效。因此同一机器的两个浏览器窗口访问服务器时会生成两个不同的 session 对象，但是由浏览器窗口内的链接、脚本等打开的新窗口除外。这类子窗口会共享父窗口的 Cookie，因此会共享一个 session 对象。

新开的浏览器窗口会生成新的 session 对象，但子窗口除外，子窗口会共用父窗口的 session 对象。例如，在链接上右击，在弹出的快捷菜单中选择"在新窗口中打开"命令时，子窗口便可以访问父窗口的 session 对象。

8. Cookie 和会话机制之间的区别与联系

(1) Cookie 数据存放在客户的浏览器上，会话数据存放在服务器上。

(2) 会话是在服务器端保存用户信息，Cookie 是在客户端保存用户信息。会话中保存的是对象，Cookie 保存的是字符串。

(3) Cookie 通常用于保存不重要的用户信息，重要的信息使用 session 对象保存。Cookie 不是很安全，别人可以分析存放在本地的 Cookie 并进行 Cookie 欺骗，考虑到安全应当使用会话。

(4) session 对象会在一定时间内保存在服务器上，当访问增多会影响服务器的性能。考虑到减轻服务器负担，应当使用 Cookie。

(5) session 对象随会话结束而关闭，Cookie 可以长期保存在客户端，单个 Cookie 保存的数据不能超过 4KB，很多浏览器都限制一个站点最多保存 20 个 Cookie。

3.6.4　任务：购物车的设计

1. 需求分析

设计一个购物车，输入商品名后将商品加入购物车，可以查看购物车中的货物。

2. 任务实施

(1) 购物车页面的代码如下。

```
<%@ page language="java" import="java.util.*" pageEncoding="GBK"%>
 <html>
   <head><title>购物车</title></head>
   <body>
    <form action="sessionloat.jsp" method="post">
    book:<input type="text" name="book"><br>
     <input type="submit" value="submit">
   </form>
   <%
    ArrayList books = (ArrayList)session.getAttribute("books");
    if (books == null){
      books = new ArrayList();
      session.setAttribute("books",books);
    }
    String book = request.getParameter("book");
    if (book!= null){
      book = new String(book.getBytes("GBK"));
      books.add(book);
    }
    %>
    购物车的商品:<br>
    <%
      out.print(books.size() + "<br>");
      for (int i = 0;i<books.size();i++)
      out.println(books.get(i) +"<br>");
    %>
   <a href="display.jsp">查看购物车</a>
  </body>
</html>
```

(2) 显示购物车内容页面的代码如下。

```
<%@ page language="java" import="java.util.*" pageEncoding="GBK" %>
  <%
    ArrayList books = (ArrayList)session.getAttribute("books");
    for (int i=0;i<books.size();i++)
        out.println(books.get(i) + "<br>");
  %>
    <a href="sessionloat.jsp">继续购物</a>
```

3.7 application 对象的原理及应用

3.7.1 什么是 application 对象

1. applicationde 对象的基本概念

当 Web 服务器启动时,Web 服务器会自动创建一个 application 对象。application 对象一旦创建,它将一直存在,直到 Web 服务器关闭。即 application 对象的生命周期从 Web 服务器启动直到 Web 服务器关闭。

一个 Web 服务器中通常有多个 Web 服务目录(网站),当 Web 服务器启动时,它自动为每个 Web 服务目录都创建一个 application 对象。这些 application 对象各自独立,而且和 Web 服务目录一一对应。访问同一个网站的客户都共享一个 application 对象,因此 application 对象可以实现多客户间的数据共享,也即所有访问该网站的客户都共享一个 application 对象。

application 对象是一个应用程序级的对象,它的作用范围是当前 Web 应用程序,所有访问当前网站的客户都共享一个 application 对象。

具体来说,不管哪个客户来访问网站 A,也不管客户访问网站 A 的哪个页面,都可以对网站 A 的 application 对象进行操作,因为所有访问网站 A 的客户都共用一个 application 对象。

因此,当在 application 对象中存储数据后,所有访问网站 A 的客户都能够对其进行访问,实现了多客户之间的数据共享。

2. application 对象的方法

application 对象的基类是 javax.servlet.ServletContext。有些 Web 服务器不直接支持使用 application 对象,必须用 ServletContext 类来声明 application 对象,再调用 getServletContext()方法来获取当前页面的 application 对象。

ServletContext 类用于表示应用程序的上下文。一个 ServletContext 类的对象表示一个 Web 应用程序的上下文。具体来说,在 Web 服务器中,提供了一个 Web 应用程序运行时的环境,专门负责 Web 应用程序的部署、编译、运行以及生命周期的管理。通过 ServletContext 类,可以获取 Web 应用程序运行时的环境信息。application 对象的常用方法如表 3-10 所示。

表 3-10　application 对象的常用方法

方　　法	描　　述
String getAttribute(String name)	根据属性名称获取属性值
Enumeration getAttributeNames()	获取所有的属性名称
void setAttribute(String name,Object object)	设置属性,指定属性名称和属性值
void removeAttribute(String name)	根据属性名称删除对应的属性
ServletContext getContext(String uripath)	获取指定 URL 的 ServletContext 对象
String getContextPath()	获取当前 Web 应用程序的根目录
String getInitParameter(String name)	根据初始化参数名称,获取初始化参数值
int getMajorVersion()	获取 Servlet API 的主版本号
int getMinorVersion()	获取 Servlet API 的次版本号
String getMimeType(String file)	获取指定文件的 MIME 类型
String getServletInfo()	获取当前 Web 服务器的版本信息
String getServletContextName()	获取当前 Web 应用程序的名称
void log(String message)	将信息写入日志文件中

3.7.2　application 对象的应用

1. 使用 application 对象存储数据

(1) 设置 application 对象的属性和获取 application 对象的属性。

【例 3-25】 利用设置 application 对象的属性来保存公司基本信息。

TestApplication.jsp 页面的代码如下。

```
<%@ page language="java" import="java.util.*" pageEncoding="utf-8"%>
<!DOCTYPE HTML PUBLIC "-//W3C//DTD HTML 4.01 Transitional//EN">
<html>
  <head><title>application 应用</title></head>
  <body>
    <h1>利用设置 application 对象的属性来保存公司基本信息</h1>
    <%
        String CompName = "ABC 公司";
        String CompTel = "01012345688";
        String CompAddr = "四川成都";
        application.setAttribute("ComName", CompName);
        application.setAttribute("CompTel", CompTel);
        application.setAttribute("CompAddr", CompAddr);
    %>
  </body>
</html>
```

在 TestCommon.jsp 页面中,通过 application 对象获取属性,注意 TestApplication.jsp 和 TestCommon.jsp 这两个页面之间没有任何关联,只要通过 application 对象设置了属性,那么就可以在任何页面通过 application 对象获取属性或者删除属性。

【例 3-26】 获取 application 对象的属性。

TestCommon.jsp 页面的代码如下。

```
<%@ page language="java" import="java.util.*" pageEncoding="utf-8"%>
<!DOCTYPE HTML PUBLIC "-//W3C//DTD HTML 4.01 Transitional//EN">
<html>
  <head><title>公司信息</title></head>
  <body>
    <h1>公司信息</h1>
    <%
        String CompName, CompTel, CompAddr;
        CompName = (String)application.getAttribute("ComName");
        CompTel = (String)application.getAttribute("CompTel");
        CompAddr = (String)application.getAttribute("CompAddr");
      %>
      <p>公司名称:<% = CompName %></p>
      <p>公司电话:<% = CompTel %></p>
      <p>公司地址:<% = CompAddr %></p>
  </body>
</html>
```

(2) 删除 application 对象的属性。

【例 3-27】 删除 application 对象的属性。

在 TestCommon.jsp 页面中,删除 application 对象的属性,代码如下。

```
<%@ page language="java" contentType="text/html; charset=GBK"%>
<% request.setCharacterEncoding("GBK"); %>
<html>
  <body>
    <%-- 删除属性 --%>
    <%
      application.removeAttribute("CompAddr");
    %>
      <p>公司名称:<% = application.getAttribute("ComName")%></p>
      <p>公司电话:<% = application.getAttribute("CompTel")%></p>
      <p>公司地址:<% = application.getAttribute("CompAddr")%></p>
  </body>
</html>
```

2. 统计网页访问的人数

【例 3-28】 统计网页访问人数。

(1) 登录页面 login.jsp 的代码如下。

```
<%@ page language="java" import="java.util.*" pageEncoding="GBK"%>
<!DOCTYPE HTML PUBLIC "-//W3C//DTD HTML 4.01 Transitional//EN">
<html>
  <head><title>登录</title></head>
```

```
<%
    integer count = (Integer)application.getAttribute("count");
    if (count != null) count = 1 + count; else count = 1;
      application.setAttribute("count",count);
  %>
  <body>
    <form name = "loginForm" method = "post" action = "showCount.jsp">
      用户名:<input type = "text" name = "userName" />
      密码:<input type = "password" name = "pwd" />
      <input type = "submit" value = "登录">
    </form>
  </body>
</html>
```

（2）显示访问人数页面 showcount.jsp 的代码如下。

```
<%@ page language = "java" import = "java.util.*" pageEncoding = "GBK" %>
<%
    integer i = (Integer)application.getAttribute("count");
    out.println("统计访问量:目前有 " + i +" 个人访问过本网站" );
%>
```

3. 使用 application 对象获取 Web 应用程序的环境信息

以下代码获取虚拟目录对应的真实路径。

```
<%@page contentType = "text/html;charset = GBK" %>
<%@page import = "java.io.*" %>
<h1><% = application.getRealPath("/") %>
```

得到的就是以下内容，即 server.xml 文件中配置的虚拟目录。

```
<Context path = "/demo"docBase = "H;\webdemo"/>
```

一般 application 表示的是上下文，但是在实际的开发中很少使用 application，而使用 getServletContent()方法来表示 application。

【例 3-29】 获取 Web 应用程序的环境信息。

GetApplictionInfo.jsp 页面的代码如下。

```
<%@ page language = "java" contentType = "text/html; charset = GBK"pageEncoding = "GBK" %>
<% request.setCharacterEncoding("GBK"); %>
<html>
  <body>
    <h2>使用 application 对象获取 Web 应用程序的环境信息</h2>
    获取当前 Web 服务器的版本信息 <% = application.getServerInfo() %><br>
    获取 Servlet API 的主版本号 <% = application.getMajorVersion() %><br>
    获取 Servlet API 的次版本号: <% = application.getMinorVersion() %>
    获得指定 URL 的 ServletContext 对象:>
    <%
      ServletContext sc = application.getContext("/ch3/TestApplication01.jsp");
```

```
        out.println(sc.getServerInfo());
      %>
      <%--需要在web.xml设置display-name元素--%>
      获取当前Web应用程序的名称:<%= application.getServletContextName() %>
      获取当前Web应用程序的上下文路径:<%= application.getContextPath() %>
    </body>
</html>
```

4. 进行 IO 操作

在知道真实路径后,就可以进行 IO 操作了。

【例 3-30】 在指定位置保存文件。

(1) 定义一个表单。

```
<%@page contentType="text/html;charset=GBK"%>
<form action="application04.jsp"method="post">
  输入文件名称:<input type="text" name="filename"><br>
  输入文件内容:<br>
  <textarea name="content"cols="30"rows="10"></textarea><br>
  <input type="submit"value="提交">
</form>
```

(2) application04.jsp 页面的代码如下。

```
<%@page contentType="text/html;charset=GBK"%>
<%@page import="java.io.*"%>
<%
  request.setCharacterEncoding("GBK");
  String fileName = this.getServletContext().getRealPath("/") + "note" + File.separator +
request.getParameter("filename");
  String content = request.getParameter("content").replaceAll("\r\n","<br>");
  PrintStream ps = new PrintStream(new FileOutputStream(new File(fileName)));
  ps.println(content);
  ps.close();
%>
```

【例 3-31】 列出目录中的全部文件。

```
<%@page contentType="text/html;charset=GBK"%>
<%@page import="java.io.*"%>
<%
  request.setCharacterEncoding("GBK");
  String fileName = this.getServletSontext().getRealPath("/") +"note";
  File f = new File(fileName);
  String files[] = f.list();
  for(int i=0;i<files.length;i++){
%>
  <h3><%=files[i]%></h3>
<%}%>
```

【例 3-32】 列出文件的全部内容。

```
<%@ page contentType = "text/html;charset = GBK" %>
<%@ page import = "java.io.*" %>
<%
  request.setCharacterEncoding("GBK");
  String fileName = this.getServletContext().getRealPath("/") + "note"
                  + File.separator + request.getParameter("filename");
  File f = new File(fileName);
  BufferedReader buf = new BufferedReader(new InputStreamReader(new FileInputStream(f)));
  String str = buf.readLine(); //读取内容
%>
  <h3><% = str %></h3>
<%
  buf.close(); //关闭
%>
```

3.7.3 session 对象和 application 对象的比较

1. 两者的作用范围不同

session 对象是用户级的对象，而 application 对象是应用程序级的对象。

一个用户一个 session 对象，每个用户的 session 对象不同，在用户所访问网站的多个页面之间共享同一个 session 对象。

一个 Web 应用程序一个 application 对象，每个 Web 应用程序的 application 对象不同，但一个 Web 应用程序的多个用户之间共享同一个 application 对象。

在同一个网站下，每个用户的 session 对象不同；所用用户的 application 对象相同。

在不同网站下，每个用户的 session 对象不同；每个用户的 application 对象不同。

2. 两者的生命周期不同

session 对象在用户首次访问网站创建，用户离开该网站（不一定要关闭浏览器）消亡。application 对象在启动 Web 服务器时创建，关闭 Web 服务器时消亡。

3. JSP 内置对象的范围

对象的范围决定了 JSP 是否可以进行对象访问，如表 3-11 表示。

表 3-11 范围的分类

名 称	说 明
page	在一个页面范围内有效，通过 pageContext 对象访问该范围内的对象
request	在一个服务器请求范围内有效，与客户端请求绑定在一起
session	在一次会话范围内容有效，在会话期间与 session 绑定的对象皆属于该范围
application	在一个应用服务器范围内有效，当应用服务启动后即创建该对象，并向所有用户所共享

【例 3-33】 JSP 内置对象范围测试。

(1) page 范围测试。

测试页面 testone.jsp 的代码如下。

```
<%@ page language = "java" contentType = "text/html; charset = UTF-8" pageEncoding = "UTF-8" %>
<html>
    <head><title>page 范围测试 -- testone.jsp</title></head>
    <body>
    <%
        String name = "page";
        pageContext.setAttribute("name",name);
    %>
        testOne:<% = pageContext.getAttribute("name") %><br/>
    <%
          pageContext.include("testTwo.jsp");
    %>
    <br/>
  </body>
</html>
```

测试页面 testTwo.jsp 的代码如下。

```
<%@ page language = "java" contentType = "text/html; charset = UTF-8" pageEncoding = "UTF-8" %>
<html>
    <head><title>page 范围测试 -- testTwo.jsp</title></head>
    <body>
        testTwo:<% = pageContext.getAttribute("name") %>
    </body>
  </html>
```

(2) request 范围测试。

测试页面 testone.jsp 的代码如下。

```
<%@ page language = "java" contentType = "text/html; charset = UTF-8" pageEncoding = "UTF-8" %>
<html>
      <head><title>request 范围测试 -- testone.jsp</title></head>
    <body>
    <%
        String name = "request";
        request.setAttribute("name",name);
    %>
    testOne:<% = request.getAttribute("name") %><br/>
    <% pageContext.include("testTwo.jsp"); %>
    </body>
</html>
```

测试页面 testTwo.jsp 的代码如下。

```
<%@ page language = "java" contentType = "text/html; charset = UTF - 8" pageEncoding = "UTF -
8" %>
<html>
  <head><title>request 范围测试 -- testTwo.jsp</title></head>
    <body>
      testTwo:<% = request.getAttribute("name") %>
    </body>
</html>
```

(3) session 范围测试。

测试页面 testone.jsp 的代码如下。

```
<%@ page language = "java" contentType = "text/html; charset = UTF - 8" pageEncoding = "UTF -
8" %>
<html>
    <head><title>session 范围测试 -- testone.jsp</title></head>
    <body>
  <%
      String req = "request", ses = "session";
      request.setAttribute("reqName",req);
      session.setAttribute("sessionName",ses);
      response.sendRedirect("testTwo.jsp");
    %>
  </body>
</html>
```

测试页面 testTwo.jsp 的代码如下。

```
<%@ page language = "java" contentType = "text/html; charset = UTF - 8" pageEncoding = "UTF -
8" %>
<html>
<head><title>session 范围测试 -- testTwo.jsp</title></head>
    <body>
        request:<% = request.getAttribute("reqName") %><br/>
        session:<% = session.getAttribute("sessionName") %>
    </body>
</html>
```

(4) application 范围测试。

测试页面 testone.jsp 的代码如下。

```
<%@ page language = "java" contentType = "text/html; charset = UTF - 8" pageEncoding = "UTF -
8" %>
<html>
<head><title>> application 范围测试 -- testone.jsp</title></head>
  <body>
  <%
    String app = "application";
```

```
    String ses = "session";
    session.setAttribute("sesName",ses);
    application.setAttribute("appName",app);
    response.sendRedirect("testTwo.jsp");
  %>
  </body>
</html>
```

测试页面 testTwo.jsp 的代码如下。

```
<%@ page language = "java" contentType = "text/html; charset = UTF - 8" pageEncoding = "UTF -
8" %>
<html>
  <head><title>application 范围测试 -- testTwo.jsp</title></head>
  <body>
    session:<% = session.getAttribute("sesName") %><br/>
    Application:<% = application.getAttribute("appName") %>
  </body>
</html>
```

3.7.4 任务：简易聊天室与网页计数器的设计

1. 任务要求

设计一个简单的网页版聊天室，实现浏览器的在线聊天功能。

2. 任务实施

一个简单的聊天室，所有用户的输入信息都保存在 application 对象中，这样所有的人都可以访问此应用。但是由于只能保持一个 application 对象，因此所有的信息只能以集合的形式保存。

(1) 设计主页面。定义一个页面框架，包括显示聊天内容以及聊天输入界面。

main.html 页面的代码如下。

```
<frameset rows = "80%,20%">
  <frame name = "top" src = "content.jsp">
  <frame name = "bottom" src = "input.jsp">
</frameset>
```

(2) 设计聊天输入界面。

input.jsp 页面的代码如下。

```
<%@ page contentType = "text/html;charset = GBK" %>
<%@ page import = "java.io.*" %>
<%@ page import = "java.util.*" %>
<form action = "input.jsp" method = "post">
  请输入内容:<input type = "text" name = "content">
  <input type = "submit" value = "说话">
</form>
```

```
<%
  request.setCharacterEncoding("GBK") ;
  if(request.getParameter("content")!= null){
    //说话的全部内容应该保存在 application
    //application 中存在一个集合用于保存所有说话的内容
    List all = null ;
    all = (List)getServletContext().getAttribute("notes") ;
    //程序必须考虑是否是第一次运行
    if(all == null){
      all = new ArrayList() ; //里面没有集合,所以重新实例化
    }
    all.add(request.getParameter("content")) ;
    //将修改后的集合重新放回到 application 之中
    getServletContext().setAttribute("notes",all) ;
  }
%>
```

(3) 设计聊天内容显示界面。

content.jsp 页面的代码如下。

```
<%@ page contentType = "text/html;charset = GBK" %>
<%@ page import = "java.util.*" %>
<%
  response.setHeader("refresh","2") ;
  request.setCharacterEncoding("GBK") ;
  List all = (List)getServletContext().getAttribute("notes") ;
  if(all == null){
%>
    <h2>没有留言!</h2>
<%
  }else{
    Iterator iter = all.iterator() ;
    while(iter.hasNext()){
%>
    <h3><% = iter.next() %></h3>
<%}
  }
%>
```

3.8　其他内置对象

3.8.1　config 对象

config 对象是在对一个 JSP 程序初始化时,JSP 引擎向它传递消息用的,此消息包括 JSP 程序初始化时所需要的参数及服务器的有关信息。config 对象对应的接口是 javax.servlet.ServletConfig,主要的功能是取得一些初始化的配置信息。SerletConfig 接口的常用方法如下。

public String getInitParameter(String name)：获取指定名称的初始化参数内容。

Public Enumeration getInitParameterNames()：获取全部的初始化参数名称。

所有的参数必须在 web.xml 中配置，即如果一个 JSP 文件要想通过初始化参数获得一些信息，则必须要在 web.xml 文件中完成映射。

【例 3-34】 读取初始化参数(init.jsp)。

```
<html>
  <head><title>初始化参数</title></head>
    <body>
        <h1><% = config.getInitParameter("driver") %></h1>
        <h1><% = config.getInitParameter("url") %></h1>
    </body>
</html>
```

init.jsp 要读取 driver 和 url 两个初始化参数，这两个参数直接在 web.xml 中配置。修改 web.xml 文件如下。

```
<?xml version="1.0" encoding="ISO-8859-1"?>
<web-app xmlns="http://java.sun.com/xml/ns/j2ee"
  xmlns:xsi="http://www.w3.org/2001/XMLSchema-instance"
  xsi:schemaLocation="http://java.sun.com/xml/ns/j2ee http://java.sun.com/xml/ns/j2ee/
web-app_2_4.xsd"
  version="2.4">
…
<!-- JSPC servlet mappings start -->
  <servlet>
    <servlet-name>demo</servlet-name>
    <jsp-file>/WEB-INF/hello.jsp</jsp-file>
    <init-param>
      <param-name>driver</param-name>
      <param-value>oracle.jdbc.driver.OracleDriver</param-value>
    </init-param>
    <init-param>
      <param-name>url</param-name>
      <param-value>jdbc:oracle:thin:@localhost:1521:MLDN</param-value>
    </init-param>
  </servlet>
  <servlet-mapping>
    <servlet-name>demo</servlet-name>
    <url-pattern>/hello.lxh</url-pattern>
  </servlet-mapping>
…
</web-app>
```

3.8.2 page 对象

page 对象代表 JSP 页面本身，是当前 JSP 页面本身的一个实例。page 对象在当前 JSP 页面中可以用 this 关键字来替代。

page 对象主要应用在 JSP 页面的 Java 程序段或 JSP 表达式中。page 对象的基类是 java.lang.Object。如果直接通过 page 对象来调用方法，就只能调用 Object 类中的那些方法。

javax.servlet.jsp.JspPage 接口继承自 javax.servlet.Servlet。可以使用 JspPage 接口对 page 对象进行强制类型转换，再调用 JspPage 接口中的各种方法。

可以使用 HttpJspPage 接口对 page 对象进行强制类型转换，再调用 HttpJspPage 接口中的各种方法。

在 JSP 页面中，this 关键字表示当前 JSP 页面。page 对象的常用方法如表 3-12 所示。

表 3-12 page 对象的常用方法

方 法	描 述
ServletConfig getServletConfig()	返回当前页面的一个 ServletConfig 对象
ServletContext getServletContext()	返回当前页面的一个 ServletContext 对象
String getServletInfo()	获取当前 JSP 页面的 Info 属性

【例 3-35】 获取 JSP 页面的 info 属性。

```
<%@ page contentType = "text/html;charset = GB2312" %>
<%@ page info = "作者:张三 版权:软件学院" %>
<html>
  <body>
    <% = this.getServletInfo() %>
</body>
</html>
```

3.8.3 pageContext 对象

pageContext 对象代表页面上下文，也就是当前页面所在的环境。pageContext 对象是 javax.servlet.jsp.PageContext 类的实例，提供了对 JSP 页面所有对象及命名空间的访问方法。pageContext 对象的方法可以访问除自身外的 8 个 JSP 内部对象，还可以直接访问绑定在 application、page 、request、session 对象上的 Java 对象。该对象主要用于访问 JSP 之间的共享数据。pageContext 对象通过以下 4 个方法来设置和获取页面上下文中的属性值，如图 3-13 所示。

表 3-13 pageContext 对象的方法

方 法	描 述
Object getAttribute(String name)	取得 page 范围内的 name 属性
Object getAttribute(String name,int scope)	取得指定范围内的 name 属性
void setAttribute(String name,Object value)	设置 page 范围内的 name 属性
void setAttribute(String name,Object value,int scope)	设置指定范围内的 name 属性

scope 可以取以下值。

```
public static final int PAGE_SCOPE          = 1; //对应于 page 范围
```

```
public static final int REQUEST_SCOPE      = 2; //对应于 request 范围
public static final int SESSION_SCOPE      = 3; //对应于 session 范围
public static final int APPLICATION_SCOPE  = 4; //对应于 application 范围
```

1. pageContext 的作用范围

【例 3-36】 pageContext 的作用范围测试。

page1.jsp 页面的代码如下。

```
<%@ page contentType="text/html; charset=GBK" language="java" errorPage="" %>
<!DOCTYPE html PUBLIC "-//W3C//DTD XHTML 1.0 Transitional//EN" "http://www.w3.org/TR/
xhtml1/DTD/xhtml1-transitional.dtd">
<html xmlns="http://www.w3.org/1999/xhtml">
<head><title>pageContext 测试</title></head>
  <body>
  <%  //使用 pageContext 设置属性,该属性默认在 page 范围内
    pageContext.setAttribute("page","hello");
  //使用 request 设置属性,该属性默认在 request 范围内
    request.setAttribute("request","hello");
  //使用 pageContext 将属性设置在 request 范围中
    pageContext.setAttribute("request2","hello",pageContext.REQUEST_SCOPE);
  //使用 session 将属性设置在 session 范围中
    session.setAttribute("session","hello");
  //使用 pageContext 将属性设置在 session 范围中
    pageContext.setAttribute("session2","hello",pageContext.SESSION_SCOPE);
  //使用 application 将属性设置在 application 范围中
    application.setAttribute("app","hello");
  //使用 pageContext 将属性设置在 application 范围中
    pageContext.setAttribute("app2","hello",pageContext.APPLICATION_SCOPE);
  //服务器重定向到 page2.jsp
    //<jsp:forward page="page2.jsp" />
    %>
    <!--- <a herf="page2.jsp">page2.jsp</a>客户端跳转 --->
</body>
  </html>
```

page2.jsp 页面的代码如下。

```
<%@ page contentType="text/html; charset=GBK" language="java" errorPage="" %>
<!DOCTYPE html PUBLIC "-//W3C//DTD XHTML 1.0 Transitional//EN" "http://www.w3.org/TR/
xhtml1/DTD/xhtml1-transitional.dtd">
<html xmlns="http://www.w3.org/1999/xhtml">
<head><title>pageContext 测试</title></head>
  <body>
      <h3>下面获取各属性所在的范围</h3>
      <p>page 变量所在范围:"<%=pageContext.getAttributesScope("page")%></p>
      <p>request 变量所在范围:"<%pageContext.getAttributesScope("request")%></p>
      <p>request2 变量所在范围:"<%pageContext.getAttributesScope("request2")%></p>
      <p>session 变量所在范围:"<%pageContext.getAttributesScope("session")%></p>
      <p>session2 变量所在范围:"<%pageContext.getAttributesScope("session2")%></p>
```

```
        <p>application 变量所在范围:"<% pageContext.getAttributesScope("app") %></p>
        <p>application2 变量所在范围:"<% pageContext.getAttributesScope("app2") %></p>
    </body>
</html>
```

分别采用客户端跳转(超链接)方式和服务器端重定向方式,观察不同作用域下值的变化。

2. 页面跳转

forward()方法用于将当前 ServletRequest 和 ServletResponse 重定向或转发到应用程序中的另一个活动组件,方法原型如下。

```
abstract public void forward(String relativeUrlPath) throws ServletException,
java.io.IOException
```

页面 goto.jsp 的代码如下。

```
<%
//跳转到 abc .jsp 页面,并传递一个 info 参数,其值为 aaa
    pageContext.forward("abc.jsp?info=aaa");
%>
```

跳转目标页面 abc.jsp 的代码如下。

```
<% String info=pageContext.getRequest().getParameter("info"); %>
  参数:      <%=info %>
<% String path= pageContext.getServletContext().getRealPath("/"); %>
  路径:<%= path%>
```

3. 页面包含

include()方法的原型如下。

```
abstract public void include(String relativeUrlPath,boolean flush) throws ServletException,
java.io.IOException
```

4. 其他方法

pageContext 还可用于获取其他内置对象,pageContext 对象包含如下方法。

(1) ServletRequest getRequest():获取 request 对象。

(2) ServletResponse getResponse():获取 response 对象。

(3) ServletConfig getServletConfig():获取 config 对象。

(4) ServletContext getServletContext():获取 application 对象。

(5) HttpSession getSession():获取 session 对象。

一旦在 JSP、Servlet 编程中获取了 pageContext 对象,就可以通过它提供的方法来获取其他内置对象。

5. Page 对象与 pageContext 对象的区别

page 对象是当前页面转换后的 Servlet 类的实例。从转换后的 Servlet 类的代码中可以看到这种关系：Object page = this。在 JSP 页面中，很少使用 page 对象。

pageContext 对象是 javax. servlet. jsp. PageContext 类的实例，该对象代表该 JSP 页面上下文(环境)，使用该对象可以访问页面中的共享数据。

总之，pageContext 和 page 都是 JSP 中的隐含对象，pageContext 代表 JSP 页面的上下文关系，能够调用、存取其他隐含对象；page 代表处理当前请求时这个页面的实现类的实例。

3.8.4 exception 对象

由于用户的输入或者一些不可预见的原因，页面在运行过程中总是有一些没有发现或者是无法避免的异常现象出现。此时，可以通过 exception 对象来获取一些异常信息。exception 对象是 java. lang. Exception 类的实例。

1. 什么是 exception 对象

当 JSP 页面在执行过程中发生异常或错误时，会自动创建一个 exception 对象。在 JSP 页面中，使用 page 指令设置 isErrorPage 属性值为 true 后，就可以使用 exception 对象来查找页面出错信息。exception 对象的基类是 javax. servlet. jsp. JspException。

exception 对象的常用方法如表 3-14 所示。

表 3-14 exception 对象的常用方法

方　法	描　述
String getMessage()	返回简短的 JSP 页面的出错信息
String toString()	返回详细的 JSP 页面的出错信息

2. 如何使用 exception 对象

在有可能产生异常或错误的 JSP 页面中，使用 page 指令设置 errorPage 属性，属性值为能够进行异常处理的某个 JSP 页面。简单来说，只要在当前 JSP 页面中产生了异常，就跳转到另外一个专门处理异常的 JSP 页面。

【例 3-37】 出错处理页面。

pageWithError. jsp 页面的代码如下。

```
<%@ page contentType="text/html;charset=GB2312" %>
<%@ page import="java.util.Vector" %>
<%@ page errorPage="ErrorHand.jsp" %>
<html>
  <body>
    <%
      int [] intArray = new int [5];
      for (int i = 0;i <= 5; i++) {
        intArray[i] = i;
```

```
        }
    %>
  </body>
</html>
```

在专门负责处理异常的 JSP 页面中，使用 page 指令设置 isErrorPage 属性为 true，并使用 exception 对象来获取出错信息。ErrorHand.jsp 页面的代码如下。

```
<%@ page contentType = "text/html; charset = GB2312" %>
<%@ page isErrorPage = "true" %>
<html>
  <body>
    <%
      if (exception != null) {
    %>
    <font color = "Red">出错了,错误如下:</font>
    <p><font color = "Red">简短的错误描述:</font>
    <br><% = exception.getMessage() %>
    <p><font color = "Red">详细的错误描述:</font>
    <br><% = exception.toString() %>
    <%  }  %>
  </body>
</html>
```

3.8.5　Web 安全性

在配置 Tomcat 服务器时，Web 目录中必须存在 WEB-INF 文件夹，但该文件夹不可见也不可访问，所以其安全性高，保存在此目录中的程序安全性肯定是最高的。在各个程序开发中，基本上都将一些配置信息保存在此文件夹中。注意定义 WEB-INF 目录时一定要大写。

【例 3-38】　创建一个 JSP 文件，保存在/WEB-INF 文件夹中。

```
<%@ page language = "java" import = "java.util. * " pageEncoding = "GBK" %>
<html>
  <head><title>页面的安全性 e</title></head>
  <body>
    <% out.println("<h1>hello!</h1>"); %>
  </body>
</html>
```

该文件无法访问，必须通过一个映射进行操作。下面通过修改/WEB/web.xml 文件增加配置。

```
<?xml version = "1.0" encoding = "UTF - 8"?>
<web - app version = "2.4"
  xmlns = "http://java.sun.com/xml/ns/j2ee"
  xmlns:xsi = "http://www.w3.org/2001/XMLSchema - instance"
  xsi:schemaLocation = "http://java.sun.com/xml/ns/j2ee
```

```
http://java.sun.com/xml/ns/j2ee/web-app_2_4.xsd">
  <servlet>
    <servlet-name>he</servlet-name>
    <jsp-file>/WEB-INF/hell.jsp</jsp-file>
  </servlet>
  <servlet-mapping>
    <servlet-name>he</servlet-name>
    <url-pattern>/hello</url-pattern>
  </servlet-mapping>
</web-app>
```

上面的配置表示，将/WEB-INF/hello.jsp 文件映射成为一个/hello 的访问路径，当用户输入/hello 就会自动在<servlet-mapping>节点中配置的<serclet-name>中找到对应的<servlet>节点，并找到其中的<jsp-file>所指定的真实文件。

3.8.6 任务：初始化参数的配置

1. 任务要求

通过 web.xml 配置数据库访问参数，包括数据库名、数据库连接驱动类名、数据库访问用户名、数据库访问密码等。并通过 JSP 页面读取相关内容。

2. 任务实施

(1) 创建一个 Web 项目 test。

(2) 配置 web.xml 文件，代码如下。

```
<?xml version="1.0" encoding="UTF-8"?>
<web-app version="3.0"
    xmlns="http://java.sun.com/xml/ns/javaee"
    xmlns:xsi="http://www.w3.org/2001/XMLSchema-instance"
    xsi:schemaLocation="http://java.sun.com/xml/ns/javaee
    http://java.sun.com/xml/ns/javaee/web-app_3_0.xsd">
    <servlet>
        <servlet-name>test</servlet-name>
        <jsp-file>/TestParam.jsp</jsp-file>
            <init-param>
          <param-name>driver</param-name>
          <param-value>com.microsoft.jdbc.sqlserver</param-value>
        </init-param>
          <init-param>
          <param-name>dbName</param-name>
          <param-value>test.db</param-value>
          </init-param>
          <init-param>
            <param-name>userName</param-name>
            <param-value>sa</param-value>
          </init-param>
          <init-param>
```

```
            <param-name>userPwd</param-name>
            <param-value>123455</param-value>
        </init-param>
    </servlet>
    <servlet-mapping>
        <servlet-name>test</servlet-name>
        <url-pattern>/testparam</url-pattern>
    </servlet-mapping>
</web-app>
```

(3) 编写 JSP 页面文件，代码如下。

```
<%@ page language="java" import="java.util.*" pageEncoding="utf-8"%>
  <!DOCTYPE HTML PUBLIC "-//W3C//DTD HTML 4.01 Transitional//EN">
<html>
  <head>
      <title>初始参数</title>
  </head>
  <body>
      <p>数据库驱动:<%=config.getInitParameter("driver")%></p>
      <p>数据库名称:<%=config.getInitParameter("dbName")%></p>
      <p>用户名<%=config.getInitParameter("userName")%></p>
      <p>用户名<%=config.getInitParameter("userPwd")%></p>
  </body>
</html>
```

(4) 启动服务器，在浏览器地址栏中输入 http://localhost:8080/test/testparam，显示内容如下。

```
数据库驱动: com.microsoft.jdbc.sqlserver
数据库名称: test.db
用户名 sa
用户名 123455
```

项目 3 用户合法性访问验证

1. 项目要求

在系统中几乎都会包含用户登录和用户注销的功能，此功能可以使用 session 对象实现。具体思路是：当用户登录成功后，设置一个 session 对象的范围属性，然后在其他需要验证的页面中判断是否存在此属性。如果存在，则表示已经是正常登录的合法用户；如果不存在，则给出提示，并跳转回登录页面提示用户重新登录。用户登录后可以进行注销的操作。登录成功后显示当前访问的人数。

2. 项目实施

(1) 创建 Java Web 项目。

(2) 设计登录表单。实现用户填写登录信息,并向本页提交数据,以完成登录的验证。如果登录成功,则保存属性；如果登录失败,则显示登录失败的信息。

login.jsp 页面的代码如下。

```
<%@ page contentType = "text/html;charset = GBK" %>
<h1>系统登录</h1>
<form action = "login.jsp" method = "post">
  用户名:<input type = "text" name = "name"><br>
  密码:<input type = "password" name = "password"><br>
  <input type = "submit" value = "登录">
</form>
<%
  String name = request.getParameter("name");
  String pass = request.getParameter("password");
  if(!((""".equals(name)||name == null)&&("".equals(pass)||pass == null))){
    //假设用户名是 abc,密码是 123
    if("abc".equals(name)&&"123".equals(pass)){
      session.setAttribute("name",name); //登录成功则设置 Session
      response.setHeader("refresh","2;URL = welcome.jsp");
    //response.sendRedirect("welcome.jsp");
    }
%>
  <h1>ok<a href = "welcome.jsp">跳转按这里</a></h1>
<% else{ %>
    <h3>错误的用户名或密码</h3>
<%}
  }
%>
```

(3) 登录成功页面。当用户登录成功后此页面显示登录成功的信息。如果用户没有登录,则要给出用户没有登录的提示,同时给出一个登录的链接地址。

welcome.jsp 页面的代码如下。

```
<%@ page contentType = "text/html;charset = GBK" %>
<%
  if(session.getAttribute("name")!= null){ //合法用户
%>
    <h1>欢迎光临!</h1>
    <jsp:include page = "count.jsp"/>
    <h2><a href = "logout.jsp">注销</a></h2>
<%  }
else {  %>
    <h1>请先<a href = "login.jsp">登录</a>!</h1>
<% } %>
```

(4) 用户注销页面。此页面完成用户登录后的注销,注销后跳转到登录页面,等待用户继续登录。

logout.jsp 页面的代码如下。

```
<%@ page contentType = "text/html;charset = GBK" %>
<% session.invalidate() ; %>
```

(5) 设计网站计数器。使用 application 对象实现网站计数器。每次当有新用户登录时,就要求用户数加 1。在用户第一次登录时进行计数。在 Web 根目录中创建 count.txt 的文件,此文件的默认值是 1。

count.jsp 页面的代码如下。

```
<%@ page contentType = "text/html;charset = GBK" %>
<%@ page import = "java.io. * " %>
<%!
  //读取文件
  public int load(String path){
    int temp = 0 ;
    try{
      File f = new File(path) ;
      BufferedReader buf = new BufferedReader(new InputStreamReader(new FileInputStream(f))) ;
      String str = buf.readLine() ;
      temp = Integer.parseInt(str) ;
      buf.close() ;
    }catch(Exception e){}
    return temp ;
  }
  public synchronized void save(int c,String path){   //需要同步
    try{
      File f = new File(path) ;
      PrintStream ps = new PrintStream(new FileOutputStream(f)) ;
      ps.print(c) ;
      ps.close() ;
    }catch(Exception e){}
  }
%>
<%
  String path = getServletContext().getRealPath("/") + "count.txt" ;
  int count = load(path) ;
  if(session.isNew()){        //如果是第一次访问
    save(++count,path) ;
  }
%>
<h2>您是第<% = count %>位访问者!!</h2>
```

习题 3

1. 选择题

(1) (   )不是 JSP 的内置对象。

A. session   B. request   C. cookie   D. out

(2) 在 HTTP 中,用于发送大量数据的方法是(   )。

A. get   B. post   C. put   D. options

(3) request. getRequestDispatcher()和 forward(request,response)称为(　　)。

A. 流转　　B. 转发　　C. 重定向　　D. 导航

(4) 要在 JSP 中使用 ArrayList,以下代码正确的是(　　)。

A. <%import java. util. ArrayList%>

B. <%@import "java. util. ArrayList"%>

C. <%@ page import="java. util. ArrayList"%>

D. <%@ page package="java. util. ArrayList"%>

(5) 从 HTTP 请求中获得请求参数,应该调用(　　)方法。

A. request. getAttribute()

B. request. getParameter()

C. session. getAttribute()

D. session. getParameter()

(6) (　　)方法可用于检索 session 对象的属性 userid 的值。

A. session. getAttribute("userid")

B. session. setAttribute("userid")

C. request. getParameter("userid")

D. request. getAttribute("userid")

(7) test1. jsp 代码如下。

```
<html>
<jsp:include page="test2.jsp" flush="false">
<jsp:param name="color" value="red"/>
</jsp:include>
</html>
```

要在 test2. jsp 中输出参数 color 中的值,以下选项正确的是(　　)。

A. <%=request. getParameter("color")%>

B. <%=request. getAttribute("color")%>

C. <jsp:getParamname="color"/>

D. <jsp:includeparam="color"/>

(8) 在当前页面中包含 a. htm 的正确语句是(　　)。

A. <%@include="a. htm"%>　　B. <jsp:includefile="a. htm"/>

C. <%@ include page="a. htm"%>　　D. <%@ include file="a. htm"%>

(9) page 指令用于定义 JSP 文件中的全局属性,下列关于该指令用法的描述不正确的是(　　)。

A. <%@ page%>作用于整个 JSP 页面

B. 可以在一个页面中使用多个<%@ page%>指令

C. 为增强程序的可读性,建议将<%@ page%>指令放在 JSP 文件的开头,但不是必需的

D. <%@ page%>指令中的属性只能出现一次

(10) 在 JSP 中，request 对象的(　　)方法可以获取页面请求中一个表单组件对应多个值时的用户的请求数据。

A. String getParameter(String name)

B. String getParameterValuses(String name)

C. String[] getParameter(String name)

D. String[] getParameterValues(String name)

(11) session 对象的 getId()方法的作用是(　　)。

A. 取得 session 对象的结束时间

B. 取得客户端最后一次访问服务器的时间

C. 取得客户端在服务器唯一的标识

D. 取得 session 对象的创建时间

(12) 某 JSP 页面中有如下代码：

```
<%
  pageContext.setAttribute("a","page");
  request.setAttribute("a","request");
  session.setAttribute("a","session");
  application.setAttribute("a","application");
%>
```

则 ${a}的显示结果为(　　)。

A. page　　B. request

C. session　　D. application

(13) 要在 session 对象中保存属性，可以使用(　　)语句。

A. session.getAttribute("key","value");

B. session.setAttribute("key","value");

C. session.setAtrribute("key");

D. session.getAttribute("key");

(14) 如果只希望在多个页面间共享数据，可以使用(　　)作用域。

A. request，session　　B. application，session

C. request，application　　D. pageContext，request

(15) 某 JSP 页面中有如下代码：

```
<%
  Cookie c = new Cookie("name","admin");
  c.setMaxAge(10000);
  response.addCookie(c);
%>
```

则(　　)可以正确显示 admin。

A. ${cookie.name}　　B. ${cookie.name.value}

C. ${name}　　D. ${name.value}

2. 简答题

（1）阐述 JSP 中的 4 种会话作用域。

（2）描述 Cookie 和会话的作用和区别。

（3）简述重定向和跳转的区别。

（4）编写一个模拟用户登录页面和用户注册页面。

（5）编写一个网络考试选择题页面，提交后，显示正确答案和答题结果。

第 4 章 JavaBean 技术

搭积木是指将一堆毫不相干的小木块，经过精心的设计和合理的安排组装成想要的建筑。Java EE 程序是基于组件开发的，就和搭积木一样，一个完整的系统由若干个程序模块组成。在前面的程序设计中，在 JSP 页面中嵌入了大量的 Java 代码，JSP 页面中大量的 Java 代码与 HTML 标签混杂，导致在维护和修改上比较困难。为了分离页面中的 HTML 代码和 Java 代码，可以编写一些类来封装页面的数据和业务逻辑，这样不但能够使代码复用，还能将不同逻辑功能分开。这可以在 JSP 技术中使用 JavaBean 组件来实现。

【技能目标】 掌握使用 JavaBean 封装业务逻辑的方法。

【知识目标】 JavaBean 的实际应用。

【关键词】 接口(interface)　监听(listening)　作业域(scope)　属性(property)　可重用的组件(JavaBean Java)

4.1 JavaBean 的构建

4.1.1 JavaBean 概述

在 JSP 页面开发的初级阶段，没有框架与逻辑分层概念，对业务的处理采用将 Java 代码嵌入到 JSP 页面中的方法，其系统架构如图 4-1 所示。

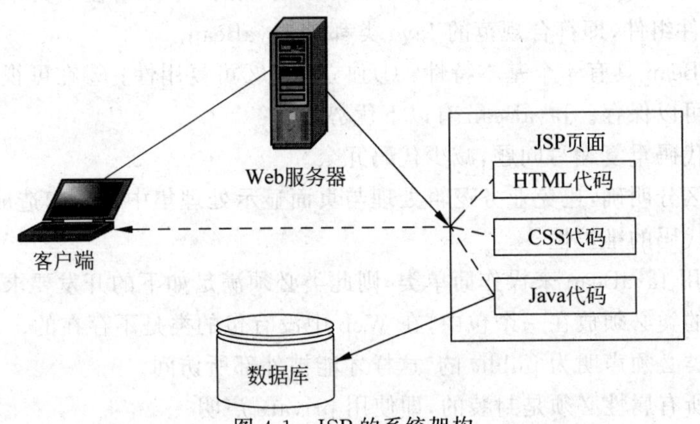

图 4-1 JSP 的系统架构

此种开发方式虽然看似流程简单，但在 JSP 页面中包含 HTML 代码、CSS 代码、Java 代码等，既不利于页面编程人员的设计，也不利于 Java 程序员对程序的开发，而且将 Java 代码写入到 JSP 页面中，不能体现面向对象的开发模式，达不到代码复用的目标，开发效率低，同时系统维护困难。

如果使 HTML 代码与 Java 代码相分离，将 Java 代码单独封装成为一个处理某种业务逻辑的类，然后在 JSP 页面中调用此类，则可以降低 HTML 代码与 Java 代码之间的耦合度，提高 Java 程序代码的复用性及灵活性。

这种与 HTML 代码相分离，而使用 Java 代码封装的类，就是一个 JavaBean 组件。可使用 JavaBean 组件来完成业务逻辑的处理。JavaBean 与 JSP 整合的开发模式如图 4-2 所示。

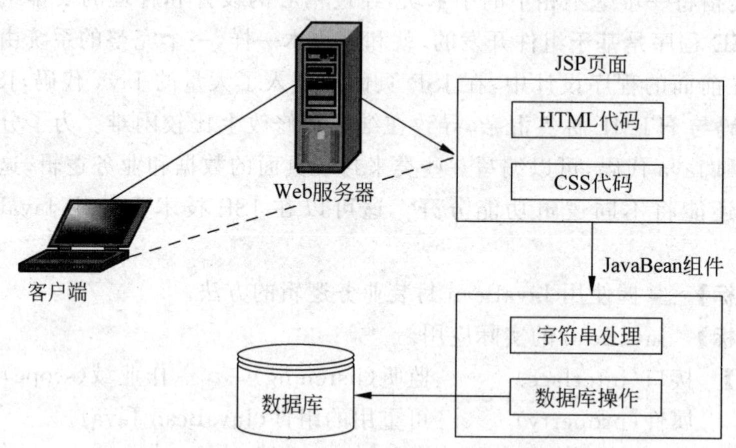

图 4-2　JavaBean 与 JSP 整合的开发模式

1. 什么是 JavaBean

在 JavaBean 标准的定义中，Bean 的正式说法是："Bean 是一个基于 Sun 公司的 JavaBean 标准的、可在编程工具中被可视化处理的可复用的软件组件。"

所以，JavaBean 是基于 Sun 公司的 JavaBean 标准的，可在编程工具中被可视化处理的可复用的软件组件，即符合规范的 Java 类都是 JavaBean。

因此 JavaBean 具有 4 个基本特性：①独立性；②可复用性；③在可视化开发工具中使用；④状态可以保存。JavaBean 有以下优势。

(1) 解决代码重复编写问题，减少代码冗余。

(2) 功能区分明确，避免业务逻辑处理与页面显示处理集中在一起造成混乱。

(3) 提高代码的维护性。

如果要应用 JavaBean 来操作简单类，则此类必须满足如下的开发要求。

(1) 所有的类必须放在一个包中，在 Web 中没有包的类是不存在的。

(2) 所有类必须声明为 public 的，这样才能被外部所访问。

(3) 类的所有属性必须是封装的，即使用 private 声明。

(4) 封装的属性如果需要外部操作，必须编写对应的 setter、getter 方法。

(5) 一个 JavaBean 中至少存在一个无参构造方法，此方法可为 JSP 中的标签所使用。

2. JavaBean 分类

JavaBean 有以下两种类型。

（1）可视化的 JavaBean：在传统的应用中，JavaBean 主要用于实现一些可视化界面，如一个窗体、按钮、文本框等，这样的 JavaBean 称为可视化的 JavaBean。

（2）非可视化的 JavaBean：主要用于实现一些业务逻辑或封装一些业务对象，由于这样的 JavaBean 并没有可视化的界面，所以称为非可视化的 JavaBean。

可视化的 JavaBean 一般应用于 Swing 的程序中，在 Java Web 开发中并不会采用，而是使用非可视化的 JavaBean，实现一些业务逻辑或封装一些业务对象，本书主要介绍非可视化的 JavaBean。

3. JavaBean 结构要素

通常一个标准的 JavaBean 组件有如下几项特性。

（1）JavaBean 类使用 public 声明。

（2）JavaBean 类中必须存在一个无参构造方法。

（3）JavaBean 类中必须包含成员变量。

（4）JavaBean 类中必须包含用于给成员变量赋值、取值的 setXXX()方法与 getXXX()方法。

（5）JavaBean 类必须实现 java.io.Serializable 接口。

实现 java.io.Serializable 接口的对象被 JVM（Java 虚拟机）转化为一个字节序列，并且能够将这个字节序列完全恢复为原来的对象，序列化机制可以弥补网络传输中不同操作系统的差异问题。作为 JavaBean，对象的序列化也是必需的。

如果在 JSP 中使用 JavaBean 组件，不必实现 java.io.Serializable 接口，JavaBean 组件仍然可以运行。

创建 JavaBean 的语法格式如下。

```
package 包名;
public class 类名 implements java.io.Serializable{
  构造方法();
  属性(Property);
  方法(Method);
}
```

【例 4-1】 第一个 JavaBean。

```
package cn.edu.mypt.bean;                                   //包
public class UserBean implements java.io.Serializable{      //实现序列化
    private String userID,userName,userPwd,userType;
    public String getUserID() {
      return userID;
    }
    public void setUserID(String userID) {
      this.userID = userID;
    }
    public String getUserName() {                           //userName 属性的 getXXX()方法
      return userName;
    }
```

```
    public void setUserName(String userName) {              //userName 属性的 setXXX()方法
      this.userName = userName;
    }
    public String getUserPwd() {
      return userPwd;
    }
    public void setUserPwd(String userPwd) {
      this.userPwd = userPwd;
    }
    public String getUserType() {
      return userType;
    }
    public void setUserType(String userType) {
      this.userType = userType;
    }
    public UserBean(){}       //创建无参构造方法
    public UserBean(String userID, String userName, String userPwd, String userType){
      this.userID = userID;
      this.userName = userName;
      this.userPwd = userPwd;
      this.userType = userType;
    }
    public UserBean(String userID, String userName, String userPwd){
      this(userID, userName, userPwd, null);
    }
    public UserBean(String userID, String userName){
      this(userID, userName, null, null);
    }
}
```

上述这个 JavaBean 具备了 JavaBean 的所有特性。声明了 4 个 Sting 类型的属性，分别为 userID、userName、userType 和 userPwd，并且分别为每个属性定义两个方法：setXXX()方法与 getXXX()方法。

4.1.2 JavaBean 的配置

1. Web 开发的标准目录结构

Java Web 应用具有固定的目录结构。开发一个 Java Web 应用，首先应该创建这个 Web 应用的目录结构，如图 4-3 和表 4-1 所示。

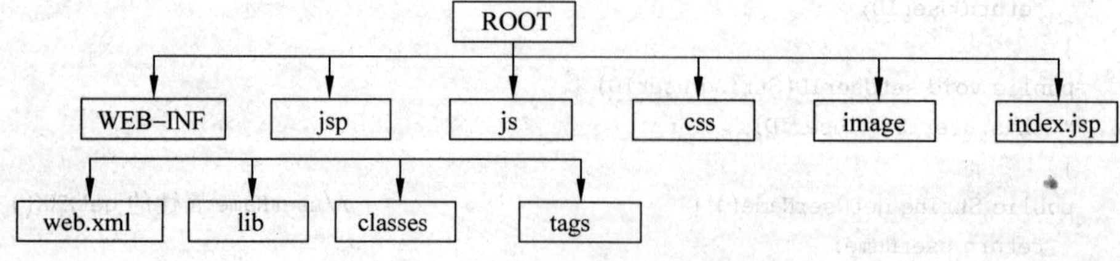

图 4-3 Java Web 应用的目录结构

表 4-1 Java Web 应用的目录结构

目录名	作 用
WEB ROOT	Web 的根目录，一般虚拟目录直接指向此文件夹，此文件夹下必然包含 WEB-INF
WEB-INF	Web 目录中最安全的文件夹，保存各种类、第三方 JAR 包、配置文件
web. xml	Web 的部署描述符
classes	保存所有的 JavaBean，如果不存在，可以手动创建
lib	保存所有的第三方 JAR 文件
tags	保存所有的标签文件
jsp	存放. jsp 文件，一般根据功能再创建子文件夹
js	存放所有需要的. js 文件
css	样式文件的保存路径
images	存放所有的图片文件

除上面的目录结构外，所有的目录基本上都会在根目录中存放一个首页文件，首页文件一般是以 index. htm、index. jsp、index. html 等命名的。

2. 手动配置 JavaBean

手动配置 JavaBean 采用如下步骤。

(1) 创建目录。在 Web 应用的 WEB-INF\classes 目录下创建 JavaBean 对应的路径，如 Tomcat 目录下的 wbapps\mysite\WEB-INF\classes\bean。

(2) 创建并编译。将编写的 JavaBean 文件 User. java 复制到新创建的目录下，进入该目录，使用 javac user. java 命令进行编译，如果编译成功，则将在当前目录生成 User. class 文件。

(3) 重启服务器。重启服务后，则在 Web 应用的 mysite 下的任何 JSP 页面中都可以使用全类名 bean. User 的 JavaBean。

3. 使用 IDE 配置 JavaBean

使用 MyEclipse 集成开发工具来配置 JavaBean 最为快捷、方便，其步骤如下。

(1) 在 Web 应用 mysite 的 src 目录中创建 bean 包。

(2) 在 bean 包中新建 User. java 类文件，编辑完成后，MyEclipse 将自动编译。

(3) 单击工具栏中的"部署"按钮，部署应用到 Tomcat 服务器。

4.1.3 JavaBean 成员

1. 属性

一个简单的属性表示为一般数据类型的变量，它的 setXXX()方法与 getXXX()方法以属性来命名，例如一个简单的属性 userName，假设它的类型为 String，可以采用如下代码定义 JavaBean。

```
private String userName;                    //定义 String 类型的简单属性
public String getUserName() {               //简单属性的 getXXX()方法
  return name;
}
public void seUsertName(String name) {      //简单属性的 setXXX()方法
  this.name = name;
}
```

如果这个简单属性的数据类型为 boolean,除了可以使用 setXXX()方法外,还可以采用 isXXX()方法来代替 getXXX()方法。例如:

```
public boolean isFlag() {                    //boolean 型属性的取值方法
  return flag;
}
public void setFlag(boolean flag) {          //boolean 型属性的 setXXX()方法
  this.flag = flag;
}
```

2. 使用索引

一个索引属性可以表示为数组或集合。对于索引属性,需要提供两套对应的 getXXX()与 setXXX()方法,一套用来设置整个数组,一套用来获得或设定数组中的某个元素。例如,对数组 userList,假设它的数据类型为 String,则关键代码如下。

```
public String[] userList = new String[6];
public String[] getUserList() {                          //返回整个数组
  return userList;
}
public void setUserList(String[] userList) {             //为整个数组赋值
  this.userList =  userList;
}
public void setUserList(int index,String value){         //为数组中的某个元素赋值
  this.userList[index] = value;
}
public String getUserList(int index){                    //返回数组中的某个值
  return userList[index];
}
```

索引属性除了可以表示数组外,也可以表示其他集合。以 Map 集合为例,同样需要使用两套 setXXX()方法与 getXXX()方法。一套用来设置整个集合,另一套是用来设置集合中某个元素,代码如下。

```
public void setMap(Object value,Object key){             //为 Map 集合中的某个键值赋值
  map.put(value, key);
}
public Object getMap(Object key){                        //返回 Map 集合中的某个键值
  return map.get(key);
}
public void setMap(Map map) {                  //为 Map 集合赋值
```

```
    this.map = map;
  }
  public Map getMap(){                         //返回整个 Map 集合
    return this.map;
  }
```

其中为 Map 集合中某个键值赋值的 setXXX()方法与以往不同,其参数有两个,这是因为 Map 接口中定义了一对数值,分别为"数值"与之关联的"键",在 setXXX()方法内使用 Map 接口中定义的 put()方法将"数值"与"键"放入集合中。返回 Map 集合中某个"键"值数据的 getXXX()方法也与以往不同,参数为"键"值,在 getXXX()方法内使用 get()方法返回与此"键"值关联的"数值"。

3. 方法

JavaBean 的方法与一般的方法没有本质区别,除了 setXXX()方法与 getXXX()方法外,在 JavaBean 中可以定义自己的方法。

【例 4-2】 定义一个类 UserModel.java,其具有用户登录方法

```
public class UserModel {
  static List<UserBean> userList = new ArrayList();
  public UserBean userLogin(String userId, String userPwd) {
    UserBean u = findById(userId);       //根据用户 ID 查找,存在则返回一个 UserBean 对象,
                                          //否则为 null
    if (u == null) return null;
    if (userPwd.equals(u.getUserPwd())) return u;
    return null;
  }
}
```

4. 监听

如果在简单属性或索引属性上添加一种监听机制,即当某个属性值发生改变时则通知监听器,这个属性就属于绑定属性。

监听器需要实现 java.beans.PropertyChangeListener 接口,负责接收由 JavaBean 组件产生的 java.beans.PropertyChangeEvent 对象,在 PropertyChangeEvent 对象中包含了发生改变的属性名称、改变前的值和改变后的值,以及每个监听器可能要访问的新属性的值。

JavaBean 还需要实现 addPropertyChangeListener()方法和 removePropertyChangeListener()方法,以便添加和取消属性变化的监听器。还可以通过 java.beans.PropertyChangeSupport 类的 firePropertyChange()方法传递属性名称、改变前的值和改变后的值等信息。

5. 约束

约束的属性与绑定属性类似,但是属性值的变化首先要被所有的监听器验证之后,值

的变化才能由 JavaBean 组件发生作用。只要有一个监听器否决了该属性的变化,该属性的值都不能被真正修改。这种添加了监听约束的属性就属于约束属性。

监听器需要实现 java. beans. VetoableChangeListener 接口,负责接收由 JavaBean 组件产生的 java. beans. PropertyChangeEvent 对象。JavaBean 组件可以通过 java. beans . VetoableChangeSupport 类激活由监听器接收的实际事件。

JavaBean 还需要实现 addVetoableChangeListener()方法和 removeVetoableChangeListener()方法,以便添加和取消属性变化的监听器。还可以通过 java. beans. VetoableChangeSupport 类的 fireVetoableChange()方法传递属性名称、改变前的值和改变后的值等信息。

4.1.4 任务:用户 JavaBean 的定义

1. 任务要求

(1) 定义用户 JavaBean,包括用户名、用户 ID、用户密码、用户类型(普通用户、系统管理员等)等基本信息。

(2) 定义一个用户管理处理业务的 JavaBean,具有用户登录、查找等方法。

2. 任务实施

(1) 新建一个 Java Web 项目 Demo。

(2) 在 src 下创建 JavaBean,包括用户名、用户 ID、用户密码、用户类型(普通用户、系统管理员等)等属性。其代码参考例 4-1。

(3) 定义一个用户管理处理业务的 JavaBean。

```
public class UserModel {
    static List < UserBean > userList = new ArrayList();
    public UserBean userLogin(String userId, String userPwd) {
        UserBean u = findById(userId);
        if (u == null) return null;
        if (userPwd.equals(u.getUserPwd())) return u;
        return null;
    }
    public UserBean findById(String userID) {
        int n = userList.size();
        int i;
            for (i = 0;i < n && !userID.equals(userList.get(i).getUserID());i++);
        if (i < n) return userList.get(i); else return null;
    }
}
```

4.2 应用 JavaBean

在 JSP 页面中应用 JavaBean 非常简单,一种方法是使用 page 指令导入 JavaBean,另一种方法是通过 JSP 动作标签< jsp:useBean >、< jsp:setProperty >、< jsp:getProperty >

来实现对 JavaBean 对象的操作，但所编写的 JavaBean 对象要遵循 JavaBean 标准，只有严格遵循 JavaBean 标准，在 JSP 页面才能够方便地调用及操作 JavaBean。

在 JSP 页面中，JavaBean 的生命周期可以自行设置，通常作用在 page、request、session、application 4 个范围内。默认的情况下，JavaBean 作用于 page 范围之内。

4.2.1　用 page 指令导入 JavaBean

通过<%@ page import＝" * " %> JSP 指令导入 JavaBean 类，例如：

```
<%@ page import = "cn.edu.mypt.bean" %>
```

【例 4-3】 导入并使用 JavaBean。

```
<%@ page contentType = "text/html" pageEncoding = "GBK" %>
<%@page import = "cn.edu.mypt.bean. * " %>
  <html>
  <head><title>JavaBean 测试</title></head>
  <body>
    <%
      UserBean u = new UserBean();
      u.setUserID("100");
      u.setUserName("张山");
       %>
      <h2>编号<% = c.getUserID() %></h2>
      <h2>姓名<% = c.getUserName() %></h2>
  </body>
</html>
```

当程序运行时，程序将所需的开发包导入 JSP 文件中，然后产生 User 的实例化对象，并调用其中的 setter 和 getter 方法。

4.2.2　用标签访问 JavaBean

采用 JSP 标签访问 JavaBean，可以减少 JSP 页面中的 Java 程序代码，使 HTML 与 Java 程序分离，增加 JSP 页面代码的可读性与维护性。

1. 创建一个 JavaBean 实例

语法格式如下。

```
<jsp:useBean id = "beanName" scope = "page|request|session|application" class = "className" />
```

或者

```
<jsp:useBean id = "beanName" scope = "page|request|session | application" class = "className ">
//初始化代码
</jsp:useBean>
```

(1) id：用于在所定义的范围中确认 JavaBean 的变量，使之能在后面的程序中使用此变量名来分辨不同的 JavaBean，这个变量名对大小写敏感。如果 JavaBean 已经在别的<jsp:useBean>标签中创建,则当使用这个 JavaBean 时,id 的值必须与原来的那个 id 值一致;否则则意味着创建了同一个类的两个不同的对象。

(2) scope:用于设置 Bean 存在的范围以及 id 的有效范围,有 page、request、session、application 4 种选择,默认值是 page。

(3) class:使用 new 关键字以及 class 的构造方法从一个类实例化一个 Bean。这个类不能是抽象的,必须有一个公用的、没有参数的构造方法。

【例 4-4】 使用 JSP 中的标签指令调用 JavaBean。

```
<%@ page contentType = "text/html" pageEncoding = "GBK" %>
  <jsp:useBean id = "u" scope = "page" class = "cn.edu.mypt.bean.UserBean"/>
  <html>
  <head><title>标签指令调用 javaBean</title></head>
  <body>
    <%
      u.setUserid("100");
      u.setUserName("张山");
    %>
    <h2>编号<% = u.getUserID() %></h2>
    <h2>姓名<% = u.getUserName() %></h2>
  </body>
</html>
```

2. 设置属性

<jsp:setProperty>标签有 4 种使用方法。

1) 自动匹配

语法格式如下。

```
<jsp:setProperty name = "实例化对象名" property = " * "/>
```

【例 4-5】 JavaBean 与表单。

表单代码如下。

```
<%@ page contentType = "text/html" pageEncoding = "GBK" %>
  <html>
    <head><title>JavaBean 与表单</title></head>
    <body>
    <form action = "input_bean.jsp" method = "post">
      编号:<input type = "text" name = "userID"><br>
      姓名:<input type = "text" name = "userName"><br>
       <input type = "submit" value = "submit"/>
       <input type = "reset" value = "reset"/>
     </form>
  </body>
</html>
```

其中,表单中的 userID 和 userName 对应 JavaBean 中的属性,以达到自动匹配的目的。

接收表单数据页面 input_bean.jsp 的代码如下。

```
<%@ page language="java" import="java.util.*" pageEncoding="GBK"%>
  <html>
    <head><title>接收表单数据的程序</title></head>
    <body>
      <% request.setCharacterEncoding("GBK"); %>
      <jsp:useBean id="u" scope="page" class="cn.edu.mypt.bean.UserBean"/>
      <jsp:setProperty name="u" property="userName"/>
      <h2>编号<%=u.getUserID() %></h2>
      <h2>姓名<%=u.getUserName() %></h2>
  </body>
</html>
```

在进行自动匹配时，参数的名称是通过表单控件指定的，如果符合则会自动调用对应的 setter 方法进行内容的设置。

2）设置属性

语法格式如下。

```
<jsp:setProperty name="实例化对象名"property="属性名称"/>
```

如果在使用<jsp:setProperty>设置属性时没有指定"*"，而指定了具体的属性，那么表示只为具体的属性设置相应的内容。

【例 4-6】 设置 userName 属性。

```
<%@ page language="java" import="java.util.*" pageEncoding="GBK"%>
  <html>
    <head><title>设置属性</title></head>
    <body>
      <% request.setCharacterEncoding("GBK"); %>
      <jsp:useBean id="u" scope="page" class="cn.edu.mypt.bean.UserBean"/>
     <jsp:setProperty name="u" property="userName"/>
     <h2>编号<%=u.getUserID() %></h2>
     <h2>姓名<%=u.getUserName() %></h2>
  </body>
</html>
```

在这里只设置了 userName 属性的内容，所以在输出时只有 userName。

3）指定参数

语法格式如下。

```
<jsp:setProperty name="实例化对象名" property="属性名称"param="参数名称"/>
```

可以通过 param 来指定属性的具体参数。

【例 4-7】 将指定的参数赋值给指定的属性。

```
<%@ page language="java" import="java.util.*" pageEncoding="GBK"%>
  <html>
    <head><title>测试</title></head>
```

```
    <body>
        <% request.setCharacterEncoding("GBK"); %>
        <jsp:useBean id = "u" scope = "page" class = "cn.edu.mypt.bean.UserBean"/>
        <jsp:setProperty name = "u" property = "userName" param = "uid"/>
        <jsp:setProperty name = "u" property = "userID" param = "uName"/>
        <h2>编号<% = u.getUserID() %></h2>
        <h2>姓名<% = u.getUserName() %></h2>
    </body>
</html>
```

4）指定内容

语法格式如下。

```
<jsp:setProperty name = "实例化对象名"property = "属性名称" value = "内容"/>
```

【例 4-8】 给对象属性赋值。

```
<%@ page language = "java" import = "java.util.*" pageEncoding = "GBK" %>
  <html>
    <head><title>给对象属性赋值</title></head>
    <body>
      <% request.setCharacterEncoding("GBK"); %>
      <% String id = "10110"; %>
      <jsp:useBean id = "user" scope = "page" class = "cn.edu.mypt.bean"/>
      <jsp:setProperty name = "user" property = "userName" value = "张山"/>
      <jsp:setProperty name = "user" property = "userID" param = <% = id %>/>
      <h2>编号<% = user.getUserID() %></h2>
      <h2>姓名<% = user.getUserName() %></h2>
  </body>
</html>
```

3. 获取属性

语法格式如下。

```
<jsp:getProperty name = "beanInstanceName" property = "propertyName" />
```

其中，name 为 JavaBean 的名称，由<jsp:useBean>指定；

property 为 JavaBean 的属性名。

【例 4-9】 获取对象属性。

```
<%@ page language = "java" import = "java.util.*" pageEncoding = "GBK" %>
  <html>
    <head><title>获取对象属性值</title></head>
    <body>
      <% request.setCharacterEncoding("GBK"); %>
      <jsp:useBean id = "user" scope = "page" class = "cn.edu.mypt.bean.USerBean"/>
      <jsp:setProperty name = "user" property = "userName"/>
      <h2><jsp:getProperty name = "user" property = "userID" /></h2>
      <h2>姓名<% = c.getUName() %></h2>
```

```
  </body>
</html>
```

4.2.3 JavaBean 的移除

JavaBean 使用结束后,如果需要释放其占用的资源,可以对 JavaBean 做移除操作。

对于作用域为 page 或 request 的 JavaBean,并不需要手动移除,因为这两个作用域的有效范围很小,如果必须使用代码移除,可以使用如下代码。

```
pageContext.removeAttribute(String name);
request.removeAttribute(String name);
```

对于作用域为 application 的 JavaBean,通常情况下都不能被移除,因为处于该作用域的 JavaBean 都是需要被永久保存的,例如网站的浏览次数等。

对于作用域为 session 的 JavaBean,有些需要在 Session 范围内永久保存,如用户的登录信息。而有些并不需要永久保存,针对这样的 JavaBean,在使用结束后,要及时通过 removeAttribute(String name)方法将其移除,释放其占用的资源,否则如果在 Session 作用域中保存大量的垃圾 JavaBean,将严重影响服务器的性能。

可以使用如下代码将作用域为 session 的 JavaBean 对象移除。

```
session.removeAttribute(String name);
```

4.2.4 任务:显示用户所有信息

1. 任务要求

在页面中显示用户的所有信息。

2. 任务实施

(1) 新建一个 Java Web 项目 mytest。

(2) 在 src 下创建 JavaBean,包括用户名、用户 ID、用户密码、用户类型(普通用户、系统管理员等)等属性,代码如下。

```
package cn.edu.mypt.bean;
public class UserBean {
  private String userID,userName, userPwd,userType;
  //getXxx,setXxx 略
  public UserBean(){}
  public UserBean(String userID,String userName,String userPwd,String userType){
    this.userID = userID;
    this.userName = userName;
    this.userPwd = userPwd;
    this.userType = userType;
  }
  public UserBean(String userID,String userName,String userPwd){
```

```
        this(userID,userName,userPwd,null);
    }
    public UserBean(String userID,String userName){
        this(userID,userName,null,null);
    }
}
```

(3) 定义一个 JavaBean,用于获取用户的信息。

```
package cn.edu.mypt.model;
import java.util.List;
import cn.edu.mypt.bean.UserBean;
public class UserModel {
    static List<UserBean> userList = new ArrayList();
    public List<UserBean> getUser() {
        return userList;
    }

    public void load(){
        userList.add(new UserBean("1","zhangsan"));
        userList.add(new UserBean("2","lishi"));
        userList.add(new UserBean("3","wang"));
        userList.add(new UserBean("4","chen"));
    }
}
```

(4) 设计显示用户信息的页面 userList.jsp。

```
<%@ page language = "java" import = "java.util.*" pageEncoding = "utf-8"%>
<!DOCTYPE HTML PUBLIC "-//W3C//DTD HTML 4.01 Transitional//EN">
<html>
    <head><title>显示用户信息</title></head>
    <body>
    <jsp:useBean id = "userModel" class = "cn.edu.mypt.model.impl.UserModel" scope =
"application"/>
    <jsp:useBean id = "user" class = "cn.edu.mypt.bean.UserBean" scope = "page"/>
        <%
            userModel.load();
            List userList = userModel.getUser();
            if (userList!= null){
            int n = userList.size();
            for (int i = 0;i<n;i++)
                user = (cn.edu.mypt.bean.UserBean)userList.get(i);
            %>
                <% = user.getUserID() %>------<% = user.getUserName() %><br>
            <% }
                }
                else {
            %>
            <h1>没有数据</h1>
            <% } %>
```

(5) 启动服务器，部署项目，在浏览器中浏览 userList.jsp 页面。

4.3　JavaBean 的保存范围

在<jsp:useBean>标签中有一个 scope 属性，表示 JavaBean 的保存范围，保存的范围共有 4 种。

(1) page：保存在一页的范围内，跳转之后此 JavaBean 无效。

(2) request：JavaBean 对象可以保存在一次服务器跳转范围内。

(3) session：在一个用户的操作范围中保存，重新打开浏览器的时候才会产生新的 JavaBean。

(4) application：在整个服务器上保存，服务器关闭时才会消失。

下面基于网站计数器来讨论 JavaBean 的作用域范围。计数器的定义如下。

```
package cn.edu.mypt.bean;
public class Counter {
  static private int count;
    public int getCount() {
      return ++count;
    }
    public void setCount(int count) {
        this.count = count;
    }
}
```

4.3.1　page 范围的 JavaBean

page 范围的 JavaBean 可以在此文件以及文件中的所有静态包含文件中使用，直到页面执行完毕向客户端发回响应或转到另一个文件为止。

【例 4-10】　page 范围的 JavaBean。

PageDemo01.jsp 页面的代码如下。

```
<%@ page language = "java" import = "java.util.*" pageEncoding = "utf-8" %>
<!DOCTYPE HTML PUBLIC "-//W3C//DTD HTML 4.01 Transitional//EN">
<html>
  <head><title>>定义 page 范围的 JavaBean</title></head>
  <body>
    <jsp:useBean id = "count" class = "cn.edu.mypt.bean.Counter" scope = "page"/>
      <h1>访问了
      <font color = "red"><jsp:getProperty name = "count" property = "count"/></font>次!
      </h1>
  <!-- <jsp:forward page = "PageDemo02.jsp"/> 测试注释 --->
  </body>
</html>
```

PageDemo02.jsp 页面的代码如下。

```
<%@ page language="java" import="java.util.*" pageEncoding="utf-8"%>
<!DOCTYPE HTML PUBLIC "-//W3C//DTD HTML 4.01 Transitional//EN">
<html>
    <head><title>跳转后的页面</title></head>
  <body>
    <jsp:useBean id="count" class="cn.edu.mypt.bean.Counter" scope="page"/>
    <h1>访问了<font color="red">
    <jsp:getProperty name="count" property="count"/>
    </font>次!</h1>
  </body>
</html>
```

4.3.2 request 范围的 JavaBean

request 范围的 JavaBean 可以在任何执行相同请求的 JSP 文件中使用，直到页面执行完毕向客户端发回响应或转到另一个文件为止。可以使用 request 对象访问这个 Bean，例如：

```
request.getAttribute(beanInstanceName)
```

【例 4-11】 request 范围的 JavaBean。

RequestDemo01.jsp 页面的代码如下。

```
<%@ page contentType="text/html;charset=GBK"%>
<jsp:useBean id="count" class="cn.edu.mypt.bean.Counter" scope="request"/>
<h1>访问了<font color="red">
<jsp:getProperty name="count" property="count"/>
</font>次!</h1>
<jsp:forward page="RequestDemo02.jsp"/>
```

RequestDemo02.jsp 页面的代码如下。

```
<%@ page contentType="text/html;charset=GBK"%>
<jsp:useBean id="count" class="cn.edu.mypt.bean.Counter" scope="request"/>
<h1>访问了<font color="red">
<jsp:getProperty name="count" property="count"/>
</font>次!</h1>
```

4.3.3 session 范围的 JavaBean

session 范围的 JavaBean 自其创建开始，就能在任何使用同一 session 对象的 JSP 文件中使用。这个 JavaBean 存在于整个 Session 生存周期内，任何分享此 session 对象的 JSP 文件都能使用同一 JavaBean。

【例 4-12】 session 范围的 JavaBean。

```
<%@ page contentType="text/html;charset=GBK"%>
<jsp:useBean id="count" class="cn.edu.mypt.bean.Counter" scope="session"/>
<h1>访问了<font color="red">
```

```
    <jsp:getProperty name="count" property="count"/></font>次!
</h1>
```

4.3.4 application 范围的 JavaBean

application 范围的 JavaBean 自其创建开始,就能在任何使用同一 application 对象的 JSP 文件中使用。这个 JavaBean 存在于整个 application 生存周期内,任何分享此 application 对象的 JSP 文件都能使用同一 Bean。

【例 4-13】 application 范围的 JavaBean。

application.jsp 的代码如下。

```
<%@ page contentType="text/html;charset=GBK"%>
<jsp:useBean id="count" class="cn.edu.mypt.bean.Counter" scope="application"/>
<h1>访问了<font color="red">
    <jsp:getProperty name="count" property="count"/>
</font>次!</h1>
```

4.3.5 任务:用户登录权限的控制

1. 任务要求

只有当用户登录后才能访问 success.jsp 页面。

2. 任务实施

(1) 新建一个 Java Web 项目 mytest。

(2) 在 src 下创建 JavaBean,包括用户名、用户 ID、用户密码、用户类型(普通用户、系统管理员等)等属性。其代码参考 4.2 节的任务。

(3) 定义一个 JavaBean,用于获取用户的信息以及用户登录。

```
package cn.edu.mypt.model.impl;
//导入包,略
public class UserModel {
    static List<UserBean> userList = new ArrayList();
    //用户登录,根据用户 ID 和密码登录,登录成功返回用户信息,否则返回 null
    public UserBean userLogin(String userId, String userPwd) {
        UserBean u = findById(userId);
        System.out.println("--login-1--");
        if (u == null) return null;
        if (userPwd.equals(u.getUserPwd())) return u;
        return null;
    }
    //根据用户 ID 查找用户,若找到返回用户信息,否则返回 null
    public UserBean findById(String userID) {
        int n = userList.size();
        int i;
        for (i = 0;i < n && !userID.equals(userList.get(i).getUserID());i++)
```

```
      ;
    if (i<n) return userList.get(i); else return null;
  }
  //获取所有用户信息
  public List<UserBean> getUserList() {
    return userList;
  }
  //模拟数据
  public void load() {
    userList.add(new UserBean("1","zhangsan","123456"));
    userList.add(new UserBean("2","lishi","123456"));
    userList.add(new UserBean("3","wang","123456"));
    userList.add(new UserBean("4","chen","123456"));
  }
}
```

(4) 设计用户登录页面 login.jsp。

```
<%@ page language="java" import="java.util.*" pageEncoding="utf-8"%>
<!DOCTYPE HTML PUBLIC "-//W3C//DTD HTML 4.01 Transitional//EN">
<html>
  <head><title>用户登录</title></head>
  <body>
  <%
    String uID,uPwd,msg;
    msg=(String)request.getAttribute("message");
    uID=(String)request.getAttribute("userID");
    uPwd=(String)request.getAttribute("userPwd");
    uID=(uID==null)?"":uID;
    uPwd=(uPwd==null)?"":uPwd;
    msg=(msg==null)?"":msg;
   %>
  <%=msg%>
    <form action="dologin.jsp" method="post">
      用户 ID:<input type="text" name="userID" value="<%=uID%>"/>
      用户名:<input type="text" name="userPwd" value="<%=uPwd%>"/>
      <input type="submit" value="登录"/>
    </form>
  </body>
</html>
```

(5) 设计用户登录处理页面 dologin.jsp。

```
<%@ page language="java" import="java.util.*" pageEncoding="utf-8"%>
  <jsp:useBean id="userModel" class="cn.edu.mypt.model.impl.UserModel" scope="application"/>
  <jsp:useBean id="user" class="cn.edu.mypt.bean.UserBean" scope="page"/>
  <%
  //userModel.load(); //初始化用户信息
  String uID,uPwd;
```

```
  request.setCharacterEncoding("utf-8");
  uID = request.getParameter("userID");
  uPwd = request.getParameter("userPwd");
  user = userModel.userLogin(uID, uPwd);
  if (user!= null) {
    request.setAttribute("userID",uID);
    request.setAttribute("userPwd",uPwd);
    session.setAttribute("USER", user);
    response.sendRedirect("success.jsp");
  }
  else {
    request.setAttribute("message", "用户名或密码有错");
    request.getRequestDispatcher("login.jsp").forward(request, response);
  }
%>
```

(6) 设计用户登录成功页面 success.jsp。

```
<%@ page language = "java" import = "java.util.*" pageEncoding = "utf-8" %>
<!DOCTYPE HTML PUBLIC "-//W3C//DTD HTML 4.01 Transitional//EN">
<html>
  <head><title>用户登录成功</title></head>
    <body>
      <jsp:useBean id = "user" class = "cn.edu.mypt.bean.UserBean" scope = "page"/>
        <%
          user = (cn.edu.mypt.bean.UserBean)session.getAttribute("USER");
          if (user == null) {
            request.setAttribute("message", "请登录");
            request.getRequestDispatcher("login.jsp").forward(request, response);
          }
          else {
        %>
        <h1>欢迎你!<% = user.getUserName() %></h1>
            <% } %>
  </body>
</html>
```

项目 4 用户管理系统业务逻辑设计

1. 项目要求

利用 JavaBean 实现用户的登录、注册、删除、修改、查找等业务逻辑。

2. 项目实施

(1) 创建一个 Java Web 项目。

(2) 创建一个用户 JavaBean,包括用户名、用户 ID、用户密码、用户类型(普通用户、系统管理员等)等基本信息。

```
package cn.edu.mypt.bean;
public class UserBean {
    private String userID, userName,userPwd, userType;
    //getXxx setXxx 略
    public UserBean(){}
    public UserBean(String userID,String userName,String userPwd,String userType){
      this.userID = userID;
      this.userName = userName;
      this.userPwd = userPwd;
      this.userType = userType;
    }
    public UserBean(String userID,String userName,String userPwd){
      this(userID,userName,userPwd,null);
    }
    public UserBean(String userID,String userName){
      this(userID,userName,null,null);
    }
}
```

(3) 定义接口 IUserModel 。

```
package cn.edu.mypt.model;
//导入包,略
public interface IUserModel {
public boolean userReg(UserBean userr); //用户注册,若注册成功返回 true,否则返回 false
public UserBean userLogin(String userId,String userPwd); //用户登录,若登录成功返回用户信
                                                        //息,否则返回 null
public boolean UserModify(UserBean user); //修改用户信息,若修改成功返回 true,否则返回
                                         //false
public boolean UserDelete(String userID); //删除用户,若删除成功返回 true,否则返回 false
public UserBean findById(String userID); //根据用户 ID 查找用户,若找到返回用户信息,否则
                                        //返回 null
public List<UserBean> getUser(); //获取所有用户信息
public boolean ModifyPwd(String userID,String oldPwd,String newPwd);
        //根据用户 ID 和旧密码修改用户密码,若修改成功返回 true,否则返回 false
}
```

(4) 创建模型层 JavaBean,模拟用户的登录、注册、删除、修改、查找等操作。

```
package cn.edu.mypt.model.impl;
//导入包,略
public class UserModel implements IUserModel {
    static List<UserBean> userList = new ArrayList();
    public boolean userReg(UserBean user) {
        UserBean u = findById(user.getUserID());
        if (u == null) {
            userList.add(user);
            return true;
        }
        return false;
```

```
    }
    @Override
    public UserBean userLogin(String userId, String userPwd) {
        UserBean u = findById(userId);
        if (u == null) return null;
        if (userPwd.equals(u.getUserPwd())) return u;
        return null;
    }
    @Override
    public boolean UserModify(UserBean user) {
        UserBean u = findById(user.getUserID());
        if (u == null) return false;
        u.setUserName(user.getUserName());
        u.setUserPwd(user.getUserPwd());
        u.setUserType(user.getUserType());
        return true;
    }
    @Override
    public boolean UserDelete(String userID) {
        UserBean u = findById(userID);
        if (u == null) return false;
        userList.remove(u);
        return true;
    }
    @Override
    public UserBean findById(String userID) {
          int n = userList.size();
          int i;
        for (i = 0;i < n && !userID.equals(userList.get(i).getUserID());i++)
              ;
        if (i < n) return userList.get(i); else return null;
    }
    @Override
    public List < UserBean > getUser() {
        return userList;
    }
    public void setUserList(List < UserBean > userList) {
        this.userList  =  userList;
    }
    @Override
    public boolean ModifyPwd(String userID, String oldPwd, String newPwd) {
        UserBean u = userLogin(userID,oldPwd);
        if (u == null) return false;
        u.setUserPwd(newPwd);
        return true;
    }
    public void load() {
        userList.add(new UserBean("1","zhangsan","123456"));
        userList.add(new UserBean("2","lishi","123456"));
    }
}
```

(5) 用户登录页面参考 4.3 节的任务。

(6) 设计用户注册页面 register.jsp。

```
<%@ page language = "java" import = "java.util.*" pageEncoding = "utf-8" %>
<!DOCTYPE HTML PUBLIC "-//W3C//DTD HTML 4.01 Transitional//EN">
<html>
  <head><title>用户注册</title></head>
  <body>
  <%
  String uID,uPwd,uName,msg;
  msg = (String)request.getAttribute("message");
  uID = (String)request.getAttribute("userID");
  uPwd = (String)request.getAttribute("userPwd");
  uName = (String)request.getAttribute("userName");
  uID = (uID == null)?"":uID;
  uPwd = (uPwd == null)?"":uPwd;
  uName = (uName == null)?"":uName;
  msg = (msg == null)?"":msg;
  %>
  <a href = "index.jsp">返回</a>
  <% = msg %>
    <form action = "doreg.jsp" method = "post">
        userID:<input type = "text" name = "userID" value = "<% = uID %>"/>
        userName:<input type = "text" name = "userName" value = "<% = uName %>"/>
        userPwd:<input type = "text" name = "userPwd" value = "<% = uPwd %>"/>
        <input type = "submit" value = "reg"/>
    </form>
  </body>
</html>
```

(7) 设计用户注册处理程序 doreg.jsp。

```
<%@ page language = "java" import = "java.util.*" pageEncoding = "utf-8" %>
<jsp:useBean id = "userModel" class = "cn.edu.mypt.model.impl.UserModel" scope =
"application"/>
  <jsp:useBean id = "user" class = "cn.edu.mypt.bean.UserBean" scope = "page"/>
  <%
    String uID,uPwd,uName;
    request.setCharacterEncoding("utf-8");
    uID = request.getParameter("userID");
    uPwd = request.getParameter("userPwd");
    uName = request.getParameter("userName");
    user.setUserID(uID);
    user.setUserName(uName);
    user.setUserPwd(uPwd);
    if (userModel.userReg(user)){
      response.sendRedirect("index.jsp");
    }
    else{
      request.setAttribute("userID",uID);
```

```
        request.setAttribute("userID",uPwd);
        request.setAttribute("userName",uName);
        session.setAttribute("USER", user);
        request.setAttribute("message", "用户名 ID 已经存在");
        request.getRequestDispatcher("reg.jsp").forward(request, response);
    }
%>
```

(8) 设计用户列表页面 list.jsp。

```
<%@ page language = "java" import = "java.util.*" pageEncoding = "utf-8" %>
<!DOCTYPE HTML PUBLIC "-//W3C//DTD HTML 4.01 Transitional//EN">
<html>
  <head><title>用户列表</title></head>
    <body>
  <jsp:useBean id = "userModel" class = "cn.edu.mypt.model.impl.UserModel"
              scope = "application"/>
  <jsp:useBean id = "user" class = "cn.edu.mypt.bean.UserBean" scope = "page"/>
    <table>
    <tr><th>序号<th>用户 ID<th>用户名<th>删除</th><th>修改 </th></tr>
    <%
      List userList = userModel.getUser();
      if (userList!= null) {
      int n = userList.size();
      for (int i = 0;i<n;i++) {
        user = (cn.edu.mypt.bean.UserBean)userList.get(i);
      %>
      <tr>
        <td><% = i+1 %></td><td><% = user.getUserID() %></td><td>
              <% = user.getUserName() %></td>
        <td><a href = "del.jsp?userID = <% = user.getUserID() %>">删除</a></td>
        <td><a href = "modify.jsp?userID = <% = user.getUserID() %>">修改</a></td>
      </tr>
       <%   }
      }
       %>
    </table>
   <a href = "index.jsp">返回</a>
  </body>
</html>
```

(9) 设计用户删除处理页面 del.jsp。

```
<%@ page language = "java" import = "java.util.*" pageEncoding = "utf-8" %>
  <jsp:useBean id = "userModel" class = "cn.edu.mypt.model.impl.UserModel" scope =
"application"/>
  <%
    String uID,uPwd,uName;
    request.setCharacterEncoding("utf-8");
    uID = request.getParameter("userID");
```

```
    userModel.UserDelete(uID);
    response.sendRedirect("userList.jsp");
  %>
```

（10）设计修改页面 modify.jsp。

```
<%@ page language = "java" import = "java.util.*" pageEncoding = "utf-8" %>
<!DOCTYPE HTML PUBLIC "-//W3C//DTD HTML 4.01 Transitional//EN">
<html>
  <head><title>修改用户信息</title></head>
  <body>
  <%
  String uID,uPwd,msg;
  msg = (String)request.getAttribute("message");
  uID = (String)request.getParameter("userID");
  uPwd = (String)request.getAttribute("userPwd");
  uID = (uID == null)?"":uID;
  uPwd = (uPwd == null)?"":uPwd;
  msg = (msg == null)?"":msg;
   %>
   <a href = "index.jsp">返回</a>
   <% = msg %>
   用户 ID:<% = uID %>
    <form action = "domodify.jsp" method = "post">
      <input type = "hidden" name = "userID" value = "<% = uID %>"/>
      用户名:<input type = "text" name = "userName" value = "<% = uPwd %>"/>
      用户密码:<input type = "text" name = "userPwd" value = "<% = uPwd %>"/>
      <input type = "submit" value = "reg"/>
    </form>
  </body>
</html>
```

（11）设计修改处理页面 domodify.jsp。

```
<%@ page language = "java" import = "java.util.*" pageEncoding = "utf-8" %>
  <jsp:useBean id = "userModel" class = "cn.edu.mypt.model.impl.UserModel" scope =
    "application"/>
  <jsp:useBean id = "user" class = "cn.edu.mypt.bean.UserBean" scope = "page"/>
  <%
    String uID,uPwd,uName;
    request.setCharacterEncoding("utf-8");
    uID = request.getParameter("userID");
    uPwd = request.getParameter("userPwd");
    uName = request.getParameter("userName");
    user.setUserID(uID);
    user.setUserName(uName);
    user.setUserPwd(uPwd);
    userModel.UserModify(user);
    request.getRequestDispatcher("userList.jsp").forward(request, response);
%>
```

(12) 设计首页 index.jsp。

```
<%@ page language = "java" import = "java.util.* " pageEncoding = "utf-8" %>
<!DOCTYPE HTML PUBLIC " - //W3C//DTD HTML 4.01 Transitional//EN">
<html>
  <head><title>用户管理</title></head>
    <body>
      <jsp:useBean id = "userModel" class = "cn.edu.mypt.model.impl.UserModel" scope =
"application"/>
        <a href = login.jsp>login</a><br>
        <a href = reg.jsp>reg</a><br>
        <a href = userList.jsp>List</a><br>
    </body>
</html>
```

(13) 启动服务器,部署项目并浏览。

习　题　4

1. 选择题

(1) 在 JSP 中调用 JavaBean 时不会用到的标签是(　　)。

A. <javabean>　　B. <jsp:useBean>

C. <jsp:setProperty>　　D. <jsp:getProperty>

(2) (　　)是错误的设置 Bean 属性值的方法。

A. <jsp:setProperty name="beanInstanceName" property=" * "/>

B. <jsp:setProperty name="beanInstanceName" property="propertyName" value="123"/>

C. <jsp:setProperty name="beanInstanceName" property="propertyName" param="parameterName" />

D. <jsp:setProperty name="beanInstanceName" property=" * " value="{string | <%= expression%>}"/>

(3) test.jsp 文件中有如下代码,要使 user 对象可以作用于整个应用程序,下画线中应填入(　　)。

```
<jsp:useBean id = "user" scope = "____" class = "com.UserBean">
```

A. page　　B. request

C. session　　D. application

(4) 在 JSP 页面中,保存数据的范围由小到大依次是(　　)。

A. page,request,application,session

B. page,application,session,request

C. page,request,session,application

D. page,session,request,application

(5) 在 JSP 页面中,正确引入 JavaBean 的方法是(　　)。

A. <%jsp: useBean id="myBean" scope="page" class="pkg. MyBean"%>

B. <jsp: useBean name="myBean" scope="page" class="pkg. MyBean">

C. <jsp: useBean id="myBean" scope="page" class="pkg. MyBean"/>

D. <jsp: useBean name="myBean" scope="page" class="pkg. MyBean"/>

(6) JavaBean 中的属性命名标准是(　　)。

A. 全部字母小写

B. 第一个字母小写,之后每个字母单词的首字母大写

C. 每个单词的首字母大写

D. 全部字母大写

(7) 在 JSP 中引用 JavaBean 应该使用(　　)。

A. page 指令　　B. include 指令　　C. include 动作　　D. useBean 动作

(8) (　　)是 JavaBean 作用域对象。

A. page　　B. session　　C. application　　D. cookie

(9) 关于 JavaBean 正确的说法是(　　)。

A. 被引用的 JavaBean 文件的扩展名为.java

B. JavaBean 文件放在任何目录下都可以被引用

C. Java 文件与 JavaBean 所定义的类名可以不同,但一定要注意区分字母的大小写

D. 在 JSP 文件中引用 JavaBean,要使用<jsp:useBean>

(10) JavaBean 文件的扩展名是(　　)。

A. .html　　B. .jsp　　C. .java　　D. .class

2. 简答题

(1) 阐述 JavaBean 的功能特征、分类和作用。

(2) 阐述 JavaBean 的作用域范围。

第5章　JSP的数据访问

动态网页设计涉及数据库的访问技术。本章介绍如何通过JSP实现对数据库数据的增、删、改、查等基本操作。

【技能目标】 掌握JDBC的数据库访问；实现JSP页面与数据库的访问。

【知识目标】 JDBC的数据库连接技术；利用JSP＋JDBC技术实现对数据库数据的编辑。

【关键词】 开放式数据库互连(ODBC)　Java数据库互连(JDBC)
缓冲池(pool)　Java命名和目录接口(JNDI)
数据源(data source)　连接(connection)
容器(container)

5.1　JDBC技术

商业应用的后台数据一般存放在数据库中,如果要连接到数据库,可以采用数据库接口技术。目前常用的有两种数据库接口ODBC(open database connectivity,开放式数据库互连)和JDBC(Java database connectivity,Java数据库互连)。

5.1.1　ODBC简介

微软公司数据库系统应用程序接口标准,支持应用程序以标准的ODBC函数和SQL语言访问不同数据库。由于ODBC是微软公司的产品,因此它几乎可以连接到所有在Windows平台下运行的数据库,由它连接到特定的数据库,不需要去各自的数据库厂商网站下载具体的驱动包。

1. ODBC体系结构

ODBC是依靠分层结构来实现的,这样可以保证其标准性和开放性。图5-1所示为ODBC的体系结构,它共分为四层:应用程序、驱动程序管理器、驱动程序和数据源。微软公司为应用层的开发者和用户提供标准的函数、语法和错误代码等,还提供了驱动程序管理器,它在Windows中是一个动态链接库即ODBC.DLL。驱动程序层由微软、DBMS厂商或第三开发商提供,它必须符合ODBC标准。

1) 应用程序

应用程序执行ODBC方法的调用和处理,提交SQL语句并检索结果。使用ODBC接口的应用程序可执行以下任务。

(1) 请求与数据源的连接和会话。

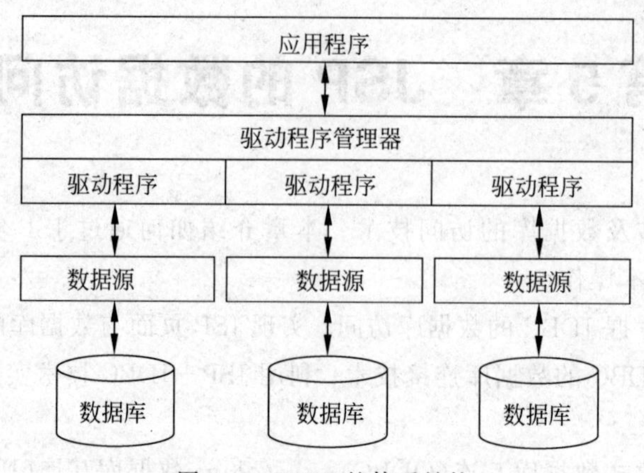

图 5-1　ODBC 的体系结构

（2）向数据源发送 SQL 请求。

（3）对 SQL 请求的结果定义存储区和数据格式。

（4）请求结果。

（5）处理错误。

（6）如果需要，把结果返回给用户。

（7）对事务进行控制，请求执行或回退操作。

（8）终止对数据源的连接。

2）驱动程序管理器

驱动程序管理器为应用程序装载驱动程序提供支持。由微软提供的驱动程序管理器是带有输入库的动态链接库 odbc.dll，其主要目的是装入驱动程序，此外还执行以下工作。

（1）处理几个 ODBC 初始化调用。

（2）为每一个驱动程序提供 ODBC 方法入口点。

（3）为 ODBC 调用提供参数和次序验证。

3）驱动程序

驱动程序建立与数据源的连接，实现和执行 ODBC 方法调用，向数据源提交 SQL 请求，并把结果返回给应用程序。当应用程序调用相关方法时，驱动程序管理器装入相应的驱动程序，它对来自应用程序的 ODBC 方法调用进行应答，按照其要求执行以下任务。

（1）建立与数据源的连接。

（2）向数据源提交请求。

（3）在应用程序需要时，转换数据格式。

（4）返回结果给应用程序。

（5）将运行错误格式化为标准代码返回。

（6）在需要时说明和处理游标。

4）数据源

数据源由存取的数据和与之相连的操作系统、DBMS 及网络平台组成。

2. 建立 ODBC 数据源

在使用 ODBC 之前，需要配置 ODBC 的数据源，让 ODBC 知道连接的具体数据库。ODBC 支持连接到各种数据库，如 Oracle、MySQL、SQL Server 等。在 Windows 操作系统中，可以在控制面板中选择"管理工具"→"数据源(ODBC)"，按照向导指引即可创建 ODBC 数据源，具体操作如下。

(1) 在控制面板中选择"管理工具"，双击"数据源(ODBC)"图标。

(2) 在"ODBC 数据源管理器"的"系统 DSN"选项卡中单击"添加"按钮，如图 5-2 所示。

图 5-2　选择驱动程序

(3) 从弹出的"创建新数据源"对话框的数据源名称列表中选择相应的 DBMS，如 SQL Server 并单击"完成"按钮，如图 5-2 所示。

(4) 在弹出的对话框的"名称"文本框中输入自定义的数据源名称，输入服务器名称或 IP 地址，然后单击"下一步"按钮，如图 5-3 所示。

(5) 在弹出的对话框中根据数据库登录的方式，设置数据库访问的用户名和用户密码，如图 5-4 所示。单击"下一步"按钮。

(6) 选择数据源的数据库名，单击"下一步"按钮，如图 5-5 所示。

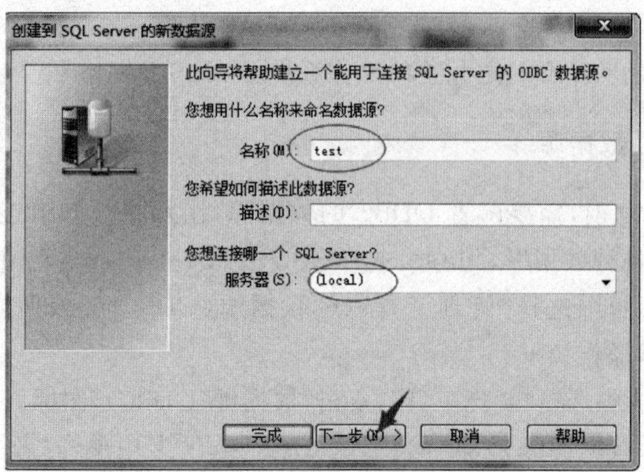

图 5-3　选择数据库服务器

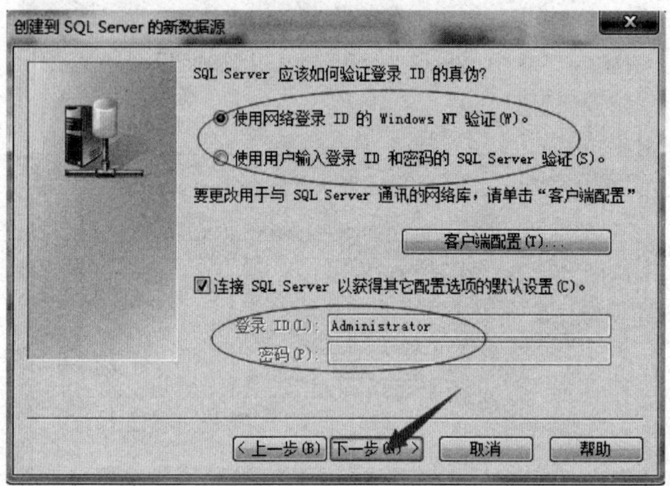

图 5-4　输入数据库访问用户名和密码

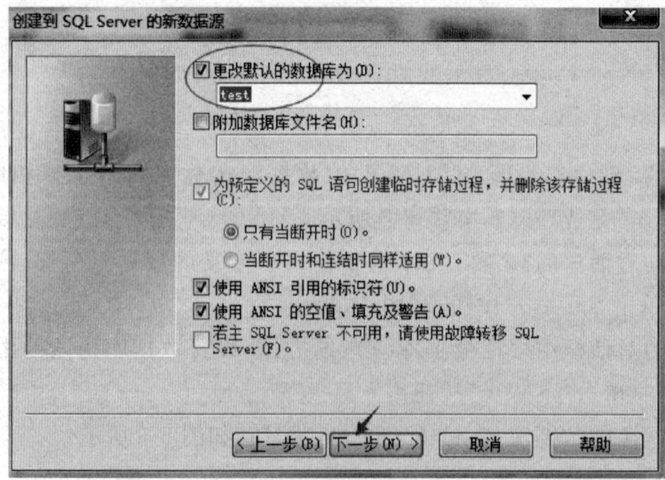

图 5-5　选择数据库

(7) 单击"测试数据源"按钮,测试数据库源配置是否正确,如图 5-6 所示,完成数据源的配置。

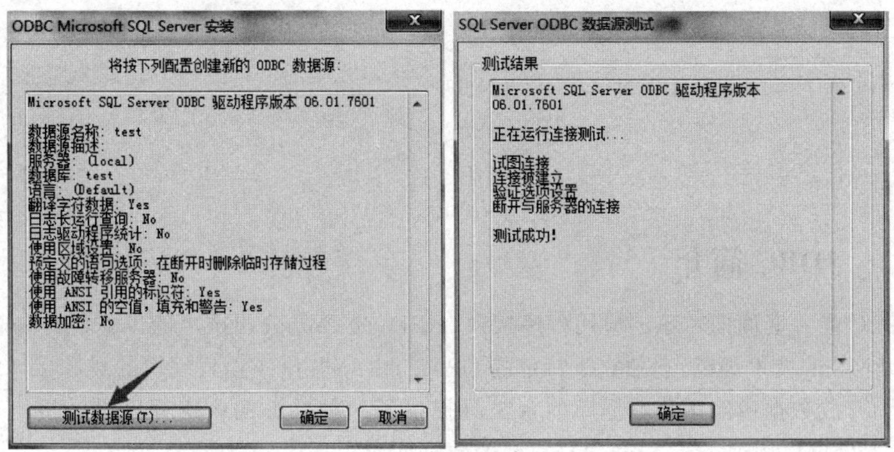

图 5-6 测试数据源

3. JDBC-ODBC 桥技术

JDBC-ODBC 桥是一个 JDBC 驱动程序,类似于一个应用程序,完成从 JDBC 操作到 ODBC 操作之间的转换工作,为所有 ODBC 操作可用的数据库实现了 JDBC 操作。由于 ODBC 技术被广泛地使用,使得 Java 可以利用 JDBC-ODBC 桥访问几乎所有的数据库。JDBC-ODBC 桥作为 sun.jdbc.odbc 包与 JDK 一起自动安装,无须特殊配置。

JDBC-ODBC 桥作为一款 Java 连接数据库的过渡性技术而产生,现在并不被 Java 技术广泛使用,而被广泛使用的是 JDBC 技术,但是并不证明它已经被淘汰。在开发经济实用的单机软件时,底层数据库通常采用 Access 数据库主要有两个原因:①它能够满足软件对数据库的要求;②不用为了安装软件而单独去安装数据库。

下面介绍一种通过 JDBC-ODBC 桥连接数据库的方法,通过这种方法不用手动配置 ODBC 数据源,而是采用默认的 ODBC 数据源。首先加载 JDBC-ODBC 桥连接的数据库驱动,代码如下。

```
Class.forName("sun.jdbc.odbc.JdbcOdbcDriver");
```

然后是建立数据库连接,这里给出的 URL 不需要手动配置 ODBC 数据源,如果想通过手动配置的 ODBC 数据源连接指定的数据库,则可以将 URL 设为手动配置的 ODBC 数据源的名称,代码如下。

```
String url = "driver = {Microsoft Excel Driver( * .xls)};DBQ = D:/books.xls";
Connection conn = DriverManager.getConnection("jdbc:odbc:" + url);
```

如果想连接其他类型的数据库,只须将 URL 中{}内的默认驱动类型更换为相应的类型即可,URL 中"DBQ="后面跟随的内容为数据库文件的存放路径。之后就是在操作数据库之前获得数据库连接状态,然后执行数据库操作,操作结束后关闭数据库连接状态，并提交事务，代码如下。

```
Statement stmt = conn.createStatement();
ResultSet rs = stmt.executeQuery("select * from [sheet1$]");
while (rs.next()) {
  System.out.println(rs.getString("name"));
}
stmt.close();
conn.close();
```

5.1.2 JDBC 简介

JDBC 是一套面向对象的应用程序接口(API)，是 Sun 公司推出的以 OOP 方式连接不同 DBMS 的技术规范，它独立于特定的 DBMS，由一些 Java 语言编写的类和接口组成。JDBC 为开放数据库应用和数据库前台工具提供了一种标准的应用程序设计接口，使程序开发人员可以用纯 Java 语言编写完整的数据库应用程序。

JDBC 与数据库连接的方式有两种：①使用客户端直接与数据库连接，如图 5-7 所示；②通过应用服务器与数据库连接，如图 5-8 所示。

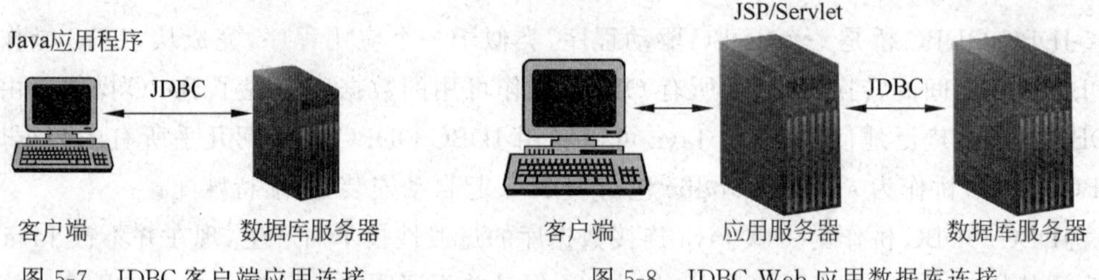

图 5-7 JDBC 客户端应用连接　　图 5-8 JDBC Web 应用数据库连接

JDBC 是一种底层 API，在访问数据库时需要在业务逻辑中直接嵌入 SQL 语句，由于 SQL 语句是面向关系的，依赖于关系模型，所以 JDBC 传承了简单直接的优点，特别是对于小型应用程序十分方便。需要注意的是，JDBC 不能直接访问数据库，必须依赖于数据库厂商提供的 JDBC 驱动程序。JDBC 完成以下 3 步工作。

(1) 同数据库建立连接。

(2) 向数据库发送 SQL 语句。

(3) 处理从数据库返回的结果。

5.1.3 JDBC 的结构

JDBC 有 4 个组件，如图 5-9 所示：应用程序、JDBC API、JDB 驱动程序管理器和数据源。JDBC API 通过一个 JDBC 驱动程序管理器和为各种数据库定制的 JDBC 驱动程序，提供与不同数据库的透明连接。

JDBC API 的作用就是屏蔽不同的数据库间 JDBC 驱动程序之间的差别，使程序员有一个标准的、纯的 Java 数据库程序接口，为在 Java 中访问任意类型的数据库提供技术支持。JDBC 驱动程序管理器为应用程序装载数据库驱动程序。JDBC 驱动程序与具体的

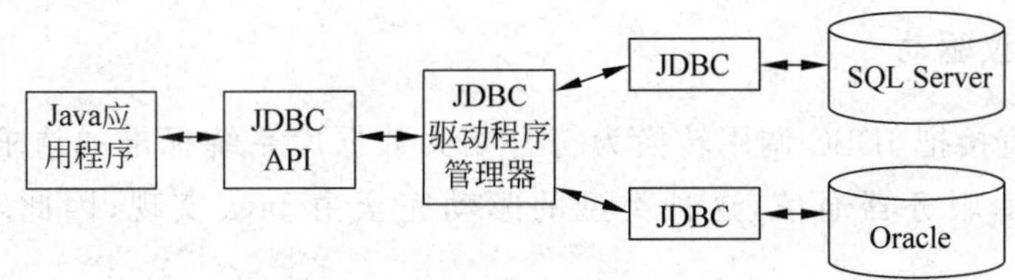

图 5-9　JDBC 的总体结构

数据库相关,用于建立与数据源的连接,向数据库提交 SQL 请求。JDBC 的驱动程序由具体的数据库厂商提供。

5.1.4　JDBC 驱动程序

JDBC 驱动程序可以分为 4 种类型,包括 JDBC-ODBC 桥、本地 API 驱动、网络协议驱动和本地协议驱动。

1. JDBC-ODBC 桥

JDBC-ODBC 桥是 JDK 提供的标准 API。这种类型的驱动实际是把所有 JDBC 的调用传递给 ODBC,再由 ODBC 调用本地数据库驱动代码。

由于 JDBC-ODBC 先调用 ODBC 再由 ODBC 去调用本地数据库接口访问数据库,所以执行效率比较低,对于那些大数据量存取的应用是不适合的,而且这种方法要求客户端必须安装 ODBC 驱动,所以对于基于 Internet、Intranet 的应用也是不合适的,因为不可能要求所有客户端都能安装 ODBC 驱动程序。

2. 本地 API 驱动

本地 API 驱动直接把 JDBC 调用转变为数据库的标准调用再去访问数据库,这种方法需要本地数据库驱动代码。

这种驱动比起 JDBC-ODBC 桥执行效率大大提高了,但是它仍然需要在客户端加载数据库厂商提供的代码库,这样就不适合基于 Internet 的应用,并且执行效率还是不够高。

3. 网络协议驱动

这种驱动实际上是根据三层结构建立的,JDBC 先把对数据库的访问请求传递给网络上的中间件服务器。中间件服务器再把请求转换为符合数据库规范的调用,再把这种调用传给数据库服务器。

由于这种驱动是基于服务器的,所以它不需要在客户端加载数据库厂商提供的代码库。而且它在执行效率和可升级性方面是比较好的,因为大部分功能实现都在服务器端,所以这种驱动可以设计的很小,可以非常快速地加载到内存中。但是,这种驱动在中间层仍然需要配置有其他数据库驱动程序,并且由于多了一个中间层传递数据,它的执行效率并不高。

4. 本地协议驱动

这种驱动直接把 JDBC 调用转换为符合相关数据库系统标准的请求，由于应用可以直接和数据库服务器通信，这种类型的驱动完全由 Java 实现，因此实现了平台独立性。

由于这种驱动不需要先把 JDBC 的调用传给 ODBC 或本地数据库接口或者是中间层服务器，所以它的执行效率非常高。而且，它根本不需要在客户端或服务器端装载任何的软件或驱动，这种驱动程序可以动态地下载，但是对于不同的数据库需要下载不同的驱动程序。

5.1.5 任务：使用 JDBC-ODBC 桥实现对数据库的访问

1. 任务要求

利用 JDBC-ODBC 桥实现对数据库的访问，假设数据库 test 中有数据表 userTable，访问数据库，获取用户信息并显示。

2. 任务实施

具体实施步骤如下。

(1) 创建连接 SQL Server 的数据源 DSN，命名为 testDSN。

(2) 编写如下代码。

```
<%@ page language = "java" import = "java.sql.*" pageEncoding = "gb2312" %>
<html>
  <head><title>ODBC 连接</title></head>
  <body>
  <%
    String driverName = "sun.jdbc.odbc.JdbcOdbcDriver";
    String dbURL = "jdbc:odbc:testDSN";
    try {
      Class.forName(driverName);
      Connection con= DriverManager.getConnection(dbURL);
      if(!conn.isClosed())
        out.println("<h1>成功链接到数据库</h>");
      con.close();
    }
    catch (Exception e) {
      out.print("出现例外!" + e.getMessage());
      e.printStackTrace();
    }
  %>
  </body>
</html>
```

5.2　JDBC 常用接口

JDBC 数据库访问步骤如图 5-10 所示。针对不同类型的数据库,JDBC 机制中提供了"驱动程序"的概念,对于不同的数据库,应用程序只须使用不同的驱动驱动。JDBC 中提供以下类或接口：DriverManager 类、Connection 接口、Statement 接口、ResultSet 接口。

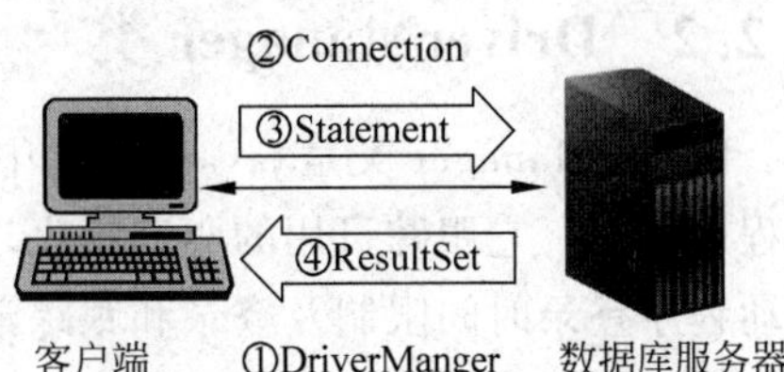

图 5-10　JDBC 数据库访问步骤

java. sql. DriverManager：依据数据库的不同,管理 JDBC 驱动。

java. sql. Connection：连接数据库传送数据。

java. sql. Statement：由 Connection 接口产生,负责执行 SQL 语句。

java. sql. ResultSet：保存 Statement 执行后所产生的查询结果。

5.2.1　Driver 接口

每种数据库的驱动程序都应该提供一个实现 java. sql. Driver 接口的类,简称 Driver 类,在加载某一驱动程序的 Driver 类时,应该创建自己的实例并向 java. sql . DriverManager 类注册该实例。

通常情况下通过 java. lang. Class 类的静态方法 forName(String className)加载要连接数据库的 Driver 类,该方法的入口参数为 Driver 类的完整路径。成功加载后,会将 Driver 类的实例注册到 DriverManager 类中,如果加载失败,将抛出 ClassNotFoundException 异常,即未找到指定 Driver 类的异常。不同数据库驱动的驱动类名称是不一样的。例如,Oracle 驱动类名称为 oracle. jdbc. driver. OracleDriver; JDBC-ODBC 桥驱动类名称为 sun. jdbc. odbc. jdbcodbcdriver; java. lang. Class 类的 forName()方法可以根据类名初始化类实例,从而实现 JDBC 驱动类注册。

主流数据库的 JDBC 驱动加裁注册的代码如下。

```
//Oracle8/8i/9iO 数据库(thin 模式)
Class.forName("oracle.jdbc.driver.OracleDriver").newInstance();
//SQL Server 2012 数据库
Class.forName("com.microsoft.sqlserver.jdbc.SQLServerDriver").newInstance();
//DB2 数据库
Class.froName("com.ibm.db2.jdbc.app.DB2Driver").newInstance();
//Informix 数据库
Class.forName("com.informix.jdbc.IfxDriver").newInstance();
//Sybase 数据库
Class.forName("com.sybase.jdbc.SybDriver").newInstance();
//MySQL 数据库
Class.forName("com.mysql.jdbc.Driver").newInstance();
//PostgreSQL 数据库
Class.forNaem("org.postgresql.Driver").newInstance();
```

JDBC 驱动包添加完成后，在程序中注册驱动的方法如下。

```
String driver = "com.microsoft.jdbc.sqlserver.SQLServerDriver";
Class.forName(driver):
```

5.2.2 DriverManager 类

DriverManager 类是 java.sql 包中用于数据库驱动程序管理的类，作用于用户和驱动程序之间。它跟踪可用的驱动程序，并在数据库和相应驱动程序之间建立连接，也处理驱动程序登录时间限制及登录和跟踪消息的显示等事务。DriverManager 类直接继承自 java.lang.Object，其常用方法如表 5-1 所示。

表 5-1　DriverManager 类的常用方法

方　法	描　述
static void deregister Driver(Driver driver)	从数据驱动程序列表中删除指定的数据驱动程序
static Connection getConnection(Strig url)	通过指定的数据 URL 创建数据连接
static Connection getConnection(String url, properties info)	通过指定的 URL 及属性信息创建数据连接
static Connection getConnection(String url, String username, String password)	通过指定的 URL、用户名、密码创建数据连接
static Driver getDriver(String URL)	通过指定的 URL 获取数据驱动程序
static Emmmeration getDriver()	获得驱动程序的枚举
static int getLoginTimeout()	获取数据连接的超时时间
static int setLoginTimeout()	设置数据连接的超时时间
static void register Driver(Driver driver)	注册数据驱动程序
static PrintWriter getLocWriter()	获取数据日志输出项
static void setLocWriter(PrintWriter out)	设置数据日志输出

成功加载 Driver 类并在 DriverManager 类中注册后，DriverManager 类即可用来建立数据库连接。当调用 DriverManager 类的 getConnection()方法请求建立数据库连接时，DriverManager 类将试图定位一个适当的 Driver 类，并检查定位到的 Driver 类是否可以建立连接，如果可以则建立连接并返回，如果不可以则抛出 SQLException 异常。其格式如下。

```
Connection conn = DriverManager.getConnerction(url,user,password);
```

其中：

url——连接数据库的字符串。不同数据库的连接 URL 有一些差异，可查询相应 JDBC 驱动文档。通常 URL 中包含数据库的 IP、端口、库名以及其他参数。例如 MySQL 数据库的连接 URL 格式为：jdbc:mysql://192.168.1.1:3306/db-customer。

user——连接数据库的用户名。

password——连接数据库的密码。

主流数据库的连接参数如下。

(1) Oracle 8/8i/9i 数据库(thin 模式)。

```
String url = "jdbc:oracle:thin:@localhost:1521:orcl";
String user = "uName";
String password = "uPwd";
Connection conn = DriverManager.getConnection(url,user,password);
```

(2) SQL Server 数据库。

```
String url = "jdbc:microsoft:sqlserver://localhost:1433;DatabaseName = pubs";
String user = "sa";
String password = "";
Connection conn = DriverManager.getConnection(url,user,password);
```

(3) MySQL 数据库。

```
String url = "jdbc:mysql://localhost:3306/testDB?user = root&password = root&useUnicode =
true&characterEncoding = gb2312";
Connection conn = DriverManager.getConnection(url);
```

5.2.3 Connection 接口

java.sql.Connection 接口代表与特定数据库的连接,在连接的上下文中可以执行 SQL 语句并返回结果,还可以通过 getMetaData()方法获得由数据库提供的相关信息,如数据表、存储过程和连接功能等。

Connection 接口的常用方法如表 5-2 所示。

表 5-2 Connection 接口的常用方法

方　法	描　述
void clearWarnings()	清除连接的所有警告信息
Statement createStatement()	创建一个 Statement 对象
Statement createStatement(int resultSetType,int resultSetConcurrency)	创建一个 Statement 对象,它将生成具有特定类型和并发性的结果集
void commit()	提交对数据库的改动并释放当前连接持有的数据库的锁
void rollback()	回滚当前事务中的所有改动并释放当前连接持有的数据库的锁
String getCatalog()	获取连接对象的当前目录名
boolean isClosed()	判断连接是否已关闭
boolean isReadOnly()	判断连接是否为只读模式
void setReadOnly()	设置连接的只读模式
void close()	立即释放连接对象的数据库和 JDBC 资源

注意: 上述方法都会抛出 SQLException 异常。

5.2.4 Statement 接口

java.sql.Statement 接口用来执行静态的 SQL 语句，通过 Connection 的 Createment() 方法可以创建 Statement 对象，发送 SQL 语句，并返回结果。

创建 Statement 对象的方法如下。

```
Statement stmt = con.createStatement();
```

Statement 接口的常用方法如表 5-3 所示。

表 5-3 Statement 接口的常用方法

方法	描述
void addBatch(String sql)	在 Statement 中增加用于数据库操作的 SQL 批处理语句
void cancel()	取消 Statement 中的 SQL 语句指定的数据库操作命令
void clearBatch()	清除 Statement 中的 SQL 批处理语句
void clearWarnings()	清除 Statement 语句中的操作引起的警告
void close()	关闭 Statement 语句指定的数据库连接
boolean execute(String sql)	执行 SQL 语句
int[] executeBatch()	执行多个 SQL 语句
resultSet executeQuery(String sql)	进行数据库查询，返回结果集
int executeUpdate(String sql)	进行数据库更新
Connection getConnection()	获取对数据库的连接
int getFetchDirection()	获取从数据库表中获取行数据的方向
int getFetchSize()	获取返回的数据库结果集行数
int getMaxFieldSize()	获取返回的数据库结果集最大字段数
int getMaxRows()	获取返回的数据库结果集最大行数
boolean getMoreResults()	获取 Statement 的下一个结果
int getQueryTimeout()	获取查询超时设置
resultSet getResultSet()	获取结果集
int getUpdateCount()	获取更新记录的数量
void setCursorName(String name)	设置数据库 Cursor 的名称
void setFetchDirection(int dir)	设置数据库表中获取行数据的方向
void setFetchSize(int rows)	设置返回的数据库结果集行数
void setMaxFieldSize(int max)	设置最大字段数
void setMaxRows(int max)	设置最大行数
void setQueryTimeout(int seconds)	设置查询超时时间

值得注意的是，Statement 接口提供了 3 个执行 SQL 语句的方法：executeQuery()、executeUpdate()和 execute()。使用哪一个方法由 SQL 语句所产生的内容决定。

(1) executeQuery()方法用于产生单个结果集的 SQL 语句，如 SELECT 语句。

(2) executeUpdate()方法用于执行 INSERT、UPDATE、DELETE 及 DDL(数据定义语言)语句，例如 CREATE TABLE 和 DROP TABLE。executeUpdate()方法的返回值是一个整数，表示它执行的 SQL 语句所影响的数据库中的表的行数(更新计数)。

execute()方法用于执行返回多个结果集或多个更新计数的语句。

5.2.5 PreparedStatement 接口

java. sql. PreparedStatement 接口继承自并扩展了 Statement 接口，用来执行动态的 SQL 语句，即包含参数的 SQL 语句，但 PreparedStatement 中包含了经过预编译的 SQL 语句，因此可以获得更高的执行效率。在 PreparedStatement 中可以包含多个用"?"代表的字段，在程序中可以利用 setXXX()方法设置该字段的内容，从而增强了程序设计的动态性。PreparedStatement 接口的常用方法如表 5-4 所示。

表 5-4　PreparedStatement 接口的常用方法

方　　法	描　　述
void addBatch(String sql)	在 Statement 语句中增加用于数据库操作的 SQL 批处理语句
void clearparameters()	清除 PreparedStatement 中的设置参数
ResultSet executeQuery(String sql)	执行 SQL 查询语句
ResultSetMetaData getMetaData()	进行数据库查询，获取数据库元数据
void setArray(int index, Array x)	设置为数组类型
void setAsciiStream(int index, InputStream stream, int length)	设置为 ASCII 输入流
void setBigDecimal(int index, BigDecimal x)	设置为十进制长类型
void setBinaryStream(int index, InputStream stream, int length)	设置为二进制输入流
void setCharacterStream(int index, InputStream stream, int length)	设置为字符输入流
void setBoolean(int index, boolean x)	设置为布尔类型
void setByte(int index, byte b)	设置为字节类型
void setBytes(int byte[] b)	设置为字节数组类型
void setDate(int index, Date x)	设置为日期类型
void setFloat(int index, float x)	设置为浮点类型
void setInt(int index, int x)	设置为整数类型
void setLong(int index, long x)	设置为长整数类型
void setRef(int index, int ref)	设置为引用类型
void setShort(int index, short x)	设置为短整数类型
void setString(int index, String x)	设置为字符串类型
void setTime(int index, Time x)	设置为时间类型

PreparedStatement 与 Statement 的区别在于它构造的 SQL 语句不是完整的语句，而需要在程序中进行动态设置。这一方面增强了程序设计的灵活性；另一方面由于 PreparedStatement 语句是经过预编译的，因此它构造的 SQL 语句的执行效率比较高。所以对于某些使用频繁的 SQL 语句，用 PreparedStatement 比用 Statement 具有明显的优势。

PreparedStatement 对象的创建方法如下。

```
PreparedStatement pstmt = con.prepareStatement("update User set reward = ? where userId = ?");
```

在该语句中,包括两个可以进行动态设置的字段：reward 和 userId。可以用下面的方法设置空字段的内容：

```
pstmt.setInt(1, 5);
pstmt. setInt (2, 1);
```

如果想给前 50 条记录设置数据,则可以用循环语句对空字段进行设置。

```
pstmt.setInt(1, 5000);
for (int i = 0; i < 50; i++){
    pstmt.setInt(2,i);
    int rowCount = pstmt.executeUpdate();
}
```

如果传递的数据量很大,可以通过将 IN 参数设置为 Java 输入流来完成。当语句执行时,JDBC 驱动程序将重复调用该输入流,读取其内容并将它们当作实际参数数据传输。JDBC 提供了 3 种将 IN 参数设置为输入流的方法：setBinaryStream()用于含有未说明字节的流；setAsciiStream()用于含有 ASCII 字符的流；setUnicodeStream()用于含有 Unicode 字符的流。这些方法比其他的 setXXX()方法要多一个用于指定流的总长度的参数,因为一些数据库在发送数据之前需要知道它传送的数据的大小。

下面是一个使用流作为 IN 参数发送文件内容的例子。

```
java.io.File file = new java.io.File("/tmp/data");
int fileLength = file.length();
java.io.InputStream fin = new java.io.FileInputStream(file);
String sql = "update table set stuff = ? where index = 4";
java.sql.PreparedStatement pstmt = con.prepareStatement(sql);
pstmt.setBinaryStream (1, fin, fileLength);
pstmt.executeUpdate();
```

当语句执行时,将反复调用输入流 fin 以传递其数据。

5.2.6 CallableStatement 接口

java.sql.CallableStatement 接口继承自并扩展了 PreparedStatement 接口,用来执行 SQL 的存储过程。

JDBC API 定义了一套存储过程 SQL 转义语法,该语法允许对所有 RDBMS 通过标准方式调用存储过程。该语法定义了两种形式,分别是包含结果参数和不包含结果参数,如果使用包含结果参数,则必须将其注册为 OUT 型参数,参数是根据定义位置按顺序引用的,第一个参数的索引为 1。

为参数赋值的方法使用从 PreparedStatement 中继承来的 setXXX()方法。在执行存储过程之前,必须注册所有 OUT 参数的类型,它们的值是在执行后通过 getXXX()方法获得的。

CallableStatement 可以返回一个或多个 ResultSet 实例。处理多个 ResultSet 对象

的方法是从 Statement 中继承来的。

CallableStatement 接口的常用方法如表 5-5 所示。

表 5-5 CallableStatement 接口的常用方法

方 法	描 述
Array getArray(int i)	获取数组
BigDecimal getBigDecimal(int index) BigDecimal getBigDecimal(int index,int scale)	获取十进制小数
boolean getBoolean(int index)	获取逻辑类型
byte getByte(int index)	获取字节类型
Date getDate(int index) Date getDate(int index,Calendar cal)	获取日期类型
double getDouble(int index)	获取日期类型双精度类型
float getFloat(int index)	获取日期类型浮点类型
int getint(int index)	获取日期类型整数类型
long getLong(int index)	获取日期类型长整数类型
Object getObject(int index) Object getObject(int index,Map map)	获取对象类型
Ref getRef(int I)	获取日期类型 Ref 类型
short getShort(int index)	获取日期类型短整数类型
String getString(int index)	获取日期类型字符串类型
Time getTime(int index) Time getTime(int index,Calendar cal)	获取时间类型
void registerOutputParameter(int index) void registerOutputParameter(int index,int type) void registerOutputParameter(int index,int type,int scale)	注册输出参数

调用存储过程的语法如下。

```
{call procedure_name} //过程不需要参数
{call procedure_name[(?,?,?,...)]} //过程需要若干个参数
{? = call procedure_name[(?,?,?,...)]} //过程需要若干个参数并返回一个参数
```

其中 procedure_name 为存储过程的名字,方括号中的内容是可选的多个用于存储过程执行的参数。

CallableStatement 的创建方法如下。

```
CallableStatement cstmt = con.prepareCall("{call getData(?, ?)}");
```

向存储过程传递执行需要参数的方法是通过 setXXX()方法完成的。例如,可以将两个参数设置如下。

```
cstmt.setByte(1, 25);
cstmt.setInt(2,64.85);
```

如果需要存储过程返回运行结果,则需要调用 registerOutParameter 方法设置存储

过程的输出参数,然后调用 getXXX()方法来获取存储过程的执行结果。例如:

```
cstmt.registerOutParameter(1,java.sql.Types.TINYINT);
cstmt.registerOutParameter(1,java.sql.Types.INTEGER);
cstmt.executeUpdate();
byte a = cstmt.getByte(1);
int b = cstmt.getInt(2);
```

从上面的程序可以看出,Java 的基本数据类型和 SQL 中支持的数据类型有一定的对应关系。这种对应关系如表 5-6 所示。

表 5-6 SQL 数据类型与 Java 数据类型的对应关系

SQL 数据类型	Java 数据类型
CHAR	String
VARCHAR	String
LONGVARCHAR	String
NUMERIC	java.math.BigDecimal
DECIMAL	java.math.BigDecimal
BIT	boolean
TINYINT	byte
SMALLINT	short
INTEGER	int
BIGINT	long
REAL	float
FLOAT	double
DOUBLE	double
BINARY	byte[]
VARBINARY	byte[]
LONGVARBINARY	byte[]
DATE	java.sql.Date
TIME	java.sql.Time
TIMESTAMP	java.sql.Timestamp

5.2.7 ResultSet 接口

java.sql.ResultSet 接口类似于一个数据表,通过该接口的实例可以获得检索结果集以及对应数据表的相关信息,如列名和类型等,ResultSet 实例通过执行查询数据库的语句生成。

使用 Statement 的 excuteQuery()方法执行 SQL 查询语句,可以获得返回的结果集。

```
Resuleset rs = Statement,executequery("select * from userinfo")
```

ResultSet 实例具有指向当前数据行的指针(游标),最初指针指向第一行记录之前(头部),通过 next()方法可以将指针移动到下一行,如果存在下一行该方法则返回 true,否则返回 false,所以可以通过 while 循环来遍历 ResultSet 结果集。默认情况下

ResultSet 实例不可以更新，只能向前移动指针，所以只能遍历一次，并且只能按从前到后的顺序。

ResultSet 用来暂时存放数据库查询操作获得的结果。它包含了符合 SQL 语句中条件的所有行，并且它提供了一套 get 方法对这些行中的数据进行访问。

通常使用如下的结构遍历结果集。

```
while(rs.net())
{
  …
}
```

ResultSet 接口的常用方法如表 5-7 所示。

表 5-7　ResultSet 接口的常用方法

方　法	描　述
boolean absolute(int row)	将指针移动到结果集对象的某一行
void afterLast()	将指针移动到结果集对象的末尾
void beforeFirst()	将指针移动到结果集对象的头部
boolean first()	将指针移动到结果集对象的第一行
Array getArray(int row)	获取结果集中的某一行并将其存入一个数组
boolean getBoolean(int columnIndex)	获取当前行某一列的值，返回一个布尔型值
byte getByte(int columnIndex)	获取当前行某一列的值，返回一个字节型值
short getShort(int columnIndex)	获取当前行某一列的值，返回一个短整型值
int getInt(int columnIndex)	获取当前行某一列的值，返回一个整型值
long getLong(int columnIndex)	获取当前行某一列的值，返回一个长整型值
double getDouble(int columnIndex)	获取当前行某一列的值，返回一个双精度型值
float getFloat(int columnIndex)	获取当前行某一列的值，返回一个浮点型值
String getString(int columnIndex)	获取当前行某一列的值，返回一个字符串
Date getDate(int columnIndex)	获取当前行某一列的值，返回一个日期型值
Object getObject(int columnIndex)	获取当前行某一列的值，返回一个对象
Statement getStatement()	获得产生该结果集的 Statement 对象
URL getURL(int columnIndex)	获取当前行某一列的值，返回一个 java. net. URL 型值
boolean isBeforeFirst()	判断指针是否在结果集的头部
boolean isAfterLast()	判断指针是否在结果集的末尾行
boolean isFirst()	判断指针是否在结果集的第一行
boolean isLast()	判断指针是否在结果集的最后一行
boolean last()	将指针移动到结果集的最后一行
boolean next()	将指针移动到当前行的下一行
boolean previous()	将指针移动到当前行的前一行

使用 rs. getXXX(argument)方法可以取得当前行的字段值。其中，XXX 表示数据类型，例如：

(1) getInt——以 int 型返回字段值。

(2) getString——以 java. lang. string 类型返回字段值。

(3) getFloat——以 float 类型返回字段值。

(4) getDate——以 java. sal. date 类型返回字段值。

argument 参数有两类值。

(1) 使用字段序号(下标从 1 开始): int id=rs. getint(1)。

(2) 使用字段名称: string name=rs. getsting("username")。

5.2.8 任务: 实现数据库连接

1. 任务要求

设计一个通用类来实现数据库的连接。

2. 任务实施

(1) 在 Tomcat 服务器的 lib 目录下加入数据库连接的驱动 JAR 包。

(2) 在 src 根目录下创建一个资源文件 DBParm. Properties,其内容为连接数据库相关参数。

```
dbName = Test
uName = sa
uPwd = 123456
dbURL = 127.0.0.1:1433
driverName = com.microsoft.sqlserver.jdbc.SQLServerDriver
```

(3) 定义一个类 DBcconfig,用于获取资源文件的相关信息。

```
package com.cykj.db;
//导入包
public class DBConfig{
  private String propertyFileName;
  private ResourceBundle resourceBundle;
  public DBConfig() {
    propertyFileName = "DBParam";
    resourceBundle = ResourceBundle.getBundle(propertyFileName);
  }

  public DBConfig(String filename) {
    propertyFileName = filename;
    resourceBundle = ResourceBundle.getBundle(propertyFileName);
  }
  public String getString(String key) {
    if (key == null || key.equals("") || key.equals("null")) {
      return "";
    }
    String result = "";
    try {
      result = resourceBundle.getString(key);      //根据 key 获取 value
    } catch (MissingResourceException e) {
```

```
        e.printStackTrace();
      }
      return result;
    }
}
```

(4) 定义一个类,用于获取数据库连接。

```
package com.cykj.db;
//导入包
public class ConnectionFactory{
  static public String driverName,dbName,dbURL,uName,uPwd;
  //获取数据库连接信息
  static public void init()
  {
    DBConfig param = new DBConfig("DBParam");
    driverName = param.getString("driverName");
    dbName = param.getString("dbName");
    dbURL = param.getString("dbURL");
    uName = param.getString("uName");
    uPwd = param.getString("uPwd");
  }
  public static Connection getConnection(String driverName,String dbName,
                                          String dbURL,String uName,String uPwd) {
      Connection con = null;
      dbURL = "jdbc:sqlserver://" + dbURL + ";DatabaseName = " + dbName;
      try {
          Class.forName(driverName);
          con = DriverManager.getConnection(dbURL,uName,uPwd);
      }
      catch (SQLException e) {
          e.printStackTrace();
      }
      catch (ClassNotFoundException e) {
          e.printStackTrace();
      }
      return con;
  }
  public static Connection getConnection() {
      Init();   //获取数据库连接参数
      Connection con = null;
      dbURL = "jdbc:sqlserver://" + dbURL + ";DatabaseName = " + dbName;
      try {
          Class.forName(driverName);
          con = DriverManager.getConnection(dbURL,uName,uPwd);
      }
      catch (SQLException e) {
          e.printStackTrace();
      }
      catch (ClassNotFoundException e) {
```

```
            e.printStackTrace();
        }
        return con;
    }

    public static void Close( ResultSet rs ,Statement stmt,Connection conn) {
        try{
            if (rs!= null ) rs.close();
            if (stmt!= null)stmt.close();
            if (conn!= null )conn.close();
        } catch (SQLException e) {
        e.printStackTrace();
        }
    }
}
```

（5）设计测试页面。

```
<%@ page language = "java" import = "java.sql.*" pageEncoding = "gb2312" %>
<%@ page import = "com.cykj.db.*" %>
<html>
  <head><title> JDBC 连接</title></head>
  <body>
  <%
  try {
  /*
    String driverName,dbName,dbURL,uName,uPwd;
    driverName = "com.microsoft.sqlserver.jdbc.SQLServerDriver";
    dbURL = "127.0.0.1:1433";
    dbName = "TestDB";
    uName = "sa";
    uPwd = "123456";
    Connection conn = ConnectionFactory.getConnection(driverName,dbName,dbURL,
                                                      uName,uPwd);
*/
    Connection conn = ConnectionFactory.getConnection();
    if(!conn.isClosed())
      out.println("<h1>成功连接到数据库</h>");
    ConnectionFactory.Close(null, null, conn);
  }
  catch (Exception e) {
      out.print("出现异常!" + e.getMessage());
      e.printStackTrace();
  }
%>
  </body>
</html>
```

（6）启动服务器，部署并浏览。

5.3　连接池技术

如果用户每次请求都向数据库获得连接，则通常需要消耗相对较大的资源，创建时间也较长。假设网站一天有 10 万访问量，数据库服务器就需要创建 10 万次连接，极大地浪费了数据库的资源，并且极易造成数据库服务器内存溢出、宕机，如图 5-11 所示。

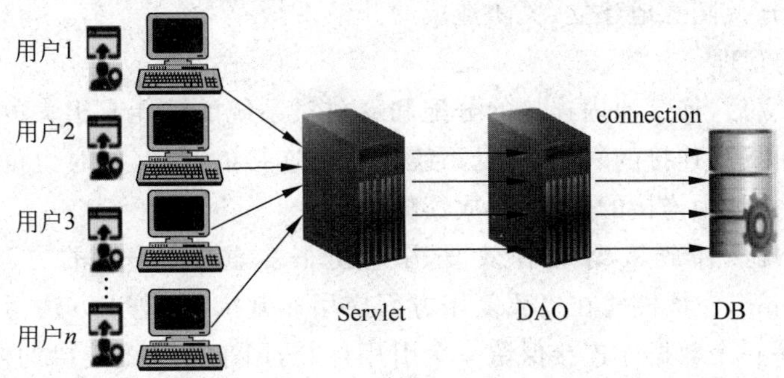

图 5-11　数据库连接

5.3.1　连接池简介

有了池，就不用自己来创建连接，而是通过池来获取 Connection 对象。当使用完连接后，调用 Connection 对象的 close()方法也不会真的关闭连接，而是把连接“归还”给池。池就可以再利用这个 Connection 对象。

1. 连接池原理

连接池技术的核心思想是连接复用，通过建立一个数据库连接池以及一套连接使用、分配、治理策略，使得该连接池中的连接可以得到高效、安全的复用，避免了数据库连接频繁建立、关闭的开销，如图 5-12 所示。另外，由于对 JDBC 中的原始连接进行了封装，从而方便了数据库应用对于连接的使用，提高了开发效率，也正是因为这个封装层的存在，隔离了应用本身的处理逻辑和具体数据库访问逻辑，使应用本身复用成为可能。

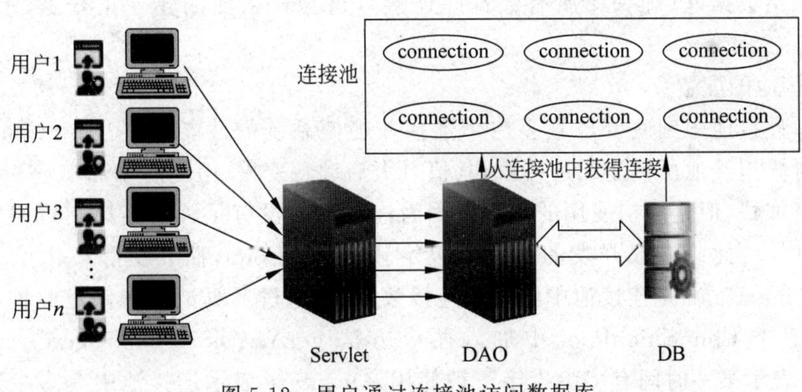

图 5-12　用户通过连接池访问数据库

1）连接池的建立

应用程序中建立的连接池其实是静态的。所谓静态连接池是指连接池中的连接在系统初始化时就已分配好，且不能随意关闭。

Java 中提供了很多容器类可以方便地构建连接池，如 Vector、Stack、Servlet、Bean 等，通过读取连接属性文件 Connections. properties 与数据库实例建立连接。在系统初始化时，根据相应的配置创建连接并放置在连接池中，以便需要使用时能从连接池中获取，这样就可以避免连接随意地建立、关闭造成的开销。

2）连接池的管理

当连接池建立后，连接池内连接的分配和释放对系统的性能有很大的影响。连接的合理分配、释放可提高连接的复用，降低系统建立新连接的开销，同时也可加速用户的访问。下面介绍连接池中连接的分配、释放策略。

连接池的分配、释放策略对有效复用连接非常重要，采用的方法是引用计数（reference counting）。该模式在资源复用方面应用非常广泛，把该方法运用到对连接的分配释放上，为每一个数据库连接保留一个引用计数，用来记录该连接的使用者的个数。具体的实现方法如下。

当客户请求数据库连接时，首先查看连接池中是否有空闲连接（指当前没有分配出去的连接）。假如存在空闲连接，则把连接分配给客户并做相应的处理（即标记该连接为正在使用，引用计数加 1）。假如没有空闲连接，则查看当前的连接数是不是已经达到 maxConn（最大连接数），假如没达到就重新创建一个连接给请求的客户；假如达到就按设定的 maxWaitTime（最大等待时间）进行等待，假如等待 maxWaitTime 后仍没有空闲连接，就抛出无空闲连接的异常给用户。

当客户释放数据库连接时，先判定该连接的引用次数是否超过了规定值，假如超过就删除该连接，并判定当前连接池内总的连接数是否小于 minConn（最小连接数），若小于就将连接池布满；假如没超过就将该连接标记为开放状态，可供再次复用。可以看出正是这套策略保证了数据库连接的有效复用，避免频繁地建立、释放所带来的系统资源开销。

3）连接池的关闭

当应用程序退出时，应关闭连接池，此时应把在连接池建立时向数据库申请的连接对象统一归还给数据库（即关闭所有数据库连接），这与连接池的建立正好是一个相反的过程。

4）连接池的配置

数据库连接池中到底要放置多少个连接才能使系统的性能更佳？minConn 是当应用启动的时候连接池所创建的连接数，其值过大启动将变慢，但是启动后响应更快；其值过小启动将加快，但是最初使用的用户将因为连接池中没有足够的连接不可避免地延缓了执行速度。因此应该在开发的过程中设定较小 minConn，而在实际应用中设定较大 minConn。maxConn 是连接池中的最大连接数，可以通过反复试验来确定此饱和点。为此在连接池类 ConnectionPool 中加入两个方法 getActiveSize( ) 和 getOpenSize( )，ActiveSize 表示某一时间有多少连接正被使用，OpenSize 表示连接池中有多少连接被打开,反映了连接池使用的峰值。将这两个值在日志信息中反映出来,minConn 的值应该小于平均 ActiveSize,而 maxConn 的值应该在 activeSize 和 OpenSize 之间。

2. 连接池的优缺点

连接池的优点如下。

(1) 创建一个新的数据库连接所耗费的时间主要取决于网络的速度以及应用程序和数据库服务器的(网络)距离,而且这通常是一个很耗时的过程,而采用数据库连接池后,数据库连接请求则可以直接通过连接池满足,从而节省了时间。

(2) 提高了数据库连接的重复使用率。

(3) 解决了数据库对连接数量的限制。

连接池的缺点如下。

(1) 连接池中可能存在多个与数据库保持连接但未被使用的连接,在一定程度上浪费了资源。

(2) 要求开发人员和使用者准确估算系统需要提供的最大数据库连接的数量。

5.3.2 Tomcat 配置连接池

自己实现一个连接池太复杂,要考虑的东西太多,比如要考虑连接对象的 close()方法、物理连接中断、数据库连接池耗尽之后如何处理等问题。在实际项目中,一般直接使用一些成熟的开源数据库连接池,常见的有 c3p0、dbcp 等。

在 Tomcat 服务器上就直接增加了数据源的配置选项,直接在服务器上配置好数据源连接池即可。在 J2EE 服务器上保存着一个数据库的多个连接,每一个连接通过 DataSource 可以找到。DataSource 被绑定在 JNDI 树上(为每一个 DataSource 提供一个名称),客户端通过名称找到在 JNDI 树上绑定的 DataSource,再由 DataSource 找到连接,如图 5-13 所示。

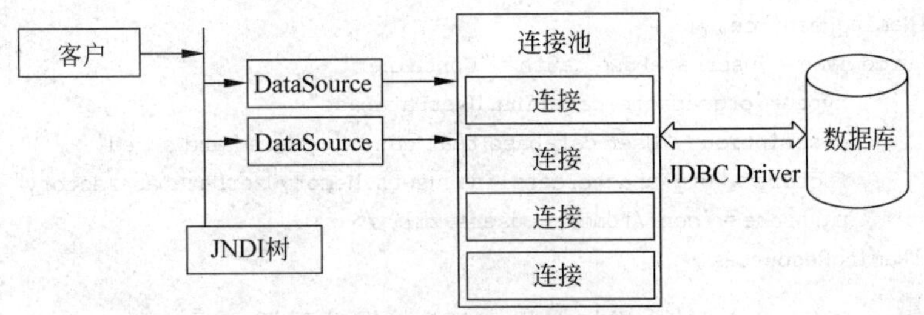

图 5-13　利用 JNDI 获取 DataSource

在以后的操作中,除数据库的连接方式不一样外,其他的所有操作都一样,只是关闭的时候不是彻底地关闭数据库,而是把数据库的连接放回连接池中。

如果要想使用数据源的配置,则必须配置虚拟目录,因为此配置是在虚拟目录之上起作用的。需要注意的是,如果要想完成以上的功能,在 Tomcat 服务器上一定要有各个数据库的驱动程序。

1. JNDI 技术简介

JNDI(Java naming and directory interface,Java 命名和目录接口),对应于J2SE 中的javax. naming 包,这套 API 的主要作用在于:它可以把 Java 对象放在一个容器中(JNDI 容器),并为容器中的 Java 对象取一个名称,以后程序想获得 Java 对象,只需通过名称检索即可。其核心 API 为 Context,它代表 JNDI 容器,其 lookup()方法用于检索容器中对应名称的对象。

Tomcat 服务器创建的数据源是以 JNDI 资源的形式发布的,所以说在 Tomcat 服务器中配置一个数据源实际上就是在配置一个 JNDI 资源,即配置数据源 DataSource。

在 DataSource 中预先建立了多个数据库连接,这些数据库连接保存在数据库连接池中,当程序访问数据库时,只需从连接池中取出空闲的连接,访问结束后,再将连接归还给连接池。DataSource 对象由容器(如 Tomcat)提供,不能通过创建实例的方法获得DataSource 对象,需要利用 Java 的 JNDI 来获得 DataSource 对象的引用。JNDI 是一种将对象和名字绑定的技术,对象工厂负责生产对象,并将其与唯一的名称绑定,在程序中可以通过名称来获得对象的引用。

2. 配置 JNDI 资源

Tomcat 服务器创建数据源的配置主要有两种方式:全局 JNDI 配置和非全局 JNDI 配置。

1) 全局 JNDI 配置

在 server. xml 中配置数据源的步骤如下。

(1) 在 Tomcat 服务器的 lib 目录下加入数据库连接的驱动 JAR 包。

(2) 修改 Tomcat 服务器的 conf 目录下的 server. xml 配置文件。打开 server. xml 配置文件,可以看到里面自带的一个全局 JNDI 配置。

```
<GlobalNamingResources>
  <Resource name = "UserDatabase" auth = "Container"
            type = "org.apache.catalina.UserDatabase"
            description = "User database that can be updated and saved"
            factory = "org.apache.catalina.users.MemoryUserDatabaseFactory"
            pathname = "conf/tomcat-users.xml" />
</GlobalNamingResources>
```

(3) 编辑 server. xml 文件,添加全局 JNDI 数据源配置。

```
<GlobalNamingResources>
     <Resource name = "UserDatabase" auth = "Container"
         type = "org.apache.catalina.UserDatabase"
         description = "User database that can be updated and saved"
         factory = "org.apache.catalina.users.MemoryUserDatabaseFactory"
         pathname = "conf/tomcat-users.xml" />
  <!-- 配置 Oracle 数据库的 JNDI 数据源 -->
  <Resource name = "jdbc/oracle"
```

```
            auth = "Container"
            type = "javax.sql.DataSource"
            maxActive = "100"
            maxIdle = "30"
            maxWait = "10000"
            username = "lead_oams"
            password = "p"
            driverClassName = "oracle.jdbc.driver.OracleDriver"
            url = "jdbc:oracle:thin:@192.168.1.229:1521:lead"/>
    <!-- 配置 MySQL 数据库的 JNDI 数据源 -->
    <Resource name = "jdbc/mysql"
            auth = "Container"
            type = "javax.sql.DataSource"
            maxActive = "100"
            maxIdle = "30"
            maxWait = "10000"
            username = "root"
            password = "root"
            driverClassName = "com.mysql.jdbc.Driver"
            url = "jdbc:mysql://192.168.1.144:3306/leadtest?useUnicode = true&amp;
    characterEncoding = utf - 8
    <!-- 配置 SQL Server 数据库的 JNDI 数据源 -->
    <Resource
            name = "jdbc/sqlserver"
            auth = "Container"
            type = "javax.sql.DataSource"
            maxActive = "100"
            maxIdle = "30"
            maxWait = "10000"
            username = "sa"
            password = "p@ssw0rd"
            driverClassName = "com.microsoft.sqlserver.jdbc.SQLServerDriver"
            url = "jdbc:sqlserver://192.168.1.51:1433;DatabaseName = demo"/>
  </GlobalNamingResources>
```

经过以上步骤，全局 JNDI 数据源就配置好了。在 server.xml 文件中，分别配置了 Oracle、MySQL 和 SQL Server 三种数据库的全局 JNDI 数据源。在配置数据源时需要配置的<Resource>元素的属性如表 5-8 所示。

表 5-8　<Resource>元素的属性

属　　性	描　　述
name	设置数据源的 JNDI 名称，表示以后要查找的名称。通过此名称可以找到 DataSource。此名称可任意更换，但是程序中最终要查找的就是此名称，为了不与其他的名称混淆，比如使用 jdbc/oracle 表示配置的是一个 JDBC 的关于 Oracle 的命名服务
type	设置数据源的类型，如 javax.sql.DataSource
auth	设置数据源的管理者，有两个可选值：Container 和 Application。Container 表示由容器来创建和管理数据源，Application 表示由 Web 应用来创建和管理数据源

续表

属　性	描　述
driverClassName	设置连接数据库的 JDBC 驱动程序
url	设置连接数据库的路径
username	设置连接数据库的用户名
password	设置连接数据库的密码
maxActive	设置连接池中处于活动状态的数据库连接的最大数目,0 表示不受限制
maxIdle	设置连接池中处于空闲状态的数据库连接的最大数目,0 表示不受限制
maxWait	设置当连接池中没有处于空闲状态的连接时,数据库连接请求的最长等待时间(单位为 ms),如果超出该时间将抛出异常,-1 表示无限期等待

2）非全局 JNDI 配置

非全局 JNDI 数据源是针对某一个 Web 项目配置的,具体的配置步骤如下。

(1) 在 Tomcat 服务器的 lib 目录下加入数据库连接的驱动 JAR 包。

(2) 针对具体的 Web 项目映射虚拟目录,然后在虚拟目录映射的配置文件中配置 JNDI 数据源。例如,对 JNDITest 项目,需要在 tomcat\conf\Catalina\localhost 目录下创建一个 JNDITest. xml 文件。

也可以在 Web 项目目录下的 META-INF\context. xml 文件中配置 JNDI,因为这样配置的数据源更有针对性。配置数据源的具体代码如下。

```
<?xml version = "1.0" encoding = "UTF - 8"?>
<!-- 将此文件放置在 tomcat\conf\Catalina\localhost 下(没有目录就新建) -->
<!-- 设置虚拟目录 -->
<Context docBase = "D:/JNDITest/WebRoot" debug = "0" reloadable = "false">
<!-- 配置 MySQL 数据库的 JNDI 数据源 -->
  <Resource name = "mysqlDataSource" //资源名称,也就是引用资源的名称
            auth = "Container"
            type = "javax. sql. DataSource"
            maxActive = "100"
            maxIdle = "30"
            maxWait = "10000"
            username = "root"
            password = "root"
            driverClassName = "com. mysql. jdbc. Driver"
            url = "jdbc:mysql://192.168.1.144:3306/leadtest?useUnicode = true&amp;
            characterEncoding = utf - 8
</Context>
```

(3) 在 Web 项目的 web. xml 文件中引用配置好的 JNDI 数据源。

```
<!--
  JNDI 配置的资源引用:
     res - ref - name:表示引用资源的名称
     res - type:此资源对应的类型为 javax. sql. DataSource
     res - auth:容器授权管理
-->
```

```
<!-- Oracle 数据库 JNDI 数据源引用 -->
<resource-ref>
        <description>Oracle DB Connection</description>
        <res-ref-name>oracleDataSource</res-ref-name>
        <res-type>javax.sql.DataSource</res-type>
        <res-auth>Container</res-auth>
</resource-ref>

<!-- MySQL 数据库 JNDI 数据源引用 -->
<resource-ref>
        <description>MySQL DB Connection</description>
        <res-ref-name>mysqlDataSource</res-ref-name>
        <res-type>javax.sql.DataSource</res-type>
        <res-auth>Container</res-auth>
</resource-ref>

<!-- SQLServer 数据库 JNDI 数据源引用 -->
<resource-ref>
        <description>SQLServer DB Connection</description>
        <res-ref-name>sqlserverDataSource</res-ref-name>
        <res-type>javax.sql.DataSource</res-type>
        <res-auth>Container</res-auth>
</resource-ref>
```

5.3.3　获取 JNDI 的资源

JDBC 的 javax.sql.DataSource 接口负责与数据库建立连接,在应用时不需要编写连接数据库代码,可以直接从数据源中获得数据库连接。其处理的步骤如下。

(1) 获得对数据源的引用。

```
Context ctx = new InitalContext();
DataSource ds = (DataSource)ctx.lookup("java:comp/env/jdbc/testDb");
```

(2) 获得数据库连接对象。

```
Connection con = ds.getConnection();
```

(3) 返回数据库连接到连接池。

```
con.close();
```

【例 5-1】　连接池测试文件。

```
<%@ page language="java" import="java.sql.*" pageEncoding="gb2312"%>
<%@ page import="javax.sql.*"%>
<%@ page import="javax.naming.*"%>
<%@ page import="com.cykj.db.*" %>
<html>
  <head><title>缓冲池连接</title></head>
  <body>
```

```
<%
   try {
     Context initCtx = new InitialContext(); //1. 创建 JNDI 的上下文对象
     Context envCtx = (Context) initCtx.lookup("java:comp/env"); ///2 查询出入口
     //3. 再进行二次查询,找到资源,使用的名称与< Resource >元素的 name 对应
     //DataSource ds = (DataSource) envCtx.lookup("jdbc/test");
     DataSource ds = (DataSource)initCtx .lookup("java:comp/env/jdbc/test");
     Connection conn = ds.getConnection();
     conn.close();
   }
   catch (Exception e) {
     out.print("出现例外!" + e.getMessage());
     e.printStackTrace();
   }
%>
  </body>
</html>
```

编辑 Web 项目目录下的 META-INF\context.xml 文件。

```
<?xml version='1.0' encoding='utf-8'?>
<Context>
    <Resource
        name="jdbc/test"
        auth="Container"
        type="javax.sql.DataSource"
        maxActive="100"
        maxIdle="30"
        maxWait="10000"
        username="sa"
        password="123456"
        driverClassName="com.microsoft.sqlserver.jdbc.SQLServerDriver"
        url="jdbc:sqlserver://127.0.0.1:1433;DatabaseName=test"/>
</Context>
```

注意：项目最好发布在 Tomcat 服务器进行调试。

5.3.4 任务：连接池的应用

1. 任务要求

设计一个通用类来获取 JNDI 资源。

2. 任务实施

(1) 在 Tomcat 服务器的 lib 目录下加入数据库连接的驱动 JAR 包。

(2) 定义一个 ConnectionFactory 类，用于获取连接。

```
package com.cykj.db;
import java.sql.Connection;
```

```
import java.sql.DriverManager;
import java.sql.SQLException;

import javax.naming.Context;
import javax.naming.InitialContext;
import javax.naming.NamingException;
import javax.sql.DataSource;
public class ConnectionFactory {
  public static Connection getConnection() {
    try {
      Context initCtx = new InitialContext();
      String JNDIname = (String) initCtx.lookup("java:comp/env/JNDIname");
      return ConnectionFactory.getConnection(JNDIname);
    }
    catch (NamingException e) {
      System.out.print("数据库未启动或连接池未配置!");
    }
      return null;
  }
  /**
   * 通过指定的 JNDI 名称获取数据库连接,如 JDBC/SURVEY
   * @param JNDIname
   * @return Connection
   */
  public static Connection getConnection(String JNDIname) {
    try {
        Context initCtx = new InitialContext();
        Context envCtx = (Context) initCtx.lookup("java:comp/env");
        DataSource ds = (DataSource) envCtx.lookup(JNDIname);
        return ds.getConnection();
    } catch (NamingException e1) {
        System.out.print("数据库未启动或连接池未配置!");
        return null;
    } catch (SQLException e) {
        System.out.print("数据库未启动或连接池未配置 2!");
        //e.printStackTrace();
        return null;
      }
    }
    //关闭连接
    public static void Close ( ResultSet rs , Statement stmt, Connection conn) throws
    SQLException
    {
        if (rs!= null ) rs.close();
        if (stmt!= null) stmt.close();
        if(conn!= null ) conn.close();
    }
}
```

(3) 编辑 Web 项目目录下的 META-INF\context.xml 文件。

```
<?xml version='1.0' encoding='utf-8'?>
<Context>
    <Resource
        name="jdbc/test"
        auth="Container"
        type="javax.sql.DataSource"
        maxActive="100"
        maxIdle="30"
        maxWait="10000"
        username="sa"
        password="123456"
        driverClassName="com.microsoft.sqlserver.jdbc.SQLServerDriver"
        url="jdbc:sqlserver://127.0.0.1:1433;DatabaseName=test"/>
</Context>
```

(4) 编辑 Web.xml 文件。

```
<web-app>
    //略
  <env-entry>
      <env-entry-name>JNDIname</env-entry-name>
      <env-entry-type>java.lang.String</env-entry-type>
      <env-entry-value>jdbc/test</env-entry-value>
    </env-entry>
</web-app>
```

(5) 设计测试页面参考 5.2 节的任务。

(6) 启动服务器,部署并浏览。

5.4 JDBC 数据库访问

5.4.1 JDBC 访问数据库的步骤

JDBC 程序的基本结构如下。

```
import java.sql.*;                                        //引入 java.lang.sql 包
Class.forName("driver_type");                             //加载 JDBC 驱动程序
//注册 JDBC 驱动程序已创建连接
Connection = DriverManager.getConnection (URL,"login","passwd");
Statement = Connection.createStatement( );
ResultSet = Statement.executeQuery(query);                //执行 SQL
...                                                       //结果集处理
Exception(SQLException,ClassNotFoundException)            //异常处理
Connection, Statement. ResultSet close();                 //释放资源
```

1. 引入 java.lang.sql 包

引入 Java 应用程序连接、检索和处理数据库信息所需的相关类库和接口的格式如下。

```
import java.sql.* ;
```

2. 加载 JDBC 驱动程序

在连接数据库之前，首先要加载目标数据库的驱动程序到 JVM(Java 虚拟机)，可以通过数据库厂商网站获得 JDBC 驱动程序及驱动的文档说明。驱动程序通常是一个 JAR 文包，驱动文档中通常包含了驱动使用的说明和示例程序。

如果在 Java 程序中需要使用第三方(非 JDK 内置)的软件包，必须把软件包添加到程序的 classpath(类路径)中。在 Eclipse 中添加软件包的方法与项目类型有关，直接将 JDBC 驱动的 JAR 包复制到项目的 web—inf/lib/下即可。

在应用程序中，有 3 种方法可以加载驱动程序。

(1) 利用 System 类的静态方法 setProperty()。

```
System.setProperty("jdbc.drivers","sun.jdbc.odbc.JdbcOdbcDriver");
```

(2) 利用 Class 类的静态方法 forName()(推荐)。

```
Class.forName("sun.jdbc.odbc.JdbcOdbcDriver");
```

(3) 直接创建一个驱动程序对象。

```
new sun.jdbc.odbc.JdbcOdbcDriver();
```

加载驱动程序的格式如下。

```
Class.forName("driver_type") ;
```

驱动程序由驱动程序开发商提供。各类数据库系统的驱动程序也不同，需要根据数据库产品进行设定。

例如，加载 SQL Server 数据库驱动程序的具体代码如下。

```
static {
  try {
    Class.forName("com.microsoft.sqlserver.jdbc.SQLServerDriver").newInstance();
  } catch (ClassNotFoundException e) {
    System.out.println("------ 在加载数据库驱动异常 -- ");
    e.printStackTrace();
  }
}
```

成功加载后，会将加载的驱动类注册到 DriverManager 类。如果加载失败，将抛出 ClassNotFoundException 异常，即未找到指定的驱动类，所以需要在加载数据库驱动类时捕捉可能抛出的异常。

通常将负责加载驱动的代码放在 static 块中，好处是只有 static 块所在的类第一次访问数据库时加载数据库驱动，可以避免重复加载驱动程序。

3. 注册 JDBC 驱动程序

java.sql.DriverManager(驱动程序管理器)类是 JDBC 的管理层，对加载的驱动程序进行系统注册，负责建立和管理数据库连接。

通过 DriverManager 类的静态方法 getConnection(String url, String user, String password)可以建立数据库连接，3 个入口参数依次为目标数据库的路径、用户名和密码，该方法的返回值类型为 java.sql.Connection，为连接数据库建立逻辑连接。

典型代码如下。

```
Connection conn = DriverManager.getConnection(
    "jdbc:microsoft:sqlserver://127.0.0.1:1433;DatabaseName = db_JSDQ10", "sa", "");
```

代码说明如下。

数据库类型：SQL Server 数据库。

数据库路径：服务器所在的计算机，即本机(127.0.0.1)。

数据库名称：db_JSDQ10。

用户名称：sa。

用户密码：密码为空。

4. 执行 SQL 语句

建立数据库连接的目的是与数据库进行通信，实现方式为执行 SQL 语句，但是通过 Connection 实例并不能执行 SQL 语句，还需要通过 Connection 实例创建 Statement 实例。Statement 实例可分为 3 种类型。

(1) Statement 实例：该类型的实例只能用来执行静态的 SQL 语句。

```
String sql = "select * from user "
Statement stmt = con.createStatement();
```

(2) PreparedStatement 实例：该类型的实例增加了执行动态 SQL 语句的功能。

```
//准备 SQL 语句
String sql = "select * from user where userid = ?" ;
PreparedStatement ps = con.prepareStatement(sql);
ps.setString(1,"01");
```

(3) CallableStatement 实例：该类型的实例增加了执行数据库存储过程的功能。

```
//准备 SQL 语句
String strSQL = "{? = call sp_jsptest(?,?)}"; //存储过程 sp_jsptest,1
//准备可调用语句对象
CallableStatement sqlStmt = sqlCon.prepareCall(strSQL);
//设置输入参数
sqlStmt.setString(2,strName);
//登记输出参数
sqlStmt.registerOutParameter(1,java.sql.Types.INTEGER);
sqlStmt.registerOutParameter(3,java.sql.Types.VARCHAR);
```

其中，Statement 是最基础的，PreparedStatement 继承自 Statement 并做了相应的扩展，而 CallableStatement 继承自 PreparedStatement，又做了相应的扩展，从而保证在基

本功能的基础上，各自又增加了一些独特的功能。

5. 获得查询结果

通过 Statement 接口的 executeUpdate()方法或 executeQuery()方法，可以执行 SQL 语句，同时将返回执行结果。如果执行的是 executeUpdate()方法，将返回一个 int 型数值，代表影响数据库记录的条数，即插入、修改或删除记录的条数；如果执行的是 executeQuery()方法，将返回一个 ResultSet 结果集，其中不仅包含所有满足查询条件的记录，还包含相应数据表的相关信息，如每一列的名称、类型和列的数量等。例如：

```
ResultSet rs = stmt.executeQuery(sql);
ResultSet rs = ps.executeQuery();
ResultSet rs  =  sqlStmt.executeQuery();          /执行该存储过程并返回结果集
rs.next(); //获取来自结果集中的数据
String strWelcome  =  rs.getString(1);            //获取输出参数的值
String strMyName  =  sqlStmt.getString(3);
int intReturn  =  sqlStmt.getInt(1);              //获取返回值
```

然后通过循环获取所有结果集中的数据。

```
while (rs.next()){
    String code = rs.getString(1);
    String name = rs.getString("name");
}
```

6. 关闭连接

在建立 Connection、Statement 和 ResultSet 实例时，均需占用一定的数据库和 JDBC 资源，所以每次访问数据库结束后，应该及时销毁这些实例，释放它们占用的资源，方法是通过各个实例的 close()方法。在关闭时建议按照如下的顺序：关闭所有连接，自里往外关闭。依次将 ResultSet、Statement、PreparedStatement、Connection 对象关闭，释放所占用的资源。大多数 JDBC 操作都会抛出 SQLException 异常，必须进行异常处理。

```
try{
  …
}
catch(SQLException e) {
  e.printStackTrace( );
}
finally {
  …
  resultSet.close();
  statement.close();
  connection.close();
}
```

建议按上面的顺序关闭的原因在于 Connection 是一个接口，close()方法的实现方式可能多种多样。如果通过 DriverManager 类的 getConnection()方法得到 Connection 实例,在调用 close()方法关闭 Connection 实例时会同时关闭 Statement 实例和 ResultSet 实例。但是通常情况下需要采用数据库连接池,在调用通过连接池得到的 Connection 实例的 close()方法时,Connection 实例可能并没有被释放,而是被放回到了连接池中,又被其他连接调用。在这种情况下如果不手动关闭 Statement 实例和 ResultSet 实例,它们在 Connection 中可能会越来越多。虽然 JVM 的垃圾回收机制会定时清理缓存,但是如果清理不及时,当数据库连接达到一定数量时,将严重影响数据库和计算机的运行速度,甚至导致软件或系统瘫痪。

5.4.2 操作数据库

数据库的常见操作主要是实现对数据的增加、删除、修改、查找等。下面介绍几种常见操作的编程。首先定义一个用户 JavaBean。

```
public class User{
  private String userID;
  private String userName;
  private String userPwd;
  //getXXX()、setXXX()略
}
```

1. 添加数据

添加数据分为两种情况:①一次只添加一条记录;②一次添加多条记录,即批量添加记录。因为 PreparedStatement 接口和 CallableStatement 接口均继承自 Statement 接口,所以通过这三种类型的实例均可完成添加一条记录的操作。

(1) 通过 Statement 实例完成的典型代码如下。

```
statement.executeUpdate("insert into record(id,name) values(060522,'马先生')");
```

(2) 通过 PreparedStatement 实例完成的典型代码如下。

```
preparedStatement = conn.prepareStatement("insert into record(id,name) values(?,?)");
preparedStatement.setInt(1, 060522);
preparedStatement.setString(2, "马先生");
preparedStatement.executeUpdate();
```

如果一次只添加一条记录,只有在执行 executeUpdate()方法时才能真正提交到数据库。例如,在通过 PreparedStatement 和 CallableStatement 实例添加时,必须在设置完参数后执行 executeUpdate()方法。

如果需要批量添加记录,则只能通过 PreparedStatement 或 CallableStatement 实例实现。在下面的例子中,既实现了单条添加记录,又实现了批量添加记录。

```
conn = DriverManager.getConnection(url, username, password); //创建数据库连接
stmt = conn.createStatement();
//执行静态 insert 语句
```

```
stmt.executeUpdate("insert into tb_testInsert(id,namepwd) values('" + id + "','+ '" +
  name + '" + pwd)");stmt.close();
conn.close();
```

上面的代码通过静态 insert 语句实现了单条添加记录。首先创建数据库连接,然后执行静态 insert 语句,最后关闭数据库连接。

```
//预处理动态 insert 语句
prpdStmt = conn .prepareStatement("insert into tb_usertable (id,name,pwd)
                                 values(?,?,?)");
prpdStmt.clearBatch();              //清除 Batch
for (int i = 0; i < names.length; i++) {
    prpdStmt.setString(1, ids[i]);    //为动态 SQL 语句赋值
    prpdStmt.setString(2, name[i]);   //为动态 SQL 语句赋值
    prpdStmt.setString(3, pwd[i]);    //为动态 SQL 语句赋值
    prpdStmt.addBatch();              //向 Batch 中添加 insert 语句
}
prpdStmt.executeBatch(); //批量执行 Batch 中的 insert 语句
```

上面的代码通过动态 insert 语句实现了批量添加记录。首先预处理动态 insert 语句并获得一个 PreparedStatement 实例,然后通过 clearBatch()方法清除 Batch 中的所有 SQL 语句,紧接着通过 for 循环为动态 insert 语句中的参数赋值,并将 insert 语句添加到 Batch 中,最后通过 executeBatch()方法执行 Batch 中的所有 insert 语句。

在定义动态 insert 语句时,如果有多列,各列之间用“,”隔开。在为参数赋值时,索引位置从 1 开始。

下面的代码通过存储过程实现了批量添加记录,与通过动态 insert 语句实现基本相同。

```
cablStmt = conn.prepareCall("{call userInsert(?,?,?)}");   //调用存储过程
cablStmt.clearBatch();
for (int i = 0; i < names.length; i++) {
    cablStmt.setString(1, id[i]);
    cablStmt.setString(2, name[i]);
    cablStmt.setString(3, pwd[i]);
    cablStmt.addBatch();
}
cablStmt.executeBatch();
```

在调用存储过程时必须满足{call userInsert(?)}格式,其中 userInsert 为存储过程的名称,? 代表存储过程的入口参数,如果有多个入口参数则用逗号隔开。

【例 5-2】 实现用户注册处理。

(1) 用户注册页面 userreg.jsp 的代码如下。

```
<%@ page language = "java" pageEncoding = "GBK" %>
<html>
  <head><title>用户注册 e</title></head>
  <body>
    <form name = "myform" action = "reg.jsp" method = "post">
```

```
        账号:< input type = "text" name = "id"/>< br >
        姓名:< input type = "text" name = "name"/>< br >
        密码:< input type = "password" name = "pwd"/>< br >
        < input type = "submit" value = "提交"/>
      </form >
   </body >
</html >
```

(2) 用户注册数据处理页面 reg.jsp 的代码如下。

```
<% @ page language = "java" import = "java.sql. * " pageEncoding = "GBK" %>
<% @ page import = "com.cykj.db. * " %>
<!-- 导入 5.2 节或 5.3 节任务中的 ConnectionFactory 类 -->
<%
   String pwd = "101";
   String name = "zhangsan";
   Connection con = ConnectionFactory.getConnection();
   String sql = "insert into usertable(userid,username,pwd) values(?,?,?)";
   PreparedStatement ps = con.prepareStatement(sql);
   ps.setString(1,ID);
   ps.setString(2,name);
   ps.setString(3,pwd);
   int i = ps.executeUpdate();
   ConnectionFactory.close(null,ps,con);
%>
```

2. 查询数据

在查询数据时,既可以利用 Statement 实例通过执行静态 select 语句完成,也可以利用 PreparedStatement 实例通过执行动态 select 语句完成,还可以利用 CallableStatement 实例通过执行存储过程来完成。

利用 Statement 实例通过执行静态 select 语句完成数据查询的典型代码如下。

```
ResultSet rs = statement.executeQuery("select * from tb_record where sex = '" + sex + "'");
```

利用 PreparedStatement 实例通过执行动态 select 语句完成数据查询的典型代码如下。

```
preparedStatement = connection.prepareStatement("select * from tb_record where sex = ?");
preparedStatement.setString(1, sex);
ResultSet rs = preparedStatement.executeQuery();
```

利用 CallableStatement 实例通过执行存储过程完成数据查询的典型代码如下。

```
callableStatement = connection.prepareCall("{call select_by_sex(?)}");
callableStatement.setString(1, sex);
ResultSet rs = callableStatement.executeQuery();
```

无论利用哪种方式查询记录,都需要执行 executeQuery()方法,并且该方法返回一个 ResultSet 结果集。在该结果集中不仅包含所有满足查询条件的记录,还包含相应数据表的相关信息，如每一列的名称、类型和列的数量等。在执行 executeQuery()方法时可能抛出 SQLException 类型的异常，所以需要通过 try-catch 语句进行捕获。

下面通过一个例子，详细讲解 ResultSet 结果集的使用方法。

在查询记录时，通常情况下将返回的结果集转换成一个 List 集合，在 List 集合中存放的是数组对象，每一个数组对象代表一条满足条件的记录。

通常情况下利用 while 循环遍历 ResultSet 结果集，并通过 next()方法判断是否存在满足查询条件的记录。如果存在则返回 true，并将指针移动到下一条记录上；如果不存在则返回 false。遍历 ResultSet 结果集的典型代码如下。

```
prpdStmt = conn.prepareStatement("select * from userTable where userType = ?");
prpdStmt.setString(1, "普通");
ResultSet rs = prpdStmt.executeQuery();
UserBean user;
while (rs.next()) {      //利用 while 循环遍历 ResultSet 结果集,并通过 next()方法判断是否
                         //存在下一条记录
  user = new UserBean();
  user.setUserId(rs.getString(1));
  user.setUserName(rs.getString(2));
  user.setUserPwd(rs.getString(3));
  list.add(user);   //将数组添加到 List 集合中
}
```

【例 5-3】　读取用户信息并显示(userlist.jsp)。

```
<%@ page language = "java" import = "java.sql.*" pageEncoding = "gb2312" %>
<%@ page import = "com.cykj.db.*" %>
<!-- 导入 5.2 节或 5.3 节任务中的 ConnectionFactory 类 -->
<html>
  <head><title>读取用户信息</title></head>
  <body>
  <%
    try {
      Connection conn = ConnectionFactory.getConnection();
      Statement stmt = conn.createStatement();
      ResultSet rs = stmt.executeQuery("select * from userTable");
      while (rs.next()) {
  %>
      <% = rs.getString(1) %><br>
  <%
      }
      ConnectionFactoryA.Close(rs, stmt, conn);
    }
    catch (Exception e) {
      out.print("出现异常!" + e.getMessage());
      e.printStackTrace();
    }
  %>
  </body>
</html>
```

3. 修改数据

修改数据分为两种情况：①一次只执行一条 update 语句；②一次执行多条 update 语句，即批量修改记录。因为 PreparedStatement 和 CallableStatement 均继承自 Statement，所以通过这 3 种类型的实例均可以完成非批量修改记录的操作。

通过 Statement 实例完成的典型代码如下。

```
statement.executeUpdate("update usertable set pwd = 'zhanhsan' where id = '6'");
```

通过 PreparedStatement 实例完成的典型代码如下。

```
preparedStatement = conn.prepareStatement("update userTable set pwd = ? where id = ?");
preparedStatement.setString(1, "zhangsan");
preparedStatement.setString(2, 6);
preparedStatement.executeUpdate();
```

通过 CallableStatement 实例完成的典型代码如下。

```
callableStatement = conn.prepareCall("{call update_pwd_by_id(?,?)}");
callableStatemen.setString(1, "zhangsan");
callableStatemen.setString(2, 6);
callableStatemen.executeUpdate();
```

如果需要批量修改记录，则只能通过 PreparedStatement 或 CallableStatement 实例实现，在下面的例子中主要讲解批量修改记录。

下面的代码通过动态 update 语句实现批量修改记录。

```
prpdStmt = conn.prepareStatement("update userTable set pwd = ? where id = ?");
prpdStmt.clearBatch();
Iterator < Entry < String, String >> entryIt = map.entrySet().iterator();
while (entryIt.hasNext()) {//通过遍历 Map 为动态 update 语句赋值并添加到 Batch 中
    Entry < String, String > entry = entryIt.next();
    prpdStmt.setString(1, entry.getValue());
    prpdStmt.setString(2, entry.getKey());
    prpdStmt.addBatch();
prpdStmt.executeBatch();
```

下面的代码通过存储过程实现批量修改记录，与通过动态 update 语句实现基本相同。

```
cablStmt = conn.prepareCall("{call update_pwd_by_id(?,?)}");
cablStmt.clearBatch();
Iterator < Entry < String, String >> entryIt = map.entrySet().iterator();
while (entryIt.hasNext()) {//通过遍历 Map 循环为动态 UPDATE 语句赋值并添加到 Batch 中
    Entry < String, String > entry = entryIt.next();
    cablStmt.setString(1, entry.getKey());
    cablStmt.setString(2, entry.getValue());
    cablStmt.addBatch();
}
cablStmt.executeBatch();
```

【例 5-4】 修改用户密码。

```
<%@ page language="java" import="java.sql.*" pageEncoding="GBK"%>
<%@ page import="com.cykj.db.*" %>
<!-- 导入 5.2 节或 5.3 节任务中的 ConnectionFactory 类 -->
<%
  request.setCharacterEncoding("GBK");
  String ID="101";
  String pwd="123456";
  Connection conn=ConnectionFactory.getConnection();
  String sql="insert into usertable(userid,username,pwd) values(?,?,?)";
  String sql="update set username=? from usertable where userid=?";
  PreparedStatement ps=con.prepareStatement(sql);
  ps.setString(1,ID);
  ps.setString(2,pwd);
  int i=ps.executeUpdate();
  ConnectionFactory.close(null,ps,conn);
%>
```

4. 删除数据

删除数据分为两种情况：①一次只执行一条 delete 语句；②一次执行多条 delete 语句,即批量删除记录。因为 PreparedStatement 和 CallableStatement 均继承自 Statement,所以通过这 3 种类型的实例均可完成非批量删除记录的操作。

通过 Statement 实例完成的典型代码如下。

```
statement.executeUpdate("delete from tb_userTable where id>6");
```

通过 PreparedStatement 实例完成的典型代码如下。

```
preparedStatement = conn.prepareStatement("delete from userTable where id>?");
preparedStatement.setString(1, 6);
preparedStatement.executeUpdate();
```

通过 CallableStatement 实例完成的典型代码如下。

```
callableStatement = conn.prepareCall("{call delete_by_id(?)}");
callableStatemen.setString(1, 6);
callableStatemen.executeUpdate();
```

如果需要批量删除记录,则只能通过 PreparedStatement 或 CallableStatement 实例实现,在下面的例子中将主要讲解批量删除记录。

下面的代码通过动态 delete 语句实现了批量删除记录。

```
prpdStmt = conn.prepareStatement("delete from userTablee where id=?");
prpdStmt.clearBatch();
for (int i = 0; i<ids.length; i++) {
    prpdStmt.setString(1, ids[i]);
```

```
        prpdStmt.addBatch();
    }
    prpdStmt.executeBatch();
```

下面的代码通过存储过程实现了批量删除记录，与通过动态 delete 语句实现基本相同。

```
    cablStmt = conn.prepareCall("{call delete_by_id(?)}");
    cablStmt.clearBatch();
    for (int i = 0; i < ids.length; i++) {
        cablStmt.setString(1, ids[i]);
        cablStmt.addBatch();
    }
    cablStmt.executeBatch();
```

【例 5-5】 删除用户信息。

```
<%@ page language = "java" import = "java.sql.*" pageEncoding = "GBK" %>
<%@ page import = "com.cykj.db.*" %>
<!-- 导入 5.2 节或 5.3 节任务中的 ConnectionFactory 类 -->
<%
   request.setCharacterEncoding("GBK");
   String ID = "101";
   Connection conn = ConnectionFactory.getConnection();
   String sql = "delete from usertable where userid = ?";
   PreparedStatement ps = con.prepareStatement(sql);
   ps.setString(1,ID);
   int i = ps.executeUpdate();
   ConnectionFactoryA.Close(rs, stmt, conn);
%>
```

5.4.3 JDBC 事务

1. 事务处理的基本概念

1）什么是事务

事务是指一组相互依赖的操作单元的集合，用来保证对数据库的正确修改，保持数据的完整性。如果一个事务的某个单元操作失败，将取消本次事务的全部操作。例如，银行交易、股票交易和网上购物等都需要利用事务来控制数据的完整性。例如，将账户 A 的资金转入账户 B 中，在 A 中扣除成功，在 B 中添加失败，导致数据失去平衡，事务将回滚到原始状态，即 A 中没少，B 中没多。

事务是现代数据库理论中的核心概念之一。如果一组处理步骤或者全部发生或者一步也不执行，则称该组处理步骤为一个事务。当所有的步骤像一个操作一样被完整地执行，则称该事务被提交。由于其中的一部分或多步执行失败，导致没有被提交，则事务必须回滚到最初的系统状态。

2）ACID 原则

事务必须服从 ISO/IEC 制定的 ACID 原则。ACID 是原子性（atomicity）、一致性

(consistency)、隔离性(isolation)和持久性(durability)的缩写。

(1) 原子性：事务是一个完整的操作。事务的各步操作是不可分的(原子的)；要么都执行，要么都不执行。

(2) 一致性：当事务完成时，数据必须处于一致状态。

(3) 隔离性：对数据进行修改的所有并发事务是彼此隔离的，这表明事务必须是独立的，它不应以任何方式依赖于或影响其他事务。

(4) 持久性：事务完成后，它对数据库的修改被永久保持，事务日志能够保持事务的持久性。

事务的原子性表示事务执行过程中的任何失败都将导致事务所做的任何修改失效。一致性表示当事务执行失败时，所有被该事务影响的数据都应该恢复到事务执行前的状态。隔离性表示并发事务时彼此隔离。持久性表示当系统或介质发生故障时，确保已提交事务的更新不能丢失。持久性通过数据库备份和恢复来保证。

3) 并发性

当多个用户并发访问数据库中相同的数据时，可能会出现并发问题。如果没有锁定且多个用户同时访问一个数据库，则当他们的事务同时使用相同的数据时可能会发生问题。并发问题包括以下几种。

(1) 丢失更新。当两个或多个事务选择同一行，然后基于最初选定的值更新该行时，会发生丢失更新问题。每个事务都不知道其他事务的存在，最后的更新将重写由其他事务所做的更新，这将导致数据丢失。

(2) 未确认的相关性(脏读)。当第二个事务选择其他事务正在更新的行时，会发生未确认的相关性问题。第二个事务正在读取的数据还没有确认并且可能由更新此行的事务所更改。

(3) 不一致的分析。当第二个事务多次访问同一行而且每次读取的数据不同时，会发生不一致的分析问题。不一致的分析和未确认的相关性类似，因为其他事务也是正在更改第二个事务正在读取的数据。然而，在不一致的分析中，第二个事务读取的数据是由已进行了更改的事务提交的。而且，不一致的分析涉及两次或更多次读取同一行，而且每次信息都由其他事务更改。

(4) 幻象读。当对某行执行插入或删除操作，而该行属于某个事务正在读取的行范围内时，会发生幻象读问题。事务第一次读取的行范围显示出其中一行已不复存在于第二次读取或后续的读取中，因为该行已被其他事务删除。同样，由于其他事务的插入操作，事务的第二次或后续读取显示有一行已不存在于原始读中。

2. JDBC 的事务支持

JDBC 对事务的支持体现在以下两个方面。

(1) 自动提交模式。Connection 提供了一个 Auto-Commit 属性来指定事务何时结束。当 Auto-Commit 为 true 时，每个独立 SQL 操作执行完毕，事务立即自动提交，也就是说每个 SQL 操作都是一个事务。一个独立 SQL 操作什么时候算执行完毕，JDBC 规范是如下规定的。

① 对数据操纵语言(DML,如 insert、update、delete)和数据定义语言(如 create、drop),语句执行结束就视为执行完毕。

② 对 select 语句,当与它关联的 ResultSet 对象关闭时,视为执行完毕。

③ 对存储过程或其他返回多个结果的语句,当与它关联的所有 ResultSet 对象全部关闭,所有 update count(update、delete 等语句操作影响的行数)和 output parameter(存储过程的输出参数)都已经获取之后,视为执行完毕。

当 Auto-Commit 为 false 时,每个事务都必须显示调用 commit()方法进行提交,或者显式调用 rollback()方法进行回滚。Auto-Commit 默认为 true。

(2) 事务隔离级别。JDBC 提供了 5 种不同的事务隔离(加锁)级别,在 Connection 中进行了定义。

① 禁止事务操作和加锁。

```
static int TRANSACTION_NONE = 0;
```

② 允许脏读、重复读和幻象读。

```
static int TRANSACTION_READ_UNCOMMITTED = 1;
```

③ 禁止脏读,允许重复读和幻象读。

```
static int TRANSACTION_READ_COMMITTED = 2;
```

④ 禁止脏读和重复读,允许幻象读。

```
static int TRANSACTION_REPEATABLE_READ = 4;
```

⑤ 禁止脏读、重复读和允许幻象读。

```
static int TRANSACTION_SERIALIZABLE = 8;
```

这 5 种隔离级别中,最后一种为表加锁,第 3 种和第 4 种为行加锁。

JDBC 根据数据库提供的默认值来设置事务支持及其隔离,当然也可以手动设置。

```
setTransactionIsolation(TRANSACTION_READ_UNCOMMITTED);
```

可以通过 getTransactionIsolation()方法查看数据库的当前设置。

需要注意的是,在进行手动设置时,数据库及其驱动程序必须支持相应的事务操作。

上述设置随着值的增加,其事务的独立性增加,更能有效地防止事务操作之间的冲突;同时也增加了加锁的开销,降低了用户之间访问数据库的并发性,程序的运行效率也会随之降低,因此必须平衡程序运行效率和数据一致性之间的冲突。

一般来说,若只涉及数据库的查询操作,可以采用 TRANSACTION_READ_UNCOMMITTED 方式;若数据查询远多于更新,可以采用 TRANSACTION_READ_COMMITTED 方式;若更新操作较多,可以采用 TRANSACTION_REPEATABLE_READ;若对数据一致性要求更高,可考虑最后一项,由于涉及表加锁,因此会对程序运行效率产生较大的影响。

另外,在 Oracle 中数据库驱动对事务处理的默认值是 TRANSACTION_NONE,即不支持事务操作,所以需要在程序中手动进行设置。

3. JDBC 的事务处理

1）程序基本结构

JDBC 支持事务操作，一般情况下，事务的操作需要以下步骤。

(1) 把数据库连接对象的自动提交 SQL 的操作的属性关闭(默认关闭事务，即打开自动提交。因为 DDL 和 DCL 的语句都会导致事务立即提交，而事务是由一步或几步数据操作序列组成的逻辑单元，这一系列操作要么全部执行，要么全部放弃执行，所以要将自动提交属性关闭)。

(2) 执行一系列数据库操作，如果成功就调用 Connection 的 commit()方法提交事务。

(3) 如果事务中的操作没有完全成功，就回滚事务。也就是说当 Connection 遇到一个未处理的 SQLException 异常时，系统将会非正常退出，事务也会自动回滚。但如果程序捕获到了异常，则需要在异常处理块中显式回滚事务。

关键代码如下。

```
//<%@ page import = "com.cykj.db. * " %>
<!-- 导入 5.2 节或 5.3 节任务中的 ConnectionFactory 类 -->
try {
  Connection conn = ConnectionFactory.getConnection();
  conn.setAutoCommit(false); //禁止自动提交,设置回滚点
  stmt = conn.createStatement();
  stmt.executeUpdate("alter table ..."); //数据库更新操作 1
  stmt.executeUpdate("insert into table ..."); //数据库更新操作 2
  conn.commit(); //事务提交
}
catch (Exception ex)
  {
    ex.printStackTrace();
    try {
      conn.rollback(); //操作不成功则回滚
    }catch(Exception e) {
    e.printStackTrace();
  }
}
```

2）设置事务级别

假设现在有一个 Connection 对象 con，那么设置事务级别的方法如下。

```
con.setTransactionLevel(TRANSACTION_SERIALIZABLE) ;
```

也可以使用 getTransactionLevel()方法来获取当前事务的级别。

```
con.getTransactionLevel();
```

在默认情况下，JDBC 驱动程序运行在自动提交模式下，即发送到数据库的所有命令运行在它们自己的事务中。这样做虽然方便，但代价是程序运行时的开销比较大。可以

利用批处理操作减小这种开销,因为在一次批处理操作中可以执行多个数据库更新操作。但批处理操作要求事务不能处于自动提交模式下,为此首先要禁用自动提交模式。

3) 保存点

JDBC 定义了 SavePoint 接口,提供了一个更精细的事务控制机制。当设置了一个保存点后,可以回滚到该保存点处的状态,而不是回滚整个事务。Connection 接口的 setSavepoint()和 releaseSavepoint()方法可以设置和释放保存点。

```
Savepoint sp = con.setSavepoint();       //设置保存点
con.releaseSavepoint(sp);                //释放保存点
```

JDBC 虽然定义了事务的以上支持行为,但是各个 JDBC 驱动、数据库厂商对事务的支持程度可能各不相同,如果在程序中任意设置,可能得不到想要的效果。为此,JDBC 提供了 DatabaseMetaData 接口,提供了一系列对 JDBC 特性支持情况的获取方法。比如,通过 DatabaseMetaData.supportsTransactionIsolationLevel()方法可以判断对事务隔离级别的支持情况,通过 DatabaseMetaData.supportsSavepoints()方法可以判断对保存点的支持情况。

【例 5-6】 事务处理。

```
public class TransactionTest {
  private String driver,url,user,pass;
  private Connection conn;
  private Statement stmt;
  public void initParam(String paramFile) throws Exception{
    //利用配置文件初始化参数
    Properties prop = new Properties();
    prop.load(new FileInputStream(paramFile));
    driver = prop.getProperty("driver");
    url = prop.getProperty("url");
    user = prop.getProperty("user");
    pass = prop.getProperty("pass");
    dbName = prop.getProperty("dbname");
  }
  public void insertInTranscation(String[] sqls){
    Savepoint sp = null;
    try{
      Connection conn = ConnectionFactory.getConnection(driver,dbName,url,user,pass);
        //关闭自动提交,并开启事务
        conn.setAutoCommit(false);
        System.out.println("开启了事务!");
        stmt = conn.createStatement();
        for (String sql : sqls) {
          stmt.execute(sql);
          sp = conn.setSavepoint();      //如果使用保存点
        }
        //conn.commit();                  //提交事务时如果没有使用保存点
        //System.out.println("提交了事务!");
    }catch(Exception e){
```

```
            if(sp!= null){
                try {
                    conn.rollback(sp);            //回滚到上一个保存点 但是事务还是没有结束
                } catch (SQLException e1) {
                    e1.printStackTrace();
                }
            }
            e.printStackTrace();
        }finally{
            try {
                conn.commit();                //提交事务,使用保存点后依然提交
                System.out.println("提交了事务!");
                ConnectionFactory.close(null,stmt,conn);
            } catch (Exception e2) {
            }
        }
    }
}
```

5.4.4 JDBC 批处理

executeBatch()方法返回一个更新计数的数组,每个值对应于批处理操作中的一条命令,方法原型如下。

```
int[] executeBatch() throws SQLException
```

批处理操作可能会抛出一个类型为 BatchUpdateException 的异常,这个异常表明批处理操作中至少有一条命令失败了。

将一批命令提交给数据库来执行,如果全部命令执行成功,则返回更新计数数组。该数组的 int 元素对应于批处理中的命令,批处理中的命令根据被添加到批处理中的顺序排序。方法 executeBatch()方法返回的数组中的元素如下。

(1) 大于或等于 0 的数: 指示成功处理了命令,是给出执行命令所影响数据库中行数的更新计数。

(2) SUCCESS_NO_INFO 的值: 指示成功执行了命令,但受影响的行数是未知的。

如果批量更新中的命令之一无法正确执行,则此方法抛出 BatchUpdateException 异常,并且 JDBC 驱动程序可能继续处理批处理中的剩余命令,也可能不执行。无论如何,驱动程序的行为必须与特定的 DBMS 一致。

可以用下面的代码实现 JDBC 数据的批处理。

```
try {
  Connection con = null;
  con.setAutoCommit(false);
  PreparedStatement prepStmt =
          con.prepareStatement("UPDATE DEPT SET MGRNO = ? WHERE DEPTNO = ?");
  prepStmt.setString(1, "1");
  prepStmt.setString(2, "2");
```

```
    prepStmt.addBatch();
    prepStmt.setString(1, "2");
    prepStmt.setString(2, "3");
    prepStmt.addBatch();
    int[] numUpdates = prepStmt.executeBatch();
    for (int i = 0; i < numUpdates.length; i++) {
      if (numUpdates[i] == -2)
      System.out.println("Execution " + i + ": unknown number of rows updated");
      else
      System.out.println("Execution " + i + "successful: " + numUpdates[i] + " rows updated");
    }
    con.commit();
  } catch (BatchUpdateException b) {
  } catch (SQLException e) {
    e.printStackTrace();
  }
```

【例 5-7】 实现对班级数据信息和学生数据信息的同步处理。

```
<%@ page language="java" import="java.sql.*" pageEncoding="gb2312"%>
  <%
   String claid = request.getParameter("claid");
   String claname = request.getParameter("claname");
   String stuid = request.getParameter("stuid");
   String stuname = request.getParameter("stuname");
  %>
   cal:<% = claid%>:<% = claname%>
   stu:<% = stuid%>:<% = stuname%>
  <%
   Connection con = null;
   PreparedStatement ps = null,ps2 = null;
   int m = 0,n = 0;
   try{
      Connection conn = ConnectionFactory.getConnection();
      con.setAutoCommit(false);
      String sql1 = "insert into classes(cla_id,cla_name) values(?,?)";
      String sql2 = "insert into students(code,name,cla_id) values(?,?,?)";
      ps = con.prepareStatement(sql1);
      ps.setString(1,claid);
      ps.setString(2,claname);
      m = ps.executeUpdate();
      ps2 = con.prepareStatement(sql1);
      ps2 = con.prepareStatement(sql2);
      ps2.setString(1,stuid);
      ps2.setString(2,stuname);
      ps2.setString(3,claid);
      n = ps2.executeUpdate();
      con.commit();
       %>
      <h1>成功提交 = <% = m%>n = <% = n%>/h1>
```

```
        <%
    }
    catch (Exception ex) {
        out.print(ex);
        con.rollback();
        %>
        <h1>成功失败</h1>
        <%
    }
     finally {
        ConnectionFactory.close(null,ps1,null);
        ConnectionFactory.close(null,ps2,conn);
        con.setAutoCommit(true);
    }
   %>
  </body>
</html>
```

5.4.5 任务：用 JDBC 实现数据库访问

1. 任务要求

假设数据库 test 中有数据表 userTable，利用 JDBC 访问数据库，获取用户信息并显示。数据表包含用户 ID(ID)、用户名(name)、密码(pwd)、用户类型(uType)等字段。

2. 任务实施

(1) 参考 5.2 节或 5.3 节中的任务，创建 ConnectionFactory 类以及 UserBean 类用于获取数据库连接。

(2) 定义如下的接口，实现数据库的增、删、改、查等操作。

```
package cn.edu.mypt.dao;
import java.util.List;
import cn.edu.mypt.bean.UserBean;
public interface UserDao {
  public int add(UserBean user);                    //增加用户信息
  public int delele(String userID);                 //删除用户指定用户信息
  public int updata(UserBean user);                 //修改指定用户信息
  public UserBean findByID(String userID);          //查找指定用户
  public List findByType(String uType);             //查找指定类型的用户
  public List getUsers();                           //获取所有用户信息
}
```

(3) 定义 UserDaoimpl 类实现数据库访问接口。

```
package cn.edu.mypt.dao.impl;
//导入包(略)
import com.cykj.db.ConnectionFactory;
import cn.edu.mypt.bean.UserBean;
```

```
import cn.edu.mypt.dao.UserDao;

public class UserDaoimpl implements UserDao {
  static Connection con = ConnectionFactory.getConnection();
//增加用户信息
@Override
public int add(UserBean user) {
  String sql = "insert into usertable(userid,username,pwd) values(?,?,?)";
  PreparedStatement ps = null;
  int i = 0;
  try {
    ps = con.prepareStatement(sql);
    ps.setString(1, user.getUserID());
    ps.setString(2, user.getUserName());
    ps.setString(3, user.getUserPwd());
    i = ps.executeUpdate();
    ConnectionFactory.Close(null, ps, con);
  } catch (SQLException e) {
    e.printStackTrace();
  }
  return i;
}
//删除指定用户信息
@Override
public int delele(String userID) {
  String sql = "delete from usertable where userid = ?";
  PreparedStatement ps = null;
  int i = 0;
  try {
    ps = con.prepareStatement(sql);
    ps.setString(1, userID);
    i = ps.executeUpdate();
    ConnectionFactory.Close(null, ps, con);
  } catch (SQLException e) {
    e.printStackTrace();
  }
  return i
}
//修改指定用户信息
@Override
public int update(UserBean user) {
  String sql = "update usertable set username = ?, userpwd = ? where userid = ?";
  PreparedStatement ps = null;
  int i = 0;
  try {
    ps = con.prepareStatement(sql);
    ps.setString(3, user.getUserID());
    ps.setString(1, user.getUserName());
    ps.setString(2, user.getUserPwd());
```

```
        i = ps.executeUpdate();
        ConnectionFactory.Close(null, ps, con);
    } catch (SQLException e) {
        e.printStackTrace();
    }
    return i;
}
//查找指定用户
@Override
public UserBean findByID(String userID) {
    String sql = "select userid,username,userpwd,userType from userTable where userid = ?";
    PreparedStatement ps = null;
    ResultSet rs;
    UserBean user = null;
    try {
        ps = con.prepareStatement(sql);
        ps.setString(1, userID);
        rs = ps.executeQuery();
        rs.next();
        user = new UserBean();
        user.setUserID(rs.getString(1));
        user.setUserName(rs.getString(2));
        user.setUserPwd(rs.getString(3));
    } catch (SQLException e) {
        e.printStackTrace();
    }
        return user;
}
//查找指定类型的用户
@Override
public List findByType(String userType){
    String sql = "select userid,username,userpwd,userType from userTable where userType = ?";
    PreparedStatement ps = null;
    ResultSet rs;
    UserBean user = null;
    List allUser = new ArrayList();
    try {
        ps = con.prepareStatement(sql);
        ps.setString(1, userType);
        rs = ps.executeQuery();
        while (rs.next()){
            user = new UserBean();
            user.setUserID(rs.getString(1));
            user.setUserName(rs.getString(2));
            user.setUserPwd(rs.getString(3));
            user.setUserType(rs.getString(4));
            allUser.add(user);
        }
    } catch (SQLException e) {
```

```
        e.printStackTrace();
    }
    return allUser;
}
//获取所有用户信息
@Override
    public List getUsers() {
        String sql = select userid,username,userpwd,userType from userTable userTable ";
        PreparedStatement ps = null;
        ResultSet rs;
        UserBean user = null;
        List allUser = new ArrayList();
        try {
            ps = con.prepareStatement(sql);
            rs = ps.executeQuery();
            while (rs.next()){
                user = new UserBean();
                user.setUserID(rs.getString(1));
                user.setUserName(rs.getString(2));
                user.setUserPwd(rs.getString(3));
                user.setUserType(rs.getString(4));
                allUser.add(user);
            }
        } catch (SQLException e) {
            e.printStackTrace();
        }
        return allUser;
    }
}
```

(4) 设计页面 userList.jsp 用于读取数据库信息。

```
<%@ page language = "java" import = "java.util.*" pageEncoding = "utf-8"%>
<!DOCTYPE HTML PUBLIC "-//W3C//DTD HTML 4.01 Transitional//EN">
<html>
    <head><title>用户列表</title></head>
    <body>
    <jsp:useBean id = "userDao" class = "cn.edu.mypt.dao.impl.UserDaoimpl" scope = "page"/>
    <jsp:useBean id = "user" class = "cn.edu.mypt.bean.UserBean" scope = "page"/>
    <table>
    <tr><th>序号<th>用户 ID<th>用户名<th>删除</th><th>修改 </th></tr>
    <%
     List userList = userDao.getUsers();
     if (userList!= null) {
       int n = userList.size();
       for (int i = 0;i<n;i++) {
            user = (cn.edu.mypt.bean.UserBean)userList.get(i);
     %>
     <tr>
          <td><% = i+1 %></td>
```

```
        <td><% = user.getUserID() %></td>
        <td><% = user.getUserName() %></td>
        <td><a href = "del.jsp?userID = <% = user.getUserID() %>">删除</a></td>
        <td><a href = "modify.jsp?userID = <% = user.getUserID() %>">修改</a></td>
      </tr>
      <% }
     }
     %>
    </table>
   <a href = "index.jsp">返回</a>
  </body>
</html>
```

项目5 用户管理系统的数据访问层设计

1. 项目需求

本项目实现对用户信息的管理,包括用户注册、用户登录、修改密码、修改用户信息、删除用户、查找指定用户等。

2. 项目实施

(1) 利用5.4节任务中的UserDaoimpl类实现对数据库的访问。

(2) 参照项目4编写业务处理层代码,定义一个UserModel接口,代码如下。

```
package cn.edu.mypt.model;
import java.util.List;
import cn.edu.mypt.bean.*;
public interface UserModel {
    public boolean userReg(UserBean userr); //用户注册,若注册成功返回 true,否则返回 false
    public UserBean userLogin(String userId,String userPwd);
                                    //用户登录,登录成功,返回用户信息,否则返回 null
    public boolean UserModify(UserBean user);
                                    //修改用户信息,若修改成功返回 true,否则返回 false
    public boolean UserDelete(String userID); //删除用户,若删除成功返回 true,否则返回 false
    public UserBean findById(String userID);
                                    //根据用户 ID 查找用户,找到返回用户信息,否则返回 null
    public List<UserBean> findByUType(String utype);//获取指定类型的用户信息
    public List<UserBean> getUser(); //获取所有用户信息
    public boolean ModifyPwd(String userID,String oldPwd,String newPwd);
                      //根据用户 ID 和旧密码修改用户密码,若修改成功返回 true,否则返回 false
}
```

(3) 定义UserModelImple类,它继承自UserModel类,实现用户管理的功能。

```
package cn.edu.mypt.model.impl;
import java.util.List;
import cn.edu.mypt.bean.UserBean;
```

```
import cn.edu.mypt.dao.UserDao;
import cn.edu.mypt.dao.impl.UserDaoimpl;
import cn.edu.mypt.model.UserModel;
public class UserModelImple implements UserModel {
  UserDao dao = new UserDaoimpl();
  @Override
  public boolean userReg(UserBean user) {
      UserBean u = dao.findByID(user.getUserID().trim());
      if (u == null)
      {
        if (dao.add(user)> 0);
        return true;
      }
    return false;
  }
  @Override
  public UserBean userLogin(String userId, String userPwd) {
      UserBean u = dao.findByID(userId);
      if (u!= null)
        if (u.getUserPwd().trim().equals(userPwd))
          return u;
        return null;
  }
  @Override
  public boolean UserModify(UserBean user) {
    UserBean u = dao.findByID(user.getUserID().trim());
    if (u!= null) {
      if  (dao.update(user)> 0) return true;
    }
    return false;
  }
  @Override
  public boolean UserDelete(String userID) {
     if (dao.delele(userID.trim())> 0) return true;
    return false;
  }
  @Override
  public UserBean findById(String userID) {
    return dao.findByID(userID.trim());
  }
  @Override
  public List< UserBean > findByUType(String utype) {
    return dao.findByType(utype.trim());
  }
  @Override
  public List< UserBean > getUser() {
    return dao.getUsers();
  }
  @Override
```

```
    public boolean ModifyPwd(String userID, String oldPwd, String newPwd) {
        UserBean u = userLogin(userID,oldPwd);
        if (u == null) return false;
        u.setUserPwd(newPwd);
        dao.update(u);
        return false;
    }
}
```

(4) 创建注册页面 reg.jsp。

```
<%@ page language = "java" import = "java.util.*" pageEncoding = "utf-8" %>
<!DOCTYPE HTML PUBLIC " - //W3C//DTD HTML 4.01 Transitional//EN">
<html>
    <head><title>用户注册</title></head>
    <body>
    <%
        String uID,uPwd,msg,uName;
        msg = (String)request.getAttribute("message");
        uID = (String)request.getAttribute("userID");
        uPwd = (String)request.getAttribute("userPwd");
        uName = (String)request.getAttribute("userName");
        uID = (uID == null)?"":uID;
        uPwd = (uPwd == null)?"":uPwd;
        msg = (msg == null)?"":msg;
    %>
     <% = msg %>
        <form action = "doreg.jsp" method = "post">
            用户 ID: <input type = "text" name = "userID" value = "<% = uID %>"/><br>
            用户名: <input type = "text" name = "userName" value = "<% = uName %>"/><br>
            用户密码: <input type = "text" name = "userPwd" value = "<% = uPwd %>"/><br>
            用户类型: <input name = "super" type = "radio" value = "管理员"/>
            <input name = "comm" type = "radio" value = "普通用户"/><br>
            <input type = "submit" value = "reg"/>
        </form>
    </body>
</html>
```

(5) 创建业务处理页面 doreg.jsp。

```
<%@ page language = "java" import = "java.util.*" pageEncoding = "utf-8" %>
<jsp:useBean id = "userModel" class = "cn.edu.mypt.model.impl.UserModel" scope =
"application"/>
<jsp:useBean id = "user" class = "cn.edu.mypt.bean.UserBean" scope = "page"/>
<%
    String uID,uPwd,uName;
    request.setCharacterEncoding("utf-8");
    uID = request.getParameter("userID");
    uPwd = request.getParameter("userPwd");
    uName = request.getParameter("userName");
```

```
  user.setUserID(uID);
  user.setUserName(uName);
  user.setUserPwd(uPwd);
  if (userModel.userReg(user))
  response.sendRedirect("index.jsp");
  else {
    request.setAttribute("userID",uID);
    request.setAttribute("userID",uPwd);
    request.setAttribute("userName",uName);
    session.setAttribute("USER", user);
    request.setAttribute("message", "用户名 ID 已经存在");
    request.getRequestDispatcher("reg.jsp").forward(request, response);
  }
%>
```

(6) 其他代码略。

习　题　5

1. 选择题

(1) 要执行 str="select * from customer"语句,假设 Statement 对象为 stmt,则应执行(　　)方法。

A. stmt.executeQuery(str)　　B. stmt.executeUpdate(str)

C. stmt.executeSelect(str)　　D. stmt.executeDelete(str)

(2) (　　)不属于 JDBC 的基本功能。

A. 与数据库建立连接　　B. 提交 SQL 语句

C. 处理查询结果　　D. 数据库维护管理

(3) 在进行数据库操作中,使用 Resultset 对象的 next()方法移动光标时,如果超过界限,会抛出异常,该异常通常是(　　)。

A. InterruptedExceptlon　　B. AlreadyBoundException

C. SQLException　　D. NextException

(4) 在 JDBC API 中,可通过(　　)对象执行 SQL 语句。

A. DriverManager　　B. Statement

C. ResultSet　　D. Connection

(5) (　　)对象是通过 Connection 的 createment()方法创建的,用于在已经建立的连接的基础上向数据库发送 SQL 语句。

A. request　　B. createment　　C. ResultSet　　D. ODBC

(6) cn 是 Connection 对象,则创建 Statement 对象的方法是(　　)。

A. Statement st=cn.createStatement()

B. Statement st=cn.newStatement()

C. Statement st=cn.createNewStatement()

D. Statement st=new Statement()

(7)（　　）接口专用于访问数据库中的存储过程。

A. ProcedureStatement　　B. PreparedStatement

C. Statement　　D. CallableStatement

(8) 微软公司提供的连接 SQL Server 的 JDBC 驱动程序是（　　）。

A. oracle. jdbc. driver. OracleDriver

B. com. mysql. jdbc. Driver

C. com. microsoft. jdbc. sqlserver. SQLServerDriver

D. sun. jdbc. odbc. JdbcOdbcDriver

(9)（　　）是 ResultSet 接口的方法。

A. forward()　　B. back()　　C. next()　　D. commit()

2. 简答题

(1) JDBC 中提供以下类或接口：DriverManager 类、Connection 接口、Statement 接口、ResultSet 接口，它们各自的功能是什么？

(2) 阐述 JDBC 访问数据库的基本步骤。

(3) 假设有 MySQL 数据库，有用户表 usertable={userID char(10), userName char(20), userPwd char(10), ip}，编程实现用户登录和注册的功能，其中，userID 为用户注册名，userName 为用户真实名，userPwd 为登录密码，IP 为 IP 地址。

第 6 章　Servlet 技术

JSP 程序的运行是通过浏览器访问来实现的，在这个过程中，浏览器发送访问请求，服务器接收请求，并对浏览器的请求做出相应的处理，这就是浏览器/服务器(B/S)模型，而 Servlet 就是对请求做出处理的组件，运行于支持 Java 的应用服务器中。

【技能目标】 掌握 Sevlet 的编程与部署；理解 MVC+DAO 设计模式。

【知识目标】 Servlet 的基本概念；Servlet API 的常用接口和类。

【关键词】 初始化(initialize)　文件配置(config)　上下文(context)　销毁(destroy)　服务(service)　过滤(filter)　监听器(listener)

6.1　Servlet 基础

6.1.1　Servlet 的概念

1. 什么是 Servlet

当使用 JSP 技术开发 Web 程序时，是在 JSP 中写入 Java 代码，当服务器运行 JSP 时，执行 Java 代码，动态获取数据，并生成 HTML 代码，最后显示在客户端浏览器上。其处理的过程如图 6-1 所示。

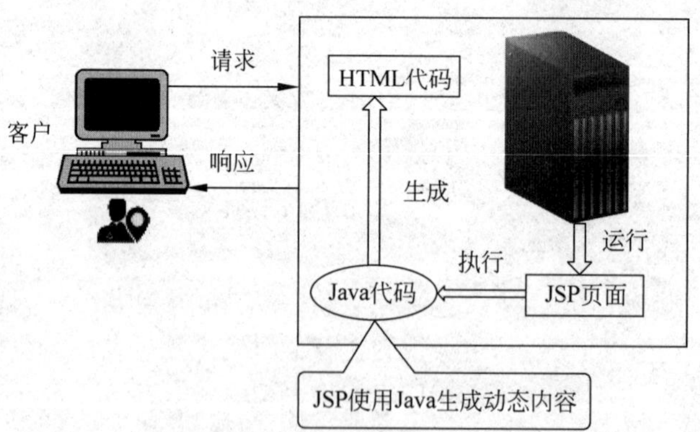

图 6-1　JSP 代码执行流程

Servlet(Java 服务器小程序)是用 Java 编写的服务器端程序，是在服务器端调用和执行的、按照 Servlet 自身标准编写的 Java 类。Servlet 带给开发人员最大的好处就是可以处理客户端传来的 HTTP 请求，并响应数据给客户端。Servlet 的特点是功能强大、可移植性强、扩展性强、使用灵活。

2. Servlet 的功能

当客户端发送请求至服务器时，服务器可以将请求信息发送给 Servlet，并让 Servlet 创建服务器返回给客户机的响应。当 Web 服务器启动或客户端第一次请求服务时，Servlet 将持续运行直到其他客户端发出请求。Servlet 的主要功能如下。

(1) 创建并返回一个包含基于客户请求性质动态内容的完整的 HTML 页面。

(2) 创建可嵌入现有 HTML 页面中的部分 HTML 页面。

(3) 与其他服务器资源进行通信。如数据库和基于 Java 的应用程序。

(4) 接收多个客户端的输入，并将结果传递到多个客户端上。

(5) 在单连接方式下传送数据时，在浏览器上打开服务器与 Applet 的新连接，并将该连接保持在打开状态。当允许客户机和服务器进行简单、高效的会话时，Applet 可以启动客户浏览器和服务器之间的连接，通过定制协议进行通信。

(6) 将定制的处理提供给所有服务器的标准程序，例如，Servlet 可以修改如何验证用户。

(7) 对特殊的处理采用 MIME 类型过滤数据。

3. Servlet 的技术特性

Servlet 技术具有以下特性。

1) 高效持久性

在服务器上仅有一个 Java 虚拟机在运行，它的优势在于当有多个来自客户端的请求，Servlet 为每个请求分配一个线程而不是进程。Servlet 只需 Web 服务器加载一次，而且可以为不同的请求进行服务。

2) 便捷性

Servlet 提供了大量的实用工具例程。例如，处理 HTML 表单数据、读取和设置 HTTP 报头、处理 Cookie 和跟踪会话等。

3) 跨平台性

Servlet 是用 Java 编写的，因此自然继承了 Java 的平台无关性。Servlet 是与平台无关的，可以在不同的操作系统平台和不同的应用服务器平台上运行。

4) 功能强大

许多使用传统 CGI 程序很难完成的任务都可以利用 Servlet 轻松地完成。例如，Servlet 能够直接和 Web 服务器交互，而普通的 CGI 程序不能。Servlet 还能够在各个程序之间共享数据，使数据库连接池之类的功能很容易实现。

5) 灵活性和可扩展性

采用 Servlet 开发的 Web 应用程序，由于 Java 类的继承性、构造方法等 Java 所能带来的所有优点，而应用灵活，可随意扩展。

6) 共享数据

Servlet 之间可通过共享数据很容易地实现数据库连接池。它能方便地实现管理用户请求，简化会话和获取前一页面信息的操作。

7）集成性

Servlet 和服务器紧密集成,它们可以密切合作完成特定的任务。

8）安全性

安全性主要体现在：①它使用 Java 的安全框架；②Servlet API 被声明为安全类型；③容器也会对 Servlet 的安全进行管理,从外界调用一个 Servlet 的唯一方法就是通过 Web 服务器,这提供了高水平的安全性保障。在 Servlet 安全策略中,既可以使用编程的安全,也可以使用声明性的安全。声明性的安全由容器进行统一管理。

6.1.2 Servlet 与 JSP 的关系

1. JSP 的执行过程

Tomcat 访问任何资源实际上都是在访问 Servlet。JSP 本身就是一种 Servlet,其实 JSP 在第一次被访问的时候会被编译为 HttpJspPage 类,该类是 HttpServlet 的一个子类。

创建一个项目 mydemo,建立一个 MyJsp.jsp 文件,其内容如下。

```
<%@ page language="java" import="java.util.*" pageEncoding="ISO-8859-1"%>
<!DOCTYPE HTML PUBLIC "-//W3C//DTD HTML 4.01 Transitional//EN">
<html>
  <head><title>test</title></head>
  <body>
    This is my test JSP page. <br>
  </body>
</html>
```

当部署并运行 MyJsp.jsp 后,在 Tomcat 的安装目录 Tomcat\work\Catalina\localhost\mydemo\org\apache\jsp 下会生成一个 MyJsp_jsp java,主要内容如下。

```
package org.apache.jsp;
//导入其他包
public final class MyJsp_jsp extends org.apache.jasper.runtime.HttpJspBase
    implements org.apache.jasper.runtime.JspSourceDependent,
               org.apache.jasper.runtime.JspSourceImports {
    //略...
  public void _jspService(final javax.servlet.http.HttpServletRequest request,
  final javax.servlet.http.HttpServletResponse response) throws java.io.IOException,
  javax.servlet.ServletException {
    //略 ...
    try {
      response.setContentType("text/html;charset=ISO-8859-1");
      pageContext = _jspxFactory.getPageContext(this, request, response, null, true,
                                                8192, true);

      //其他变量定义,略 ...
      out = pageContext.getOut();
      _jspx_out = out;
      out.write("\r\n");
```

```
        out.write("<!DOCTYPE HTML PUBLIC \"-//W3C//DTD HTML 4.01 Transitional//EN\">\r\n");
        out.write("<html>\r\n");
        out.write(" <head>\r\n");
        out.write(" <title>test</title>\r\n");
        out.write(" </head>\r\n");
        out.write(" \r\n");
        out.write(" <body>\r\n");
        out.write(" This is my firste JSP page. <br>\r\n");
        out.write(" </body>\r\n");
        out.write("</html>\r\n");
      } catch (java.lang.Throwable t) {
        //其他处理,略...
      }
      finally {
        //其他处理,略...
      }
    }
  }
```

可以看出,MyJsp.jsp 在运行时首先解析成一个 Java 类 MyJsp_jsp.java,该类继承自 org.apache.jasper.runtime HtpJspBase 类,而 HtpJspBase 又是继承自 HtpServlet 类。由此可以得出一个结论,就是 JSP 在运行时会被 Web 容器翻译为一个 Servlet。

2. Servlet 与 JSP 的区别

Servlet 的功能比较强大,但它输出 HTML 元素时是逐元素输出的,编写和修改 HTML 代码非常不方便,而 JSP 将 Java 嵌入 HTML 代码中,就大大简化和方便了网页的设计和修改。

(1) JSP 在本质上就是 Servlet,但是两者的创建方式不一样。

(2) Servlet 完全由 Java 代码构成,擅长流程控制和事务处理,通过 Servlet 来生成动态网页很不直观。

(3) JSP 由 HTML 代码和 JSP 标签构成,可以方便地编写动态网页。

在实际应用中采用 Servlet 来控制业务流程,而采用 JSP 来生成动态网页。MVC 设计模式中,JSP 位于视图层,而 Servlet 位于控制层。

(1) JSP 是 Servlet 技术的扩展,本质上是 Servlet 的简易方式。

(2) JSP 编译后是"类 Servlet"。

(3) Servlet 的应用逻辑编写在 Java 文件中,并且完全从表示层中的 HTML 代码里分离开来。而 JSP 是 Java 代码和 HTML 代码组合成一个扩展名为.jsp 的文件。

(4) JSP 侧重于视图,Servlet 主要用于控制逻辑。

6.1.3 Servlet 生命周期

Servlet 的生命周期主要由加载和实例化、初始化、服务、销毁几个过程组成,如图 6-2 所示。

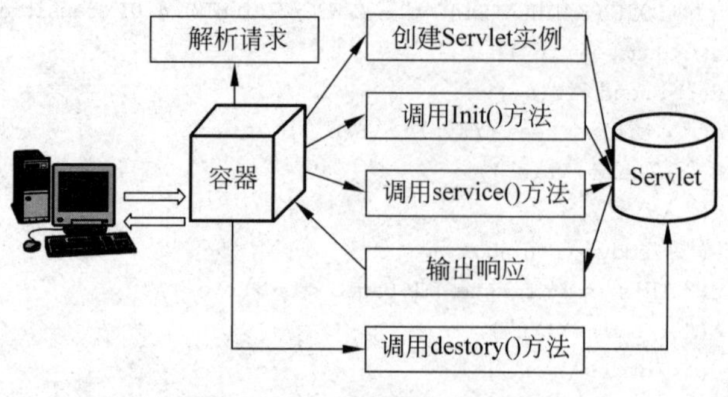

图 6-2 Servlet 的生命周期

1. 加载和实例化

Servlet 容器负责加载和实例化 Servlet，当客户端发送一个请求时，Servlet 容器会查找内存中是否存在该 Servlet 的实例，如果不存在，就创建一个 Servlet 实例。如果存在该 Servlet 的实例，就直接从内存中取出该实例来响应请求。

2. 初始化

在 Servlet 容器完成 Servlet 实例化后，Servlet 容器将调用 Servlet 的 init()方法进行初始化，初始化的目的是让 Servlet 对象在处理客户端请求前完成一些初始化工作，例如设置数据库连接参数、建立 JDBC 连接，或者是建立对其他资源的引用。init()方法在 javax.servlet.Servlet 接口中定义。对于每一个 Servlet 实例，ini()方法只被调用一次。

3. 服务

Servlet 被初始化以后，就处于能响应请求的就绪状态。当 Servlet 容器接收到客户端请求时，调用 Servlet 的 service()方法处理客户端请求。Servlet 实例通过 ServletRequest 对象获得客户端的请求。通过调用 ServletResponse 对象的方法设置响应信息。

4. 销毁

Servlet 的实例是由 Servlet 容器创建的，所以实例的销毁也是由容器来完成的。Servlet 容器判断一个 Servlet 是否应当被释放时（容器关闭或需要回收资源），容器就会调用 Servlet 的 destroy() 方法。destroy()方法指明哪些资源可以被系统回收，而不是用 destroy()方法直接进行回收。

6.1.4 Servlet 的创建

1. 手动创建 Servlet

可以通过继承 HttpServlet 类手动创建 Servlet 来建立一个项目，其目录树如图 6-3 所示。

Classes 文件夹用于存放编译得到的 servlet.class 文件，web.xml 文件说明 URL 与

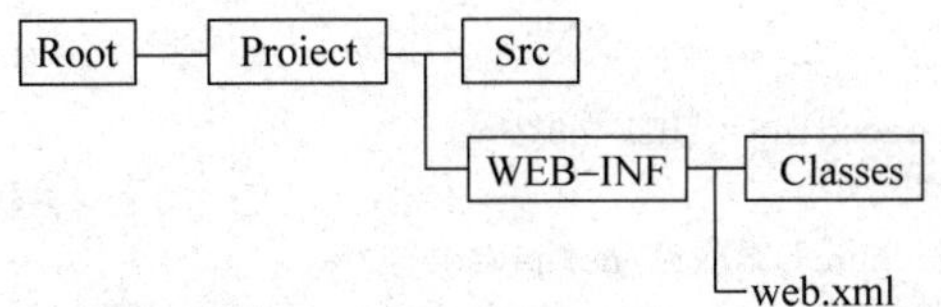

图 6-3　Servlet 项目的目录树

Servlet 之间的映射关系。Servlet 的源代码存放在 Src 文件夹下。下面手动来编写一个 Servlet，具体步骤如下。

(1) 继承 HttpServlet 抽象类，创建 HelloServlet 类。

(2) 重载适当的方法，如覆盖(或称为重写)doGet()方法或 doPost()方法。

(3) 如果有 HTTP 请求信息，则获取该信息。用 HttpServletRequest 类对象来检索 HTML 表单所提交的数据或 URL 中的查询字符串。request 对象含有特定的方法以检索客户机提供的信息，有以下 3 个方法可用：request. getParameterNames()、request. getParameter()和 request. getParameterValues()。

(4) 生成 HTTP 响应。HttpServletResponse 类的对象生成响应，并将其返回到发出请求的客户端上。它的方法允许设置请求标题和响应主体。响应对象的 getWriter()方法可以返回一个 PrintWriter 对象。使用 PrintWriter 对象的 print()方法和 println()方法可以编写 Servlet 响应以返回给客户端。或者直接使用 out 对象输出有关的 HTML 内容。关键代码如下。

```
package mypt.sevlet;
//导入相关包
public class HelloServlet extends HttpServlet {
  //略
  public void doGet(HttpServletRequest request, HttpServletResponse response)
  throws ServletException, IOException {
    doPost(request,response);
  }
  public void doPost(HttpServletRequest request, HttpServletResponse response)
  throws ServletException, IOException {
    response.setContentType("text/html");
    PrintWriter out = response.getWriter();
    out.println("<!DOCTYPE HTML PUBLIC \"-//W3C//DTD HTML 4.01 Transitional//EN\">");
    out.println("<HTML>");
    out.println("<HEAD><TITLE>A Servlet</TITLE></HEAD>");
    out.println("<BODY>");
    out.print(" Hello world");
    out.println("</BODY>");
    out.println("</HTML>");
    out.flush();
    out.close();
  }
}
```

(5) 配置 web. xml。

```
<?xml version = "1.0" encoding = "UTF - 8"?>
< web - app version = "3.0"
  xmlns = "http://java.sun.com/xml/ns/javaee"
  xmlns:xsi = "http://www.w3.org/2001/XMLSchema - instance"
  xsi:schemaLocation = "http://java.sun.com/xml/ns/javaee
  http://java.sun.com/xml/ns/javaee/web - app_3_0.xsd">
  < display - name ></display - name >
  < servlet >
    < servlet - name > HelloServlet </servlet - name >
    < servlet - class > mypt.sevlet.HelloServlet </servlet - class >
  </servlet >
  < servlet - mapping >
    < servlet - name > HelloServlet </servlet - name >
    < url - pattern >/HelloServlet </url - pattern >
  </servlet - mapping >
</web - app >
```

2. 在 Eclipse 下创建的 Servlet 实例

(1) 在项目中新建 Servlet,右击 scr,选择 New→Servlet 命令,如图 6-4 所示。

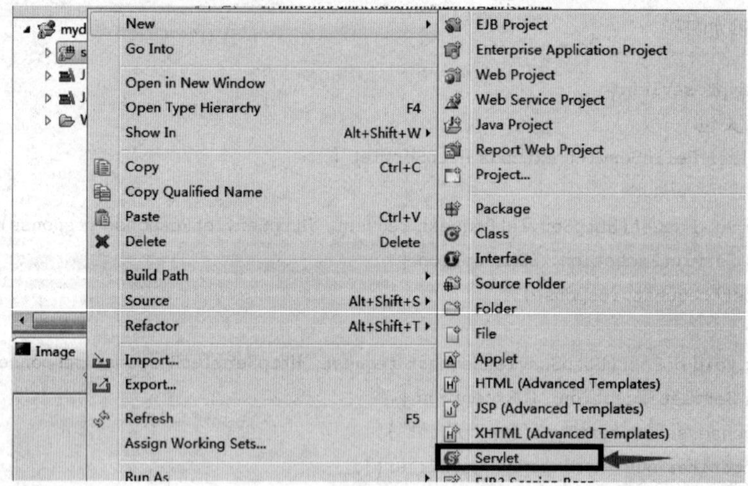

图 6-4 新建 Servlet

(2) 输入包名(Packag)和 Servlet 类名(Name),注意下边的 Supperclass 自动继承了 HttpServlet 类,如图 6-5 所示。

(3) 单击 Next 按钮,修改 Mapping URL,对应的就是 web. xml 中的以下代码,如图 6-6 所示。

```
< servlet - mapping >
    < servlet - name > HelloServlet </servlet - name >
    < url - pattern >/HelloServlet </url - pattern >
</servlet - mapping >
```

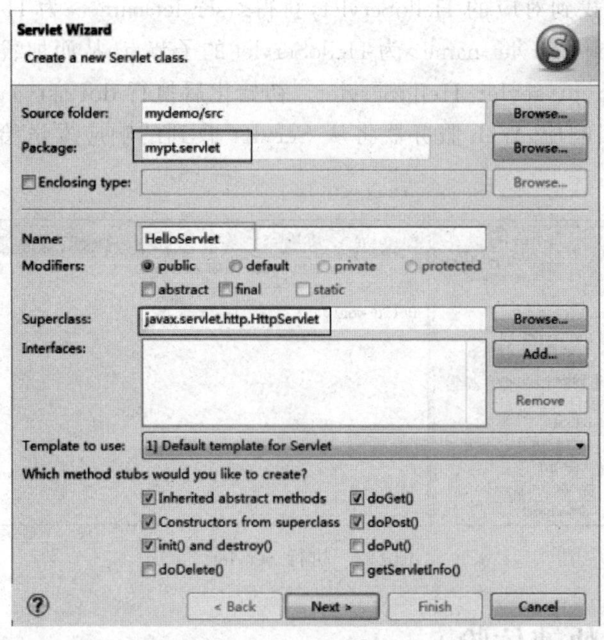

图 6-5　输入 Servlet 类名

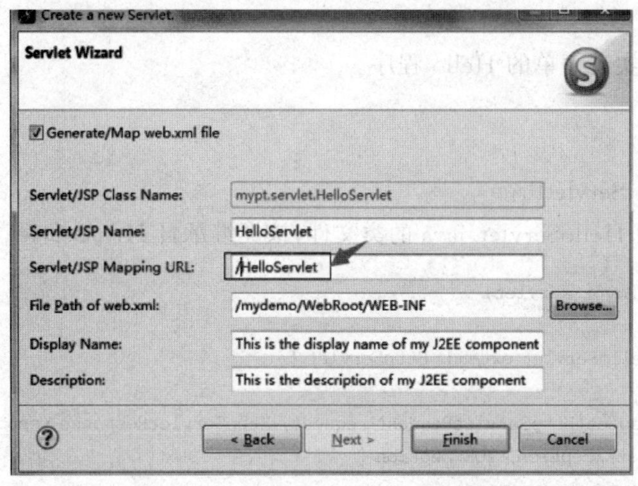

图 6-6　输入 Servlet/JSP Mappping URL

(4) 单击 Finish 按钮，创建完成后在 Servlet 的 doGet()、doPost()方法中输入前面类似代码。

(5) 启动 Tomcat，在浏览器中访问页面。

当项目部署后，在地址栏中输入 http://localhost:8080/mydemo/HelloServlet，这里 http://localhost:8080/表示服务器 IP 地址和端口号；mydemo 为项目名称或路径；HelloServlet 为资源路径。

这时 Web 服务器接收该请求，根据 web.xml 配置文件，在< servlet-mapping >节点的< url-pattern >中找到对应的 HelloServlet，获得< servlet-name >为 HelloServlet，然后在< servlet >节点中找到< servlet-name >为 HelloServlet 的子节点，从而加载< servlet-class >中指定的 Servlet 类 mypt. sevlet. HelloServlet。初始化后执行 doPost()方法，最后 Servlet 向 Web 服务器返回响应，Web 服务器将从 Servlet 收到的响应发送给客户端，如图 6-7 所示。

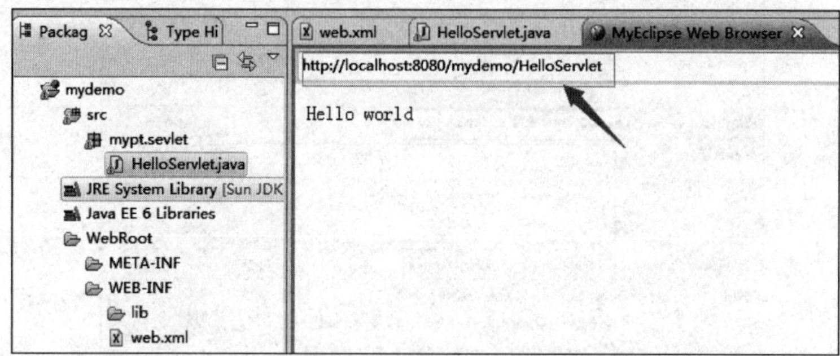

图 6-7 执行 Servlet

6.1.5 任务：快速体验 Servlet

1. 任务要求

利用 Servlet 实现简单的 Hello 程序。

2. 任务实施

(1) 创建项目 ServletPrj。

(2) 创建名为 HelloServlet. java 的类文件，该类继承自 HttpServlet 类，代码如下。

```
package cn.edu.mypt.servlet;
//导入包略
public class HelloServlet extends HttpServlet {
  //略
  public void doGet(HttpServletRequest request, HttpServletResponse response)
  throws ServletException, IOException {
      doPost(request,response);
  }
  public void doPost(HttpServletRequest request, HttpServletResponse response)
      throws ServletException, IOException {
    response.setContentType("text/html");
    PrintWriter out = response.getWriter();
    out.println("<!DOCTYPE HTML PUBLIC \"-//W3C//DTD HTML 4.01 Transitional//EN\">");
    out.println("<HTML>");
    out.println(" <HEAD><TITLE>A Servlet</TITLE></HEAD>");
    out.println(" <BODY>");
    out.print(" <h1>hello! world</h1>");
    out.println(" </BODY>");
```

```
        out.println("</HTML>");
        out.flush();
        out.close();
    }
    public void init() throws ServletException {
    }
}
```

(3) 在 web.xml 文件中,配置 HelloServlet .java 类,关键代码如下。

```
<servlet>
    <display-name>This is the display name of my J2EE component</display-name>
    <servlet-name>HelloServlet</servlet-name><!-- 指定 Servlet 的名称 -->
    <servlet-class>cn.edu.mypt.servlet.HelloServlet</servlet-class>
    <!-- 指定 Servlet 的路径 -->
</servlet>
<servlet-mapping>
    <servlet-name>HelloServlet</servlet-name>
    <url-pattern>/HelloServlet</url-pattern><!-- 指定 Servlet 的访问地址 -->
</servlet-mapping>
```

(4) 程序发布后,启动 Tomcat 服务器,在浏览器地址栏中输入 http://127.0.0.1:8080/ServletPrj/HelloServlet 并观察效果。

6.2　Servlet API

整个 Servlet 程序中最重要的就是 Servlet 接口,在此接口下定义了一个 GenericServlet 类的子类,但一般不直接继承该类,而是根据所使用的协议选择 GenericServlet 类的子类来继承。由于一般采用 HTTP 处理,所以一般用户定义的 Servlet 类都继承自 HttpServlet 类。其继承关系如图 6-8 所示。

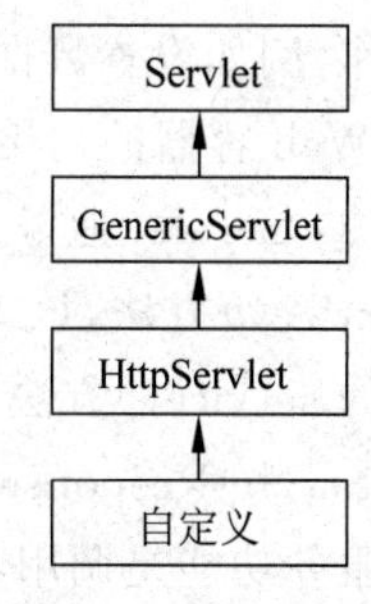

图 6-8　Servlet 接口

Servlet 使用以下两个包中的类和接口:javax.servlet 和 javax.servlet.http。包名以 javax 开头而不是 java,表示 Servlet API 是一个标准扩展。

javax.servlet 包中包含了支持通用、跨协议的 Servlet 的类。java.servlet.http 包中的类增加了 HTTP 处理功能。每个 Servlet 都必须实现 javax.servlet.Servlet 接口。大部分 Servlet 通过扩展以下其中一个特殊类来实现这一接口:javax.servlet.GenericServlet 或 javax.servlet.HttpServlet。跨协议 Servlet 应该继承 GenericServlet。HttpServlet 类是 GenericServlet 的子类,在 GenericServlet 类的基础上扩展了特定的功能。Servlet 接口只声明而不实现用于管理 Servlet 及其客户端通信的方法。

6.2.1　javax.servlet 包

javax.servlet 包中包含了 7 个接口、3 个类和 2 个异常类接口。

接口：RequestDispatcher、Servlet、ServletConfig、ServletContext、ServletRequest、ServletResponse 和 SingleThreadModel。

类：GenericServlet、ServletInputStream 和 ServletOutputStream。

异常类：ServletException 和 UnavailableException。

下面介绍常用的接口和类。

1. Servlet 接口

Servlet 编程需要引用 javax. servlet 包中的类与接口，该包中的类和接口封装了一个抽象框架，建立接收请求和产生响应的组件（即 Servlet）。其中 javax. servlet. Servlet 类是所有 Java Servlet 的基础接口类。Servlet 接口的常用方法如表 6-1 所示。

表 6-1 Servlet 接口的常用方法

方　法	描　述
init(ServeltConfig config)	用于初始化 Servlet
destroy()	销毁 Servlet
getServletInfo()	获得 Servlet 的信息
getServletConfig()	获得 Servlet 配置相关信息
service (ServletRequest req, ServletResponse res)	运行应用程序逻辑的入口，它接收两个参数，ServletRequest 表示客户端请求的信息，ServletResponse 表示对客户端的响应

(1) Servlet 初始化。在进行 Servlet 初始化时，需要使用 Servlet 编程接口提供的 init()方法。在 Servlet 生命周期中，这个方法仅会被调用一次。它可以用来设置一些准备工作，例如设置数据库连接、读取 Servlet 设置信息等。也可以通过 ServletConfig 对象获得 Web 容器的初始化变量。

(2) Servlet 业务实现。在进行 Servlet 业务实现时，需要使用 Servlet 编程接口提供的 service()方法，该方法是 Servlet 的核心。当客户向 HttpServlet 对象发出请求时，该对象的 service()方法就要被调用，而且传递给这个方法一个 ServletRequest 请求对象和一个 ServletResponse 响应对象作为参数。在 HttpServlet 中已存在 service()方法。默认的服务功能是调用与 HTTP 请求的方法相应的动作。例如，如果 HTTP 请求方法为 Get，则默认情况下就调用 doGet()。

当客户通过 HTML 表单发出一个 HTTP Post 请求时，doPost()方法被调用。与 Post 请求相关的参数作为一个单独的 HTTP 请求从浏览器发送到服务器。当需要修改服务器端的数据时，应该使用 doPost()方法。

当一个客户通过 HTML 表单发出一个 HTTP Get 请求或直接请求一个 URL 时，doGet()方法被调用。与 Get 请求相关的参数被添加到 URL 的后面，并与这个请求一起发送。当不会修改服务器端的数据时，应该使用 doGet()方法。

Servlet 的响应可以是下列两种类型之一。

① 输出流：浏览器根据它的内容类型（如 text/HTML）进行解释。

② HTTP 错误响应：重定向到另一个 URL、Servlet、JSP，并显示错误信息。

(3) Servlet 销毁。在进行 Servlet 销毁时,需使用 Servlet 编程接口提供的 destroy()方法。当一个 Servlet 被从服务中销毁时,Web 容器会调用这个方法。Servlet 可以使用这个方法完成如切断和数据库的连接、保存重要数据等操作。

当服务器卸载 Servlet 时,将在所有 service()方法调用完成后,或在指定的时间后调用 destroy()方法。一个 Servlet 在运行 service()方法时可能会产生其他线程,因此要确认在调用 destroy()方法时这些线程已终止或完成。

(4) Servlet 和 Web 容器进行通信。在 Servlet 和 Web 容器进行通信时,需要使用 Servlet 编程接口提供的 getServletConfig()方法。该方法返回 ServletConfig 对象。该对象可以使 Servlet 和 Web 容器进行通信,例如传递初始变量。

(5) 获取 Servlet 基本信息。在获取 Servlet 基本信息时,需要使用 Servlet 编程接口提供的 getServletConfig()方法。该方法返回有关 Servlet 的基本信息,如编程人员姓名和时间等。

【例 6-1】 通过一个类文件实现 Servlet 编程接口中的各个方法。

(1) 创建名为 TestServlet.java 的类文件,TestServlet 类实现了 Servlet 接口的各种方法,关键代码如下。

```
package mypt.servlet;
public class TestServlet extends HttpServlet {
  public TestServlet() {
    System.out.println("service...");
  }
  public void destroy() {
    System.out.println("destroy...");
  }
  public void doGet(HttpServletRequest request, HttpServletResponse response)
  throws ServletException, IOException {
    System.out.println("doGet...");
  }
  public void doPost(HttpServletRequest request, HttpServletResponse response)
  throws ServletException, IOException {
    System.out.println("doPost...");
  }
  public void init() throws ServletException {
    System.out.println("init");
  }
}
```

代码说明:init()、service()和 destroy()方法是 Servlet 生命周期的方法。当 Servlet 类被实例化后,容器加载 init(),以通知 Servlet 已进入服务行列。init()方法必须被加载,Servlet 才能接收请求。如果要载入数据库驱动、初始化操作等,程序员可以重写这个方法。

(2) 在 web.xml 文件中配置 TestServlet.java 类。

(3) 程序发布后,启动 Tomcat 服务器,在浏览器地址栏中输入 http://127.0.0.1:

8080/Mydemo/TestServlet。

当部署发布后，执行 TestServlet，控制台将输出以下的结果。

```
service...
init
doGet...
```

将控制台显示的信息清除后再次浏览 TestServlet，控制台将输出以下的结果。

```
doGet...
```

当服务器停止时，控制台输出以下的结果。

```
destroy...
2020 - 3 - 11 18:10:53 org.apache.coyote.http11.htt
信息:Stopping Coyote HTTP/1.1 on http - 8080
```

2. ServletConfig 接口

Servlet 运行期间需要一些辅助信息，如文件使用的编码、使用 Servlet 程序的公司等，这些信息可以在 web.xml 文件中使用一个或者多个元素进行配置。

当 Tomcat 初始化一个 Servlet 时，会将该 Servlet 的配置信息封装到一个 ServletConfig 对象中，通过调用 init(ServletConfig config)方法，将 ServletConfig 对象传递给 Servlet。ServletConfig 接口定义了一系列获取配置信息的方法，如表 6-2 所示。

表 6-2 ServletConfig 接口的常用方法

方　法	描　述
String getInitParameter(String name)	根据初始化参数名，返回对应的初始化参数值
Enumeration getInitParameterNames()	返回一个 Enumeration 对象，其中包含所有的初始化参数名
ServletContext getServletContext()	返回一个代表当前 Web 应用的 ServletContext 对象
String getServletName()	返回 Servlet 的名字，即 web.xml 中元素的值

(1) 在 web.xml 中配置 Servlet 参数信息。

```
<?xml version = "1.0" encoding = "ISO - 8859 - 1"?>
< web - app xmlns = "http://java.sun.com/xml/ns/javaee"
  xmlns:xsi = "http://www.w3.org/2001/XMLSchema - instance"
  xsi:schemaLocation = "http://java.sun.com/xml/ns/javaee
                        http://java.sun.com/xml/ns/javaee/web - app_3_0.xsd"
  version = "3.0">
< servlet >
    < servlet - name > TestServlet02 </servlet - name >
    < servlet - class > cn.itcast.chapter04.servlet.TestServlet02 </servlet - class >
    < init - param >
    < param - name > encoding </param - name >
    < param - value > UTF - 8 </param - value >
    </init - param >
</servlet >
```

```
    <servlet-mapping>
    <servlet-name>TestServlet02</servlet-name>
    <url-pattern>/TestServlet02</url-pattern>
  </servlet-mapping>
  </web-app>
```

参数信息如下。

<init-param>：要设置的参数。

<param-name>：参数的名称。

<param-value>：参数的值。

（2）编写 TestServlet02 类读取 web.xml 文件中的参数信息。

```
package mypt.servlet;
import java.io.*;
import javax.servlet.*;
import javax.servlet.http.*;
public class TestServlet extends HttpServlet{
  protected void doGet(HttpServletRequest request, HttpServletResponse response) throws
ServletException, IOException{
    PrintWriter out = response.getWriter();
    //获得 ServletConfig 对象
    ServletConfig config = this.getServletConfig();
    //获得参数名为 encoding 对应的参数值
    String param = config.getInitParameter("encoding");
    out.println("encoding = " + param);
  }
  protected void doPost(HttpServletRequest request, HttpServletResponse response) throws
ServletException, IOException{
    this.doGet(request,response);
  }
}
```

启动 Tomcat，在浏览器中输入地址 http://localhost:8080/myDedmo/TestServlet，访问 TestServlet，控制台输出结果如下。

```
encoding = UTF-8
```

3. GenericServlet 类

抽象类 GenericServlet 实现了 Servlet 接口和 ServletConfig 接口，给出了除 service()方法外的其他方法的简单实现，是一种与协议无关的 Servlet，它可以用于模拟操作系统的端口监控进程。其常用的方法如表 6-3 所示。

表 6-3　GenericServlet 类的常用方法

方　法	描　述
void init(ServletConfig config)	调用 Servlet 接口的 init()方法
String getInitParameter(String name)	返回名称为 name 的初始化参数
ServletContext getServletContext()	返回 ServletContevt 对象的引用

通常只需要重写不带参数的 init()方法,如果要编写一个通用的 Servlet,只要继承 GenericServlet 类实现 service()方法即可。

4. ServletContext 接口

ServletContext 对象表示 Web 应用的上下文。Servlet 使用 ServletContext 接口定义的方法与它的 Servlet 容器进行通信。Servlet 容器厂商负责提供 ServletContext 接口的实现。容器在应用程序加载时创建一个 ServletContext 对象,ServletContext 对象被 Servlet 容器中的所有 Servlet 共享。JSP 的隐含对象 application 是 ServletContext 的实例。ServletContext 接口的常用方法如表 6-4 所示。

表 6-4 ServletContext 接口的常用方法

方　法	描　述
Object getAttribute(String name)	获得 ServletContext 中名称为 name 的属性
ServletContext getContext(String uripath)	返回给定的 uripath 的应用的 Servlet 上下文,如 test = getContext("/text")
void removeAttribute(String name)	删除名称为 name 的属性
void setAttribute(String name,Object object)	在 ServletContext 中设置一个属性,这个属性的名称为 name,值为 object 对象
String getRealPath(String path)	返回相对路径的真实路径
void log(String message)	记录一般日志信息

5. RequestDispatcher 接口

RequestDispatcher 接口是请求的发起者。它有两个方法: forward()和 include()。客户端对于任何一个请求,可以根据业务逻辑需要选择不同的处理方法。

(1) forward(ServletRequest request,ServletResponse response): 把请求转发到服务器上的另一个资源(Servlet、JSP、HTML)。

(2) include(ServletRequest request,ServletResponse response): 把服务器上的另一个资源(Servlet、JSP、HTML)包含到响应中。

RequestDispatcher 接口有一个特点,就是浏览器上显示的 URL 是最先请求的目标资源的 URL,不会因为使用了 forward()、include()方法而改变,因此 forward()和 include()的调用对于用户来说是透明的。

6.2.2 javax.servlet.http 包

javax.servlet.http 包中可用的接口和类主要有以下几个。

HttpServletRequest 接口: 继承自 ServletRequest 接口,为 HTTPServlet 提供请求信息。

HttpServletResponse 接口: 继承自 ServletResponse 接口,为 HTTPServlet 输出响应信息提供支持。

HttpSession 接口: 为维护 HTTP 用户的会话状态提供支持。

HttpSessionBindingListener 接口：使得某对象在加入一个会话或从会话中删除时能够得到通知。

Cookie 类：用在 Servlet 中以使用 Cookie 技术。

HttpServlet 类：定义了一个抽象类，继承自 GenericServlet 抽象类。

HttpSessionBindingEvent 类：当某个实现了 HttpSessionBindingListener 接口的对象被加入会话或从会话中删除时，会收到该类对象的一个句柄。

HttpUtils 类：提供了一系列便于编写 HTTP Servlet 的方法。

1. HttpServlet 类

Servlet 的请求可能包含多个数据项，当 Web 容器接收到对 Servlet 的请求时，会把这些请求封装成一个 HttpServletRequest 对象，然后把这个对象传递给 Servlet 相应的服务方法。HttpServlet 类扩展了 GenericServlet 类，提供 Servlet 接口的 HTTP 的实现。这些服务方法实现的功能有：获取服务器信息、客户端把数据传送到服务器、客户端把文件存放到服务器、Servlet 处理 Trace 请求及客户端删除服务器端的文件。HttpServlet 的子类必须实现表 6-5 所示方法中的一个。

表 6-5　HttpServlet 子类必须实现的方法

方　法	描　述
void service (ServletRequest req, ServletResponse res)	调用 GenericServlet 类的 service()方法的实现
void doXXXX (HttpServletRequest req, HttpServletResponse res)	根据请求方式的不同，分别调用相应的处理方法，例如 doGet()、doPost()、doPut()、doDelete()等

HttpServlet 类是一个抽象类，如果需要编写 Servlet 就一定要继承 HttpServlet 类，从中将需要响应到客户端的数据封装到 HttpServletResponse 对象中。

(1) 获取服务器信息。在获取服务器端信息时，需要使用 HttpServlet 类的 doGet()方法。该方法可以获取服务器信息，并将其作为响应返回给客户端。通过 Web 浏览器、HTML、JSP 直接访问 Servlet 的 URL 时，一般使用 doGet()方法。doGet()方法在 URL 中显示正传递给 Servlet 的数据，这样可能给系统的安全方面带来一些问题，比如用户登录时，表单中的用户名和密码需要发送到服务器端，如果使用 doGet()方法，会在 URL 里显示用户名和密码，而且在使用 doGet()方法时传送的数据量也会受到限制。

(2) 客户端把数据传送到服务器。在客户端把数据传送到服务器时，需要使用 HttpServlet 类的 doPost()方法，它的好处是可以隐藏发送给服务器端的数据。doPost()方法适合于发送大批量的数据。

(3) 客户端把文件存放到服务器。在客户端把文件存放到服务器时，需要使用 HttpServlet 类的 doPut()方法。该方法与 doPost()方法调用相似，允许客户端把真正的文件存放到服务器上，而不仅仅是传送数据。

(4) Servlet 处理 Trace 请求。在 Servlet 处理 Trace 请求时，需要使用 HttpServlet 类的 doTrace()方法。该方法由容器进行调用，主要用于 Web 调试，不可以覆盖。

(5) 客户端删除服务器端的文件。在客户端删除服务器端的文件时，需要使用 HttpServlet 类的 doDelete()方法。该方法与 doPut()方法相似，允许客户端删除服务器端的文件或者 Web 页面。

【例 6-2】 把数据库驱动、URL 地址、用户名称和密码写入 web.xml 文件，通过初始化参数获取 web.xml 文件中的信息，实现数据查询的功能。

(1) 创建名为 InitServlet.java 的类文件。InitServlet 类继承自 HttpServlet 类，并实现对数据表的查询功能。

```
public class InitServlet extends HttpServlet {
  private String driver = "";
  private String URL = "";
  private String username = "";
  private String password = "";
  //初始化 web.xml 参数
  public void init() throws ServletException {
    driver = getInitParameter("driver");
    URL = getInitParameter("URL");
    username = getInitParameter("username");
    password = getInitParameter("password");
  }
  //获得数据库连接的方法
  public Connection getConnection() {
    Connection con = null;
    try {
      Class.forName(driver);
      con = DriverManager.getConnection(URL, username, password);
    } catch (Exception e) {
    }
    return con;
  }
  //通过 doGet()方法实现数据查询功能
  public void doGet(HttpServletRequest request, HttpServletResponse response)
  throws ServletException, IOException {
    response.setContentType("text/html;charset = gb2312");
    PrintWriter out = response.getWriter();
    Connection con = this.getConnection();
    try {
      Statement stmt = con.createStatement();
      ResultSet rs = stmt.executeQuery("select * from tb_user");
      while (rs.next()) {
          out.print(rs.getString("id") + "---------- " + rs.getString("name"));
          out.print("< br >");
      }
    } catch (SQLException e) {
      e.printStackTrace();
    }
  }
  public void doPost(HttpServletRequest request, HttpServletResponse response)
```

```
        throws ServletException, IOException {
        doGet(request, response);
    }
}
```

代码说明：当调用 InitServlet 类时，该类的 init()方法将被自动执行。在 init()方法体中通过 getInitParameter()方法获取 web. xml 文件中<init-name>和</init-name>之间的参数，取得<param-value>和</param-value>之间的数据。

(2) 在 web. xml 文件中配置 InitServlet 类，关键代码如下。

```
<servlet>
  <description>
    This is the description of my J2EE component
  </description>
  <display-name>
    This is the display name of my J2EE component
  </display-name>
  <servlet-name>InitServlet</servlet-name>
  <servlet-class>com.mypt.InitServlet</servlet-class>
  <init-param>
    <param-name>driver</param-name>  <!-- 指定数据库驱动参数名称 -->
    <param-value>
    com.microsoft.jdbc.sqlserver.SQLServerDriver <!-- 指定数据库驱动参数值 -->
    </param-value>
  </init-param>
  <init-param>
    <init-param>
    <param-name>URL</param-name>      <!-- 指定数据库 URL 参数名称 -->
    <param-value>                     <!-- 指定数据库 URL 参数值 -->
      jdbc:microsoft:sqlserver://localhost:1433;DatabaseName=db_database19
    </param-value>
  </init-param>
  <init-param>
    <param-name>username</param-name>    <!-- 指定数据库登录用户参数名称 -->
    <param-value>sa</param-value>        <!-- 指定数据库登录用户参数值 -->
  </init-param>
  <init-param>
    <param-name>password</param-name>    <!-- 指定数据库登录密码参数名称 -->
    <param-value>123321</param-value>    <!-- 指定数据库登录密码参数值 -->
    </init-param>
  </servlet>
    <servlet-mapping>
      <servlet-name>InitServlet</servlet-name>
      <url-pattern>/InitServlet</url-pattern>
  </servlet-mapping>
</web-app>
```

(3) 程序发布后，启动 Tomcat 服务器，在地址栏中输入 http://127.0.0.1:8080/04/InitServlet 观察运行结果。

2. ServletRequest 接口

当客户发出请求时，由 Servlet 容器创建 ServletRequest 对象来封装客户的请求信息。这个对象将被容器作为 service()方法的参数传递给 Servlet。Servlet 能够利用 ServletRequest 对象获取客户端的请求数据。ServletRequest 接口的常用方法如表 6-6 所示。

表 6-6　ServletRequest 接口的常用方法

方　法	描　述
Object getAttribut(String name)	获取名称为 name 的属性值
void setAttribute(String name,Object object)	在请求中保存名称为 name 的属性
void removeAttribute(String name)	清除请求中名字为 name 的属性

3. HttpServletRequest 接口

HttpServletRequest 接口可以获取由客户端传送的参数、客户端正在使用的通信协议、接收请求的服务器主机名及其 IP 地址等信息。JSP 的内建对象 request 是 HttpServletRequest 类的实例。HttpServlet Request 接口的常用方法如表 6-7 所示。

表 6-7　HttpServlet Request 接口的常用方法

方　法	描　述
String getContextPath()	返回请求 URL 的 Context 部分。实际是 URL 中指定 Web 程序的部分。例如：URL 为"http://localhost:8080/mingrisoft/index.jsp"。这一方法返回的是"mingrisoft"
Cookie[] getCookies()	返回客户发过来的 Cookie 对象
String getMethod()	返回客户请求的方法类型，如 Get、Post 或 Put
String getPathInfo()	返回客户请求 URL 的路径信息
String getPathTranslated()	返回 URL 中 Servlet 名称之后检索字符串之前的路径信息
String getQueryString()	返回 URL 中检索的字符串
String getRemoteUser()	返回用户名称，主要应用在 Servlet 安全机制中，用以检查用户是否已经登录
String getRequestURI()	返回客户请求使用的 URI，是 URI 中的 host 名称和端口号之后的部分。例如，若 URL 为 http://localhost:8080/mingrisoft/index.jsp，这一方法返回的是"/mingrisoft/index.jsp"
StringBuffer getRequestURL()	返回客户 Web 请求的 URL 路径
String getServletPath()	返回 URL 中 Servlet 的名称
HttpSession getSession()	返回当前会话期间的对象
String getParameter(String name)	获取请求中指定的参数名，如果请求中没有这个参数，则返回 null
String getParameterValues(String name)	返回请求中指定 name 的参数值，返回值是一个 String 类型的数组，通常这个值是通过 checkbox 或 select 表单控件提交的

4. ServletResponse 接口

Servlet 容器在接收到客户请求后，除了创建 ServletRequest 对象用于封装客户的请求信息外，还创建了一个 ServletResponse 对象，用来封装响应数据，并且同时将这两个对象一并作为参数传递给 Servlet。Servlet 利用 ServletRequest 对象获取客户端的请求数据，经过处理后由 ServletResponse 对象发送响应数据。ServletResponse 接口的常用方法如表 6-8 所示。

表 6-8 ServletResponse 接口的常用方法

方 法	描 述
PrintWriter getWriter()	返回 PrintWrite 对象，用于向客户端发送文本
String getCharacterEncodingO	返回在响应中发送的正文所使用的字符编码
void setCharacterEncodingO	设置发送到客户端的响应的字符编码
void setContentType(String type)	设置发送到客户端的响应的内容类型

5. HttpServletResponse 接口

与 HttpServletRequest 接口类似，HttpServletResponse 接口也继承自 ServletResponse 接口，用于对客户端的请求执行响应。它除了具有 ServletResponse 接口的常用方法外，还增加了新的方法，如表 6-9 所示。

表 6-9 HttpServletResponse 接口的常用方法

方 法	描 述
void addCookie(Cookie arg0)	在响应中加入 Cookie 对象
String encodeRedirectURL(String arg0)	对特定的 URL 进行加密，在 sendRedirect()方法中使用
void sendError(int arg0)	使用特定的错误代码向客户传递出错响应
void sendError(int arg0,String arg1)	使用特定的错误代码向客户传递出错响应，同时清空缓存
void sendRedirect(String arg0)	传递临时响应

6. HttpSession 接口

HttpSession 接口被 Servlet 引擎用来实现 HTTP 客户端和 HTTP 会话两者的关联。这种关联可能在多处连接和请求中持续一段时间。Session 用来在无状态的 HTTP 下越过多个请求页面来维持状态和识别用户。一个 Session 可以通过 Cookie 或重写 URL 来维持。HttpSession 的常用方法如表 6-10 所示。

表 6-10 HttpSession 接口的常用方法

方 法	描 述
long getCreationTime()	返回建立 Session 的时间，这个时间表示为自 1970-1-1 日(GMT)以来的毫秒数
String getId()	返回分配给这个 Session 的标识符。HTTP Session 的标识符是由服务器建立和维持的唯一的字符串

续表

方　法	描　述
long getLastAccessedTime()	返回客户端最后一次发出与这个 Session 有关的请求的时间，如果这个 Session 是新建立的，返回-1。这个时间表示为自 1970-1-1 日(GMT)以来的毫秒数
int getMaxInactiveInterval()	返回一个秒数，这个秒数表示客户端在不发出请求时，Session 被 Servlet 引擎维持的最长时间。在这个时间之后，Session 可能被 Servlet 引擎终止。如果这个 Session 不会被终止，这个方法返回-1。当 Session 无效后再调用这个方法会抛出 IllegalStateException 异常
Object getValue(String name)	返回以给定的名称 Session 上的对象。如果不存在这样的绑定，返回 null。当 Session 无效后再调用这个方法会抛出 IllegalStateException 异常
String[] getValueNames()	以数组形式返回绑定到 Session 上的所有数据的名称。当 Session 无效后再调用这个方法会抛出 IllegalStateException 异常
void invalidate()	这个方法会终止 Session。所有绑定在这个 Session 上的数据都会被清除。并通过 HttpSessionBindingListener 接口的 valueUnbound()方法发出通知
boolean isNew()	返回一个布尔值以判断这个 Session 是不是新的。如果一个 Session 已经被服务器建立但是还没有收到客户端的请求，这个 Session 将被认为是新的
void putValue (String name, Object value)	以给定的名称绑定对象到 Session 中。已存在的同名的绑定会被重置。这时会调用 HttpSessionBindingListener 接口的 valueBound()方法。当 Session 无效后再调用这个方法会抛出 IllegalStateException 异常
void removeValue(String name)	取消给定名称的对象在 Session 上的绑定。如果未找到给定名称绑定的对象，这个方法什么出不做。这时会调用 HttpSessionBindingListener 接口的 valueUnbound()方法。当 Session 无效后再调用这个方法会抛出 IllegalStateException 异常
int setMaxInactiveInterval (int interval)	设置一个秒数，这个秒数表示客户端在不发出请求时，Session 被 Servlet 引擎维持的最长时间

6.2.3 Servlet 的部署与配置

如果想运行 JSP，首先要把程序部署到 Web 服务器(如 Tomcat)中，这样程序才能运行起来。对于 Servlet 也是一样，要部署到 Servlet 容器中才能运行。

1. 环境设置

Servlet 包并不在 JDK 中，如果要编译和运行 Servlet，就必须把 servlet. jar 包放到 classpath 下。servlet-api. jar 包可以从 tomcat\common\lib 下获取。应该在项目的 WEB-INF 文件夹下新建 classe 文件夹用来存放所有的类。Servlet 程序的部署结构如图 6-9 所示。

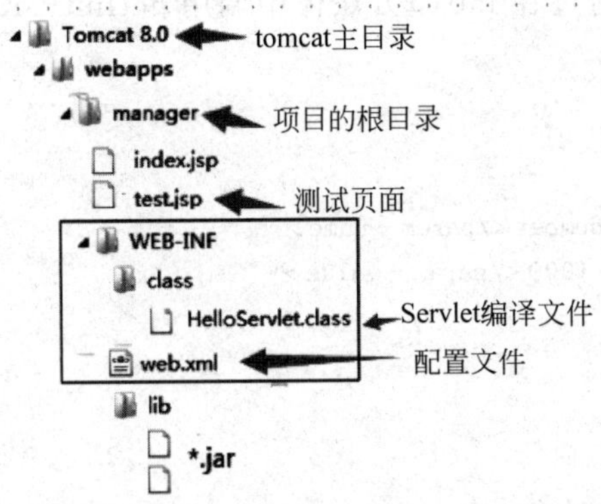

图 6-9　Servlet 程序的部署结构

2. Servlet 的配置

在 web. xml 文件中配置 Servlet 时，必须首先指定 Serlvet 的名称、Servlet 类的路径，可以有选择性地给 Servlet 添加描述信息，指定在发布时显示的名称和图标。具体代码如下。

```
<! DOCTYPE web - appPUBLIC " - //Sun Microsystems, Inc.//DTD Web Application 2. 3//EN"
"http://java.sun.com/dtd/web - app_2_3.dtd">
<web - app>
  <servlet>
<description> Simple Servlet </description>
<display - name> Servlet </display - name>
     <servlet - name> servlet 的名字</servlet - name>
     <servlet - class> servlet 包名类名</servlet - class>
</servlet>
```

<description>和</description>标签间的内容是 Serlvet 的描述信息，<display-name>和</display-name>标签间的内容是发布时 Serlvet 的名称，<servlet-name>和</servlet-name>标签间的内容是 Servlet 的名称，<servlet-class>和</servlet-class>标签间的内容是 Servlet 类的路径。

如果要配置的 Servlet 是一个 JSP 页面文件，那么可以通过下面的代码进行指定。

```
<servlet>
  <description> Simple Servlet </description>
  <display - name> Servlet </display - name>
  <servlet - name> Login </servlet - name>
  <jsp - file> login.jsp </jsp - file>
</servlet>
```

<jsp-file>和</jsp-file>标签间的内容是要访问 JSP 页面文件的名称。

3. 初始化参数

可以为 Servlet 配置一些初始化参数，例如下面的代码指定 number 的参数值为

1000,在 Servlet 中可以通过在 init()方法体中调用 getInitParameter()方法对参数进行访问。

```
<servlet>
  <init-param>
      <param-name>number</param-name>
      <param-value>1000</param-value>
  </init-param>
</servlet>
```

4. 启动装入优先权

启动装入优先权通过< load-on-startup >和</load-on-startup >间的内容进行指定,例如:

```
<servlet>
  <description>ONE</description>
  <display-name>ServletONE</display-name>
  <servlet-name>ServletONE</servlet-name>
  <servlet-class>com.ServletONE</servlet-class>
  <load-on-startup>10</load-on-startup>
  <!-- 设置 ChenLeiServletONE 载入时间 -->
</servlet>
<servlet>
  <description>TWO</description>
  <display-name>ServletTWO</display-name>
  <servlet-name>ServletTWO</servlet-name>
  <servlet-class>com.ServletTWO</servlet-class>
  <load-on-startup>20</load-on-startup>
  <!-- 设置 ChenLeiServletTWO 载入时间 -->
</servlet>
<servlet>
  <description>THREE</description>
  <display-name>ServletTHREE</display-name>
  <servlet-name>ServletTHREE</servlet-name>
  <servlet-class>com.ServletTHREE</servlet-class>
  <load-on-startup>AnyTime</load-on-startup>
</servlet>
```

ServletONE 类先被载入,ServletTWO 类后被载入,而 ServletTHREE 类可以在任何时间内被载入。

5. Servlet 的映射

```
<servlet-mapping>
  <servlet-name>Servlet 的名称(要和 servlet 标签中的相同)</servlet-name>
  <url-pattern>指定 Servlet 相对于应用目录的路径</url-pattern>
</servlet-mapping>
</web-app>
```

<servlet-mapping>标签在 Servlet 和 URL 样式之间定义了一个映射。它包含了两个子标签<servlet-name>和<url-pattern>。<servlet-name>标签给出的 Servlet 名称必须是在<servlet>标签中声明过的 Servlet 的名称。<url-pattern>标签指定对应于 Servlet 的 URL 路径,该路径是相对于 Web 应用程序上下文根的路径。

Servlet 标准允许<url-pattern>子标签在<servlet-mapping>父标签中出现多次,也就是说允许一个 Servlet 对应几个不同的 URL。

(1) 以/开始并以/*结束的字符串用来映射路径,例如:

```
<url-pattern>/admin/*</url-pattern>
```

如果没有精确匹配,那么对/admin/路径下的资源的所有请求将由映射了上述 URL 样式的 Servlet 来处理。

(2) 以*.为前缀的字符串用来映射扩展名,例如:

```
<url-pattern>*.do</url-pattern>
```

如果没有精确匹配和路径匹配,那么对具有.do 扩展名的资源的请求将由映射了上述 URL 样式的 Servlet 处理。

(3) 以一个单独的/指示这个 Web 应用程序默认的 Servlet,例如:

```
<url-pattern>/</url-pattern>
```

如果对于某个请求,没有找到匹配的 Servlet,那么将使用 Web 应用程序的默认 Servlet 来处理。

(4) 精确匹配,例如:

```
<url-pattern>/login</url-pattern>
```

6. 默认 Servlet

如果某个 Servlet 的映射路径仅仅为一个斜杠(/),那么这个 Servlet 就成为当前 Web 应用程序的默认 Servlet。

凡是在 web.xml 文件中找不到匹配的 URL,它们的访问请求都将交给默认的 Servlet 处理,也就是说,默认的 Servlet 用于处理所有其他 Servlet 都不处理的访问请求。例如:

```
<servlet>
   <servlet-name>ServletDemo</servlet-name>
   <servlet-class>mypt.servlet.ServletDemo</servlet-class>
   <load-on-startup>1</load-on-startup>
</servlet>
<!-- 将 ServletDemo 配置成缺省 Servlet -->
<servlet-mapping>
   <servlet-name>ServletDemo</servlet-name>
   <url-pattern>/</url-pattern>
</servlet-mapping>
```

在 Tomcat 安装目录下的\conf\web.xml 文件中，注册了一个名称为 org.apache.catalina.servlets.DefaultServlet 的 Servlet，并将这个 Servlet 设置为默认的 Servlet。

```
<servlet>
  <servlet-name>default</servlet-name>
  <servlet-class>org.apache.catalina.servlets.DefaultServlet</servlet-class>
    <init-param>
      <param-name>debug</param-name>
      <param-value>0</param-value>
    </init-param>
    <init-param>
      <param-name>listings</param-name>
      <param-value>false</param-value>
      </init-param>
      <load-on-startup>1</load-on-startup>
  </servlet>
<!-- The mapping for the default servlet -->
  <servlet-mapping>
    <servlet-name>default</servlet-name>
    <url-pattern>/</url-pattern>
</servlet-mapping>
```

当访问 Tomcat 服务器中的某个静态 HTML 文件和图片时，实际上是在访问这个默认的 Servlet。

6.2.4 Servlet 的线程安全

当多个客户端并发访问同一个 Servlet 时，Web 服务器会为每一个客户端的访问请求创建一个线程，并在这个线程上调用 Servlet 的 service()方法，因此 service()方法内如果访问了同一个资源的话，就有可能引发线程安全问题。例如下面的代码不存在线程安全问题。

```
package mypt.servlet.study;
public class ServletDemo extends HttpServlet {
  public void doGet(HttpServletRequest request, HttpServletResponse response) throws
ServletException, IOException {
     int i = 1;
     i++;
     response.getWriter().write(i);
  }
  public void doPost(HttpServletRequest request, HttpServletResponse response)
  throws ServletException, IOException {
     doGet(request, response);
  }
}
```

以上代码中，当多线程并发访问这个方法中的代码时，变量 i 被多个线程并发访问。i 是 doGet()方法中的局部变量，当有多个线程并发访问 doGet()方法时，每个线程中都有自己的 i 变量,各个线程操作的都是自己的 i 变量,所以不存在线程安全问题。

下面的代码则存在线程安全问题。

```
package mypt.servlet.study;
public class ServletDemo extends HttpServlet {
  int i = 1;
  public void doGet(HttpServletRequest request, HttpServletResponse response)
  throws ServletException, IOException {
    i++;
    try { Thread.sleep(1000 * 4);
    } catch (InterruptedException e) {
      e.printStackTrace();
    }
    response.getWriter().write(i + "");
  }
  public void doPost(HttpServletRequest request, HttpServletResponse response)
  throws ServletException, IOException {
    doGet(request, response);
  }
}
```

以上代码中,i 是全局变量,当多个线程并发访问变量 i 时,就会存在线程安全问题了。如同时开启两个浏览器模拟并发访问同一个 Servlet,正常情况下第一个浏览器应该看到 2,而第二个浏览器应该看到 3,结果两个浏览器都看到了 3,这就不正常。

线程安全问题只存在于多个线程并发操作同一个资源的情况下,所以在编写 Servlet 的时候,如果并发访问某一个资源(变量、集合等),就会存在线程安全问题。

下面的代码中,加了 synchronized 后,并发访问 i 时就不存在线程安全问题了。

```
package gacl.servlet.study;
public class ServletDemo extends HttpServlet {
  int i = 1;
  public void doGet(HttpServletRequest request, HttpServletResponse response)
  throws ServletException, IOException {
    //在 Java 中,每一个对象都有一把锁,这里的 this 指的就是 Servlet 对象
    synchronized (this) {
      i++;
      try {
          Thread.sleep(1000 * 4);
        } catch (InterruptedException e) {
            e.printStackTrace();
      }
      response.getWriter().write(i + "");
   }
  }
  public void doPost(HttpServletRequest request, HttpServletResponse response)
  throws ServletException, IOException {
    doGet(request, response);
  }
}
```

假如现在有一个线程访问 Servlet 对象,那么它就先拿到了 Servlet 对象的那把锁,等到它执行完之后才会把锁还给 Servlet 对象。由于是它先拿到了 Servlet 对象的那把锁,所以当有别的线程来访问这个 Servlet 对象时,只能排队等候。

这种做法虽然解决了线程安全问题,但是编写 Servlet 时最好不用这种方式处理线程安全问题,因为存在顺序排队轮流访问,效率很低。

针对 Servlet 的线程安全问题,Sun 公司提供了解决方案:让 Servlet 去实现一个 SingleThreadModel 接口,如果某个 Servlet 实现了 SingleThreadModel 接口,那么 Servlet 引擎将以单线程模式来调用其 service()方法。

让 Servlet 实现了 SingleThreadModel 接口,只要在 Servlet 类的定义中增加实现 SingleThreadModel 接口的声明即可。

对于实现了 SingleThreadModel 接口的 Servlet,Servlet 引擎仍然支持对该 Servlet 的多线程并发访问,其采用的方式是产生多个 Servlet 实例,并发的每个线程分别调用一个独立的 Servlet 实例。

实现 SingleThreadModel 接口并不能真正解决 Servlet 的线程安全问题,因为 Servlet 引擎会创建多个 Servlet 实例,而真正意义上解决多线程安全问题是指一个 Servlet 实例对象被多个线程同时调用的问题。

6.2.5 Servlet 应用

1. Sevlet 与表单

由于 Servlet 中存在 HttpServetRequest 和 HttpServletResponse 对象的声明,所以可以使用 Servlet 接收用户所提交的内容。

【例 6-3】 定义表单,由 Servlet 获取表单提交的信息。

(1) 定义表单 input.html。

```
<html>
  <head><title>获取用户信息</title></head>
  <body>
      <form action = "helloServlet" method = "post">
      编号:<input type = "text" name = "code"/><br>
      姓名:<input type = "text" name = "name"/><br>
      <input type = "submit" value = "提交"/>
    </form>
  </body>
</html>
```

(2) 编写接收用户请求类 helloServlet.java。

```
package org.myvtc.demo;
//导入包(略)
public class helloServlet extends HttpServlet {
  public void doGet ( HttpServletRequest request, HttpServletResponse response) throws
  ServletException, IOException {
```

```
    response.setContentType("text/html");
    String code = request.getParameter("code");
    String name = request.getParameter("name");
    PrintWriter out = response.getWriter()
    out.println("<!DOCTYPE HTML PUBLIC \"-//W3C//DTD HTML 4.01 Transitional//EN\">");
    out.println("<HTML>");
    out.println(" <HEAD><TITLE>用户信息</TITLE></HEAD>");
    out.println(" <BODY>");
    out.println("编号:" + code + "<br>");
    out.println("姓名:" + name);
    out.println(" </BODY>");
    out.println("</HTML>");
    out.flush();
    out.close();
  }
}
```

(3) 配置 web.xml。

```
<servlet>
  <servlet-name>Myservlet</servlet-name>
  <servlet-class>org.myvtc.demo.helloServlet</servlet-class>
</servlet>
<servlet-mapping>
  <servlet-name>Myservlet</servlet-name>
  <url-pattern>/hello</url-pattern>
</servlet-mapping>
```

2. Servlet 与 JSP 内置对象

获得内置对象的方法有以下几种。

(1) 获得 out 对象。JSP 中的 out 对象,一般可以使用 doXXX()方法中的 response 参数获得,默认情况下 out 对象无法打印中文,解决方法如下。

```
response.setContentType("text/html;charset=gb2312");
```

(2) 获得 request 和 response 对象。

```
public void doGet(HttpServletRequest request, HttpServletResponse response) throws
ServletException, IOException {
  //将 request 参数当成 request 对象使用
  //将 response 参数当成 response 对象使用
}
```

(3) 获得 session 对象。

```
HttpSession session = request.getSession();
```

(4) 获得 application 对象。

```
ServletContext application = this.getServletContext();
```

3. 在 Servlet 中取参数

web.xml 文件中有两种类型的参数设定。

（1）设置全局参数，该参数可以被所有的 Servlet 访问。

```
<context-param>
  <param-name>参数名</param-name>
  <param-value>参数值</param-value>
</context-param>
```

获取全局参数的方法如下。

```
ServletContext application = this.getServletContext();
application.getInitParameter("参数名称");
```

（2）设置局部参数，该参数只有相应的 Servlet 才能访问。

```
<servlet>
  <servlet-name>Servlet 名称</servlet-name>
  <servlet-class>Servlet 类路径</servlet-class>
  <init-param>
    <param-name>参数名</param-name>
    <param-value>参数值</param-value>
  </init-param>
</servlet>
```

获取局部参数的方法如下。

```
this.getInitParameter("参数名称");
```

此处的 this 是指 Servlet 本身。

4. 设置欢迎页面

设置欢迎页面的方法如下。

```
<welcome-file-list>
  <!-- 所要设定的欢迎页面 -->
  <welcome-file>welcom.jsp</welcome-file>
</welcome-file-list>
```

在 web.xml 中可以同时设置多个欢迎页面，Web 容器会默认第一个页面为欢迎页面。如果找不到，Web 容器将依次选择后面的页面作为欢迎页面。

在 JSP 页面或是 HTML 页面中可以通过表单或超链接跳转到 Servlet，也可以从 Sevelet 跳转到其他 Servlet、JSP 页面或其他页面。

5. Servlet 跳转

如果想进行客户端跳转，可直接使用 HttpServletResponse 接口的 sendRedirect()方法。但需要注意的是，此跳转只能传递 session 范围的属性，无法传递 request 范围的属性。

【例 6-4】 Servlet 实现客户端跳转。

(1) 客户端跳转类 clientredirect. java 的代码如下。

```
//clientredirect. java
package org. myvtc. clientredirect;
//导入包(略)
public class clientredirect extends HttpServlet {
  public void doGet(HttpServletRequest request, HttpServletResponse response)
  throws ServletException, IOException {
    request. getSession(). setAttribute("info", "myvtc");
    request. setAttribute("name", "zhangsan");
    response. sendRedirect("getinfo. jsp");
  }
  public void doPost(HttpServletRequest request, HttpServletResponse response)
  throws ServletException, IOException {
    doGet(request, response);
  }
}
```

接收属性页面 getinfo. jsp 的代码如下。

```
<%@ page language = "java" pageEncoding = "GBK" %>
  <html>
  <head><title>My JSP 'getinfo. jsp' starting page</title></head>
  <body>
  <%
    String info = (String)request. getSession(). getAttribute("info");
    String name = (String)request. getAttribute("name");
   %>
  info:<% = info %><br>
  name:<% = name %>
  </body>
</html>
```

(2) 服务器跳转。在 Sevelet 中没有<jsp:forward>之类的指令,所以如果要想执行服务器端跳转,必须依靠 RequestDispatcher 接口完成,此接口中提供了两个方法,原型如下。

```
public void forward ( Servlet Request request, SrveletResponse response ) throws
ServletExecption, IOException
public void include ( ServeletRequest request, SeveletResponse response ) throws
SeveletEception, IOException
```

使用 RequestDispatcher 接口的 forward()方法即可完成跳转功能,但如果想使用此接口还需要使用 SerletRequest 接口提供的方法进行实例化。

```
public RequestDispatcher getRequestDispatcher(String path)
```

【例 6-5】 Servlet 使用服务器端跳转。

serverredircet. java 的代码如下。

```
package org.myvtc.serverredirect;
//导入包(略)
public class serverredirect extends HttpServlet {
  public void doGet(HttpServletRequest request, HttpServletResponse response)
  throws ServletException, IOException {
    request.getSession().setAttribute("info", "myvtc");
    request.setAttribute("name", "zhangsan");
    RequestDispatcher rd = request.getRequestDispatcher("getinfo.jsp");
    rd.forward(request, response);
  }
  public void doPost(HttpServletRequest request, HttpServletResponse response)
  throws ServletException, IOException {
    doGet(request,response);
  }
}
```

6.2.6 任务：利用 Servlet 实现用户登录

1. 任务要求

利用 Servlet 实现用户登录。现有数据库 test，数据表 userTable，包括 userid、username、userPwd 等字段，编程实现用户登录，当用户登录成功，跳转到登录成功页面 success.jsp，否则要求重新登录并提示错误信息，保存原来的登录信息。

2. 任务实施

(1) 创建 userManger 项目。

(2) 创建一个 Servlet——LoginServlet，代码如下。

```
package cn.edu.mypt.servlet;
//导入包略
import cn.edu.mypt.bean.UserBean;
import cn.edu.mypt.model.UserModel;
import cn.edu.mypt.model.impl.UserModelImple;      //参见项目 3
public class LoginServlet extends HttpServlet {
  //略
  public void doGet(HttpServletRequest request, HttpServletResponse response)
  throws ServletException, IOException {
    doPost(request,response);
  }
  public void doPost(HttpServletRequest request, HttpServletResponse response)
  throws ServletException, IOException {
    request.setCharacterEncoding("utf-8");
    String uID = request.getParameter("uID");
    String uPwd = request.getParameter("uPwd");
    UserModel uModel = new UserModelImple();
    UserBean user = uModel.userLogin(uID, uPwd);
    if (user == null) {
```

```
            request.setAttribute("uID", uID);
            request.setAttribute("uPwd", uPwd);
            request.setAttribute("message", "用户名或密码有错");
            request.getRequestDispatcher("login.jsp").forward(request, response);
        }
        else {
            HttpSession ses = request.getSession();
            ses.setAttribute("USER", user);
            request.getRequestDispatcher("success.jsp").forward(request, response);
        }
    }
}
```

(3) 设计登录页面 login.jsp，参见 4.3 节中的任务。

```
<%@ page language = "java" import = "java.util.*" pageEncoding = "utf-8" %>
<!DOCTYPE HTML PUBLIC "-//W3C//DTD HTML 4.01 Transitional//EN">
<html>
  <head><title>用户登录</title></head>
  <body>
  <%
      String uID,uPwd,msg;
      msg = (String)request.getAttribute("message");
      uID = (String)request.getAttribute("userID");
      uPwd = (String)request.getAttribute("userPwd");
      uID = (uID == null)?"":uID;
      uPwd = (uPwd == null)?"":uPwd;
      msg = (msg == null)?"":msg;
  %>
    <% = msg %>
      <form action = "loginServlet" method = "post">
        用户 ID:<input type = "text" name = "uID" value = "<% = uID %>"/>
        用户名:<input type = "text" name = "uPwd" value = "<% = uPwd %>"/>
        <input type = "submit" value = "登录"/>
    </form>
  </body>
</html>
```

(4) 设计登录成功页面 success.jsp，参见 4.3 节中的任务。

(5) 配置 web.xml。

```
<?xml version = "1.0" encoding = "UTF-8"?>
<web-app version = "3.0"
  xmlns = "http://java.sun.com/xml/ns/javaee"
    xmlns:xsi = "http://www.w3.org/2001/XMLSchema-instance"
    xsi:schemaLocation = "http://java.sun.com/xml/ns/javaee
    http://java.sun.com/xml/ns/javaee/web-app_3_0.xsd">
  <servlet>
    <servlet-name>LoginServlet</servlet-name>
    <servlet-class>cn.edu.mypt.servlet.LoginServlet</servlet-class>
```

```
  </servlet>
  <servlet-mapping>
    <servlet-name>LoginServlet</servlet-name>
    <url-pattern>/LoginServlet</url-pattern>
  </servlet-mapping>
</web-app>
```

(6) 启动服务器,部署项目并运行。

6.3 Servlet 过滤器

6.3.1 过滤器的概念

1. 什么是 Servlet 过滤器

Servlet 过滤器是可通过配置文件来灵活声明的模块化的可复用组件,是 Servlet API 的组成部分。过滤器动态地处理传入请求和传出响应,可以将过滤器连接在一起,以便过滤器组可以对指定资源或资源组的输入和输出执行操作。过滤器通常包括记录过滤器、图像转换过滤器、加密过滤器和多用途 Internet 邮件扩展类型过滤器。

过滤器可使开发者直接处理请求和响应过程。过滤器可以在资源被获取前对其操作(或者在动态输出中被启动),也可以在获取资源后或被执行后立即启动。Servlet 过滤器无须修改应用代码就可以透明地添加或删除它们,独立于任何平台或者 Servlet 容器。

2. Servlet 过滤器的优点

(1) 声明方式:过滤器通过 web.xml 中的 XML 标签来声明,这样就可以添加和删除过滤器而无须改动任何应用代码或 JSP 页面。

(2) 动态:过滤器在运行时由 Servlet 容器来拦截、处理请求和响应。

(3) 灵活:过滤器在 Web 处理环境中应用很广泛,包括日志记录和安全等许多公共辅助任务。

(4) 模块化:把应用处理逻辑封装到单个类文件中,即可轻松地从请求/响应中添加或删除模块化单元。

(5) 可移植:Servlet 过滤器是跨平台、跨容器和可移植的,从而进一步支持了 Servlet 过滤器模块化和可复用的本质。

(6) 可复用:由于过滤器实现了模块化设计以及声明式的过滤器配置方式,因此可以跨越不同项目和应用使用。

(7) 透明:在请求/响应链中,过滤器可以补充(而不是以任何方式替代)Servlet 或为 JSP 页面提供核心处理业务,因而过滤器可以根据需要添加或删除而不会影响 Servlet 或 JSP 页面。

3. 过滤器的工作原理

当客户端发出 Web 资源的请求时,Web 服务器根据应用程序配置文件设置的过滤规则进行检查。若客户请求满足过滤规则，则对客户请求/响应进行拦截，对请求头和请求数据进行检查或改动，并依次通过过滤器链把请求/响应交给请求的 Web 资源处理器。请求信息在过滤器链中可以被修改，也可以根据条件使请求不发往资源处理器，并直接向客户机发回一个响应。当资源处理器完成了对资源的处理后，响应信息将逐级逆向返回。在这个过程中，用户可以修改响应信息，从而完成特定的任务。

服务器会按照 web. xml 中过滤器定义的先后顺序组装成一条链，过滤器的执行流程如图 6-10 所示。

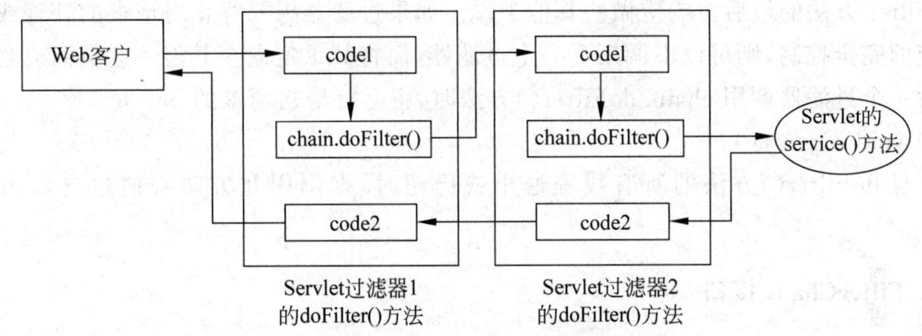

图 6-10 过滤器的执行流程

可以这样理解过滤器的执行流程：执行第 1 个过滤器的 chain. doFilter()之前的代码→执行第 2 个过滤器的 chain. doFilter()之前的代码→……→执行第 *n* 个过滤器的 chain. doFilter()之前的代码→执行请求 Servlet 的 service()方法中的代码→执行请求 Servlet 的 doGet()或 doPost()方法中的代码→执行第 *n* 个过滤器的 chain. doFilter()之后的代码→……→执行第 2 个过滤器的 chain. doFilter()之后的代码→执行第 1 个过滤器的 chain. doFilter()之后的代码。

6.3.2 Servlet 过滤器的接口

Servlet 过滤器本身生成请求对象和响应对象，它只能提供过滤功能。Servlet 过滤器能够在 Serlvet 被调用之前检查 Request 对象，修改请求头和请求内容。在 Servlet 被调用之后检查 Response，修改响应头和响应内容。要实现这样的操作，必须实现 Servlet 过滤器接口中的方法。Servlet 过滤器 API 包含了 3 个接口，它们都在 javax. servlet 包中，分别是 Filter 接口、FilterChain 接口和 FilterConfig 接口。

1. Filter 接口

所有的过滤器都必须实现 Filter 接口，该接口定义了 init()、doFilter()、destory() 3 个方法。

1) void init(FilterConfig filterConfig)

当开始使用 Servlet 过滤器时，Web 容器调用此方法一次，以准备过滤器，并负责设置 FilterConfig 对象。FilterConfig 对象将为过滤器提供初始化参数，并允许访问与之相关的 ServletContext。

2）void doFilter（ServletRequest request，ServletResponse response，FilterChain chain）

每个过滤器都接收当前的请求和响应，且 FilterChain 过滤器链中的过滤器（应该都是符合条件的）都会被执行。

在 doFilter()方法中，过滤器可以对请求和响应做任何事情，通过调用它们的方法收集数据，或者给对象添加新的行为。过滤器通过传递给此方法的 FilterChain 参数调用 chain. doFilter()方法将控制权传送给下一个过滤器。当这个方法返回后，过滤器可以在它的 Filter 方法的最后对响应做些其他工作。如果过滤器想要终止对请求的处理或得到对响应的完全控制，则可以不调用下一个过滤器，而将其重定向至其他一些页面。当链中的最后一个过滤器调用 chain. doFilter()方法时，将运行最初请求的 Servlet。

3）void destroy()

一旦 doFilter()方法的所有线程退出或已超时，则调用此方法释放过滤器占用的资源。

2. FilterChain 接口

FilterChain 接口存放在 javax. servlet 包内，通过这个接口可以把过滤的任务在不同的过滤器之间转移，它的 void doFilter（ServletRequest request，ServletResponse response）方法是 Servlet 容器提供给开发者的，用于对资源请求过滤器链的依次调用，通过 FilterChain 调用过滤器链中的下一个过滤器，如果是最后一个过滤器，则下一个就调用目标资源。

3. FilterConfig 接口

FilterConfig 接口用于建立过滤器的初始化参数、文本名称或应用程序运行下的 ServletContext。FilterConfig 接口包含以下方法。

（1）String getFilterName0：返回 web. xml 文件中定义的过滤器的名称。

（2）ServletContext getServletContext()：返回调用者所处的 Servlet 上下文。

（3）String getlnitParameter(String name)：返回过滤器初始化参数值的字符串形式。当参数不存在时，返回 nul1。name 是初始化参数名。

（4）Enumeration getlnitParameterNames()：以 Enumeration 形式返回过滤器所有初始化参数值。如果没有初始化参数，返回 null。

6.3.3 Servlet 过滤器的配置

Servlet 过滤器是一个 Web 应用组件，与 Servlet 类似，也需要在 web. xml 应用配置文件中进行部署。

1. 定义过滤器

<filter>元素包括两个必要子元素<filter-name>和<filter-class>，分别用来定义过滤器的名称和与过滤器相关的 Java 类的路径；此外还包含 4 个子元素<inti-param>、

< icon >、< display-name >和< description >。

定义 Servlet 过滤器的语法格式如下。

```
<filter>
    <filter-name>filterstation</filter-name>
    <filter-class>com.FilterStation</filter-class>
    <init-param>
        <param-name>Name1</param-name>
        <param-value>Value1</param-value>
    </init-param>
    <init-param>
        <param-name>Name2</param-name>
        <param-value>Value2</param-value>
    </init-param>
</filter>
```

说明：在名称为 filterstation 的过滤器中定义了两个参数 Name1 和 Name2，在实际应用中通过 Config 类中的 getInitParamenter()方法获得。

2. 配置过滤器的映射

可以将过滤器映射到一个或多个 Servlet 和 JSP 文件中，也可以映射到任意的 URL。

(1) 将 Servlet 映射到一个或多个 JSP 文件中，代码如下。

```
<filter-mapping>
    <filter-name>filterstation</filter-name>
    <url-pattern>/jsp/filename.jsp</url-pattern>
</filter-mapping>
```

说明：将名称为 filterstation 的过滤器映射到 jsp 目录中的 filename.jsp 文件。

(2) 映射到一个或多个 Servlet 中，代码如下。

```
<filter-mapping>
    <filter-name>FilterName</filter-name>
    <url-pattern>/FilterName1</url-pattern>
</filter-mapping>
<filter-mapping>
    <filter-name>FilterName</filter-name>
    <url-pattern>/FilterName2</url-pattern>
</filter-mapping>
```

说明：与映射到 JSP 文件的不同之处是需要提供 Servlet 名称。名为 FilterName1 和 FilterName2 的 Servlet 都被映射到 FilterName 过滤器。

(3) 映射到任意 URL，代码如下。

```
<filter-mapping>
    <filter-name>FilterName</filter-name>
    <url-pattern>/*</url-pattern>
</filter-mapping>
```

说明：只要把“/ * ”写到< url-pattern >和</url-pattern >元素之间即可。

6.3.4 过滤器的应用

通过过滤器技术，可以对 Web 服务器管理的所有 Web 资源(如 JSP、Servlet、静态图片文件或静态 HTML 文件等)进行拦截，并且添加功能，从而实现一些“共同”的功能。例如如下的应用场景。

(1) 实现 URL 级别的权限访问控制。

(2) 过滤敏感词汇。

(3) 自动登录。

(4) 压缩响应信息等功能。

(5) 流量控制。

过滤器不是用户主动调用的，而是根据规则自己执行的，编写过滤器的基本步骤如下。

(1) 编写一个 Java 类，实现 Filter 接口，并实现其中的所有方法。

(2) 在 web.xml 文件中配置过滤器。

```
<!-- 过滤器配置 -->
<filter>
  <!-- 内部名称 -->
  <filter-name>HelloFilter</filter-name>
  <!-- 类全名:包+简单类名 -->
  <filter-class>org.newboy.filter.HelloFilter</filter-class>
</filter>
<!-- 过滤器映射配置 -->
<filter-mapping>
  <!-- 内部名称,和上面的名称保持一致! -->
  <filter-name>HelloFilter</filter-name>
  <!-- 需要拦截的路径 -->
  <url-pattern>/*</url-pattern>
</filter-mapping>
```

(3) 把 Filter 部署到 Tomcat 服务器。

【例 6-6】 创建过滤器 EncodingFilter，实现对所有页面的编码转换。

```
package org.myvtc;
//导入包(略)
public class EncodingFilter implements Filter {
  String charset;
  public void destroy() {
    System.out.print("destroy");
    Thread.sleep(1000);
  }
  public void doFilter(ServletRequest arg0, ServletResponse arg1,
  FilterChain arg2) throws IOException, ServletException {
    System.out.print("执行");
```

```
        arg0.setCharacterEncoding(charset);
        arg2.doFilter(arg0, arg1);
        System.out.print("执行 2");
    }
    public void init(FilterConfig arg0) throws ServletException {
        charset = arg0.getInitParameter("charset");
        System.out.print("初始化" + charset);
    }
}
```

配置 web.xml,代码如下。

```
<filter>
    <filter-name>Encoding</filter-name>
        <filter-class>org.myvtc.EncodingFilter</filter-class>
        <init-param>
            <param-name>charset</param-name>
            <param-value>GBK</param-value>
        </init-param>
</filter>
<filter-mapping>
    <filter-name>Encoding</filter-name>
    <url-pattern>/*</url-pattern>
</filter-mapping>
```

【例 6-7】 使用过滤器进行网站流量统计。

Servlet 过滤器可以对用户提交的数据或服务器返回的数据进行更改,任何到达服务器的请求都会首先经过过滤器的处理。本例利用过滤器的这个特点编写一个专门用于流量统计的过滤器,具体开发步骤如下。

(1) 创建 FilterFlux.java 类,该类通过 Servlet 中的过滤器技术统计网站的访问量。关键代码如下。

```
public class FilterFlux extends HttpServlet implements Filter {
    //略
    private static int flux = 0;
    public synchronized void doFilter(ServletRequest request, ServletResponse response,
    FilterChain filterChain) throws ServletException, IOException {
        this.flux++;
        request.setAttribute("flux",String.valueOf(flux));
        filterChain.doFilter(request, response);
    }
}
```

(2) 创建 index.jsp 页面文件,该页面主要接收放入 request 中的对象,关键代码如下。

```
<tr align="center">
        <td><%=request.getAttribute("flux")%>次</td>
</tr>
```

(3) 发布 Servlet 过滤器时,必须在 web.xml 文件中加入<filter>和<filter-mapping>元素,<filter>元素用来定义一个过滤器。

```
<?xml version = "1.0" encoding = "UTF - 8"?>
< web - app xmlns = "http://java.sun.com/xml/ns/j2ee"
xmlns:xsi = "http://www.w3.org/2001/XMLSchema - instance"
xsi:schemaLocation = "http://java.sun.com/xml/ns/j2ee http://java.sun.com/xml/ns/j2ee/web -
app_2_4.xsd" version = "2.4">
  < display - name > web </display - name >
< filter >
    < filter - name > filterflux </filter - name >
    < filter - class > com.FilterFlux </filter - class >
  </filter >
  < filter - mapping >
    < filter - name > filterflux </filter - name >
    < url - pattern >/ * </url - pattern >
  </filter - mapping >
</web - app >
```

(4) 程序发布后,启动 Tomcat 服务器,在浏览器地址栏中输入地址 http://127.0.0.1:8080/demo 并观察运行结果。

【例 6-8】 设计一个简单的 IP 地址过滤器,根据用户的 IP 地址对网站的访问进行控制。

(1) 设计过滤器 ipfilter.java。

```
package ipf;
//导入包 略
public class ipfilter implements Filter {                    //实现 Filter 接口
  protected FilterConfig config;
  protected String rejectedlP;
  public void init(FilterConfig filterConfig)throws ServletException {
    this.config = filterConfig;                               //从 Web 服务器获取过滤器配置对象
    rejectedlP = config.getlnitParameter("RejectedlP");
    //从配置中取得过滤 lP
    if(rejectedlP == null) rejectedlP = ".";
  }
  public void doFilter(ServletRequest request,ServletResponse response,
                       FilterChain chain)throws IOException,ServletException {
    RequestDispatcher dispatcher = request.getRequestDispatcher("rejectedError.jsp");
    String remotelP = request.getRemoteAddr;                  //获取客户请求的 lP 地址
    int i = remotelP.lastlndexOf(".");
    int r = rejectedlP.lastlndexOf(".");
    String relPscope = rejectedlP.substring(0,r);             //过滤 lP 地址段
    if(relPscope.equals(remotelP.substring(0,i))){
      dispatcher.forward(request,response);                   //重定向到 rejectedError.jsp 页面
      return;                                                 //阻塞,直接返 Web 回客户端
    }
    else{chain.doFilter(request,response);                    //调用过滤链上的下一个过滤器
```

```
        }
    }
    public void destroy(){}
}
//过滤器功能完成后,由 Web 服务器调用执行,回收过滤器资源
```

注意：调用 chain.doFilter()方法以前的代码用于对客户请求的处理；以后的代码用于对响应进行处理。

(2) 配置过滤器。在应用程序 Web-INF 目录下的 web.xml 文件中添加以下代码。

```
<filter>
    <filter-name>ipfIter</filter-name>              //过滤器名称
    <filter-class>ipf.ipfiIter</filter-class>       //实现过滤器的类
    <init-param>
        <param-name>RejectedlP</param-name>         //过滤器初始化参数名 RejectedlP
        <param-value>192.168.12.*</param-value>
    </init-pamm>
</filter>
<filter-mapping>                                    //过滤器映射(规律规则)
    <filter-name>ipfiIter</filter-name>
    <url-pattem>/*</ud-pattem>
    //映射到 Web 应用根目录下的所有 JSP 文件
</filter-mapping>
```

通过以上设计与配置,就禁止了 192.168.12 网段的用户对网站的访问。

6.3.5 任务:强制登录验证

1. 任务要求

在进行用户的首次身份验证后都会在 Session 中留下相应的用户对象作为标识,在以后的操作中,只需要在进行身份验证的页面或 Servlet 中查看相应的 Session 即可。但是如果在每个页面或 Servlet 中都添加身份验证代码显然会对编程造成很大的麻烦,也会增加多余的代码。因此,可以利用过滤器对一批页面或 Servlet 统一进行身份验证,这样在设计页面或 Servlet 时不需要考虑身份验证的问题,非常方便。本任务实现通过过滤器进行用户身份验证。

2. 实施步骤

(1) 创建 long.jsp 页面文件,该文件的主要功能是填写用户的登录信息,关键代码参见 6.2 节中的任务。

(2) 创建 Servlet,其作用是直接放置一个 user 对象在 Session 中并且执行跳转另一个页面的代码,关键代码参考 6.2 节中的任务。

(3) 用户登录成功页面 success.jsp 的代码如下。

```
<%@ page language="java" import="java.util.*" pageEncoding="utf-8"%>
<!DOCTYPE HTML PUBLIC "-//W3C//DTD HTML 4.01 Transitional//EN">
```

```
<html>
  <head><title>欢迎页面</title></head>
    <body>
    <jsp:useBean id="user" class="cn.edu.mypt.bean.UserBean" scope="page"/>
    <%
        if (user!= null)user=(cn.edu.mypt.bean.UserBean)session.getAttribute("USER");
    %>
    <h1>欢迎你!<%=user.getUserName()%></h1>
  </body>
</html>
```

(4) 创建 UserInfoFilter.java 文件,该文件中的代码需要判断 Session 中是否存在 user 对象,如果 user 对象为 null 则表示用户没有登录过,这时候可以向输出流中输出错误信息,并中断过滤器,使服务器端不能得到用户请求,用户也得不到服务器传回的页面。如果 user 对象不为 null 则表示用户已经登录过,这时只需要简单地执行过滤器即可。该文件的关键代码如下。

```
public class FilterStation extends HttpServlet implements Filter {
  private FilterConfig filterConfig;
  public void init(FilterConfig filterConfig) throws ServletException {
      this.filterConfig = filterConfig;
  }
  public void doFilter(ServletRequest request, ServletResponse response,
  FilterChain filterChain) throws ServletException, IOException {
    HttpSession session=((HttpServletRequest)request).getSession();
    response.setCharacterEncoding("gb2312");
    if(session.getAttribute("user")==null){
      PrintWriter out=response.getWriter();
      out.print("<script language=javascript>alert('您还没有登录!!!');
      window.location.href='../login.jsp';</script>");
      request.getRequestDispatcher("login.jsp").forward(request, response);
    }else{
      filterChain.doFilter(request, response);
    }
  }
}
```

(5) 在 web.xml 文件中配置过滤器的 FilterStation.java 类文件。

```
<filter>
    <filter-name>filterstation</filter-name>
    <filter-class>com.FilterStation</filter-class>
</filter>
<filter-mapping>
    <filter-name>filterstation</filter-name>
    <url-pattern>/jsp/*</url-pattern>
</filter-mapping>
```

在设置需要进行身份验证的页面时,应该把登录页面排除在外;否则由于登录页面无法通过用户身份验证，而使用户无法进行登录，以致所有进行身份验证的页面都将无法访问。

(6) 程序发布后，启动 Tomcat 服务器，在址栏中输入 http://127.0.0.1:8080/demo/login.jsp 并观察运行结果。

6.4 监 听 器

6.4.1 监听器概述

Servlet 监听器用于监听 Web 容器的有效期事件，因此它是由容器管理的。利用 Listener 接口监听容器中的某个程序，并且根据其需求做出适当的响应。监听器对象在事情发生前、发生后可以做一些必要的处理。

1. 监听器的分类

按监听的对象划分，Servlet 2.4 标准中定义的事件监听器有以下 3 种。

(1) 用于监听应用程序环境对象(ServletContext)的事件监听器。

(2) 用于监听用户会话对象(HttpSession)的事件监听器。

(3) 用于监听请求消息对象(ServletRequest)的事件监听器。

按监听的事件类型划分，可以分为以下 3 种。

(1) 用于监听域对象自身的创建和销毁的事件监听器。

(2) 用于监听域对象中的属性的增加和删除的事件监听器。

(3) 用于监听绑定到 HttpSession 域中的某个对象的状态的事件监听器。

2. Servlet 监听器的原理

Servlet 监听器是当今 Web 应用开发的一个重要组成部分。它是在 Servlet 2.3 版中引入的，并且在 Servlet 2.4 标准中对其进行了较大的改进，主要是用来对 Web 应用进行监听和控制，极大地增强了 Web 应用的事件处理能力。

Servlet 监听器的功能与 Java 的 GUI 程序监听器比较接近，可以监听由于 Web 应用中状态改变而引起的 Servlet 容器产生的相应事件，然后接收并处理这些事件。

6.4.2 主要接口和对象

Servlet 2.4 和 JSP 中主要有 8 个 Listener 接口和 6 个 Event 类，如表 6-11 所示。

表 6-11 Listener 接口与 Event 类

Listener 接口	Event 类
ServletContextListener	ServletContextEvent
ServletContextAttributeListener	ServletContextAttributeEvent
HttpSessionListener	HttpSessionEvent
HttpSessionActivationListener	

续表

Listener 接口	Event 类
HttpSessionAttributeListener	HttpSessionBindingEvent
HttpSessionBindingListener	
ServletRequestListener	ServletRequestEvent
ServletRequestAttributeListener	ServletRequestAttributeEvent

在 Servlet 技术中已经定义了一些事件，并且可以针对这些事件来编写相关的事件监听器，从而对事件做出相应处理。Servlet 事件主要有 3 类：Servlet 上下文事件、会话事件与请求事件。

1. 对 Servlet 上下文进行监听

监听内容包括 ServletContext 对象的创建和删除以及属性的添加、删除和修改等操作。该监听器需要使用到如下两个接口。

1）ServletContextListener 接口

该接口存放在 javax.servlet 包内，主要用于监听 ServletContext 的创建和删除。ServletContextListener 接口提供了两个方法，也被称为“Web 应用程序的生命周期方法”。

（1）contextInitialized(ServletContextEvent event)：通知正在监听的对象，应用程序已经被加载及初始化。

（2）contextDestroyed(ServletContextEvent event)：通知正在监听的对象，应用程序已经被卸载，即关闭。

2）ServletAttributeListener 接口

该接口存放在 javax.servlet 包内，主要用于监听 ServletContext 属性的增加、删除和修改。ServletAttributeListener 接口提供了以下 3 个方法。

（1）attributeAdded(ServletContextAttributeEvent event)：如果有对象加入 Web 应用的范围，通知正在监听的对象。

（2）attributeReplaced(ServletContextAttributeEvent event)：如果在 Web 应用的范围内有对象取代另一个对象，通知正在监听的对象。

（3）attributeRemoved(ServletContextAttributeEvent event)：如果有对象从 Web 应用的范围移除时，通知正在监听的对象。

2. 监听 HTTP 会话

监听内容包括 HTTP 会话的活动情况、HTTP 会话中的属性设置情况，以及 HTTP 会话的 active、paasivate 情况等。该监听器有 4 个接口。

1）HttpSessionListener 接口

HttpSessionListener 接口用于监听 HTTP 会话的创建和销毁。该接口提供了以下两个方法。

（1）sessionCreated(HttpSessionEvent event)：通知正在监听的对象，Session 已经

被加载及初始化。

(2) sessionDestroyed(HttpSessionEvent event)：通知正在监听的对象，Session 已经被卸载(HttpSessionEvent 类的主要方法是 getSession()，可以使用该方法回传一个 Session 对象)。

2) HttpSessionActivationListener 接口

HttpSessionActivationListener 接口用于监听 HTTP 会话的 active、passivate 情况。该接口提供以下 3 个方法。

(1) attributeAdded(HttpSessionBindingEvent event)：若有对象加入 Session 的范围，通知正在监听的对象。

(2) attributeReplaced(HttpSessionBindingEvent event)：若在 Session 的范围内有对象取代另一个对象，通知正在监听的对象。

(3) attributeRemoved(HttpSessionBindingEvent event)：如果有对象被从 Session 的范围内移除，通知正在监听的对象(HttpSessionBindingEvent 类主要有 getName()、getSession()和 getValues()3 个方法)。

3) HttpBindingListener 接口

HttpBindingListener 接口用于监听 HTTP 会话中对象的绑定信息。它是唯一不需要在 web.xml 中设定的监听器。该接口提供以下两个方法。

(1) valueBound(HttpSessionBindingEvent event)：若有对象加入 Session 的范围则会被自动调用。

(2) valueUnBound(HttpSessionBindingEvent event)：若有对象被从 Session 的范围内移除则会被自动调用。

4) HttpSessionAttributeListener 接口

HttpSessionAttributeLitener 接口用于监听 HTTP 会话中属性的设置。该接口提供以下两个方法。

(1) sessinDidActivate(HttpSessionEvent event)：通知正在监听的对象，它的 Session 已经变为有效状态。

(2) sessinWillPassivate(HttpSessionEvent event)：通知正在监听的对象，它的 Session 已经变为无效状态。

3. Servlet 请求监听

在 Servlet 2.4 标准中，一旦能够在监听程序中获取客户端的请求，就可以对请求进行统一处理。要实现客户端的请求和请求参数设置的监听，需要利用以下两个接口。

1) ServletRequestListener 接口

ServletRequestListener 接口提供了以下两个方法。

(1) requestInitalized(ServletRequestEvent event)：通知正在监听的对象，ServletRequest 已经被加载及初始化。

(2) requestDestroyed(ServletRequestEvent event)：通知正在监听的对象，ServletRequest 已经被卸载，即关闭。

2）ServletRequestAttributeListener 接口

ServletRequestAttributeListener 接口提供了以下 3 个方法。

（1）attributeAdded（ServletRequestAttributeEvent event）：如果有对象加入 Request 的范围，通知正在监听的对象。

（2）attributeReplaced(ServletRequestAttributeEvent event)：如果在 Request 的范围内有对象取代另一个对象，通知正在监听的对象。

（3）attributeRemoved（ServletRequestAttributeEvent event）：如果有对象被从 Request 的范围内移除，通知正在收听的对象。

6.4.3 监听器的应用

监听器的作用是监听 Web 容器的有效事件，它由 Servlet 容器管理，利用 Listener 接口监听某个程序，并根据该程序的需求做出适当的响应。

创建监听器的步骤如下。

（1）根据应用场景，确定监听对象，编写一个继承相应接口的监听器类。

```
package cn.listen;
import javax.servlet.ServletContextEvent;
import javax.servlet.ServletContextListener;
public class MyListener implements ServletContextListener {
  public void contextDestroyed(ServletContextEvent sce) {
    //事件处理
  }
  public void contextInitialized(ServletContextEvent sce) {
    //事件处理
  }
}
```

（2）配置 web.xml。监听器的部署在 web.xml 文件中进行。在配置文件中，它的位置在过滤器的后面、Servlet 的前面。

```
<listener>
  <listener-class>cn.listen.MyListener</listener-class>
</listener>
```

【例 6-9】 在线人数计数器。

（1）设计监听器。

```
onlineCountListener.java
package onlineCountListener;
//导入包(略)
public class onlineCountListener implements HttpSessionListener {
  private int count = 0; //计数变量
  //当创建 Session 对象时，计数变量加 1
  public void sessionCreated(HttpSessionEvent se) {
    count++;
    se.getSession().getServletContext().setAttribute("onlineCount",new Integer(count));
```

```
    }
    //当销毁 Session 对象时,计数变量减 1
    public void sessionDestroyed(HttpSessionEvent se) {
        count--;
        se.getSession().getServletContext().setAttribute("onlineCount",new Integer(count));
    }
}
```

(2) 配置 web.xml。

```
<listener>
    <listener-class>onlineCountListener.onlineCountListener</listener-class>
</listener>
```

(3) 编写测试页面 onlineCount.jsp。

```
<%@ page language="java" contentType="text/html;charset=GB2312" %>
<html><head><title>监听器测试页面</title></head>
  <body>
    在线人数:<%=application.getAttribute("onlineCount") %>
  </body>
</html>
```

6.4.4 任务:在线用户的显示和用户数统计

1. 任务要求

通过监听器实现一个在线用户的显示和人数统计。

用户登录时,系统将用户名添加为 HttpSession 会话的属性。登录成功后跳转到在线首页面,显示当前登录用户的名称、所有在线用户的名称和个数。监听器能够监听 ServletContext 对象的创建和销毁。监听 HttpSession 对象相关的操作包括创建、销毁以及属性的增加、删除、修改。另外,要求把所有监听的事件日志写入 TXT 文件。

2. 任务实施

本任务中包括以下文件。

(1) UserInfoTrace.java:实现对 HTTP 会话中的属性和 ServletContext 对象的操作。

(2) UserInfoList.java:保存当前用户信息,提供对用户信息的操作。

(3) Login.jsp:用户登录页面。表单中包括一个输入用户名的文本框和一个提交按钮,提交后跳转到 showUser.jsp 页面。

(4) showUser.jsp:显示当前在线的用户。

(5) Logout.jsp:用户退出处理页面。

3. 实施步骤

(1) 创建 UserInfoList.java 类文件,用于存储在线用户和对在线用户进行具体操作。

```
package com.listener;
public class UserInfoList {
  private static UserInfoList user = new UserInfoList();
  private Vector vector = null;
  /*
    用 private 方式调用构造方法,防止被外界产生新的 instance 对象
   */
  public UserInfoList() {
    this.vector = new Vector();
  }
  /*外界使用的 instance 对象*/
  public static UserInfoList getInstance() {
    return user;
  }
  /*增加用户*/
  public boolean addUserInfo(String user) {
    if (user != null) {
      this.vector.add(user);
      return true;
    } else {
        return false;
    }
  }
  /*获取用户列表*/
  public Vector getList() {
      return vector;
  }
  /*移除用户*/
  public void removeUserInfo(String user) {
      if (user != null) {
        vector.removeElement(user);
      }
  }
}
```

(2) 创建 UserInfoTrace.java 类文件,实现 valueBound(HttpSessionBindingEvent arg0)和 valueUnbound(HttpSessionBindingEvent arg0)两个方法。当有对象加入 Session 时,valueBound()方法会自动执行。当有对象从 Session 中移除时,valueUnbound()方法会自动执行,在 valueBound()和 valueUnbound()方法中都加入了输出信息的功能,可以使用户在控制台中清楚地了解其执行过程,该文件的完整代码如下。

```
package com.listener;
import javax.servlet.http.HttpSessionBindingEvent;
public class UserInfoTrace implements javax.servlet.http.HttpSessionBindingListener {
  private String user;
  private UserInfoList container = UserInfoList.getInstance();
  public UserInfoTrace() {
    user = "";
  }
```

```
    /*设置在线监听人员*/
    public void setUser(String user) {
      this.user = user;
    }
    /*获取在线监听*/
    public String getUser() {
      return this.user;
    }
    public void valueBound(HttpSessionBindingEvent arg0) {
      System.out.println("上线" + this.user);
    }
    public void valueUnbound(HttpSessionBindingEvent arg0) {
      System.out.println("下线" + this.user);
      if (user != "") {
        container.removeUserInfo(user);
      }
    }
}
```

（3）创建 showUser.jsp 页面文件，用 setMaxInactiveInterval()方法设置 Session 的会话维持时间为 10 秒，这样可以缩短 Session 的生命周期，该页面文件的关键代码如下。

```
<%@ page contentType = "text/html; charset = gb2312" language = "java"
      import = "java.sql.*" errorPage = "" %>
<%@ page import = "java.util.*" %>
<%@ page import = "com.listener.*" %>
<html>
<head>
<meta http-equiv = "Content-Type" content = "text/html; charset = gb2312">
<title>使用监听查看在线用户</title>
<link href = "css/style.css" rel = "stylesheet" type = "text/css">
</head>
<%
  UserInfoList list = UserInfoList.getInstance();
  UserInfoTrace ut = new UserInfoTrace();
  String name = request.getParameter("user");
  ut.setUser(name);
  session.setAttribute("list",ut);
  list.addUserInfo(ut.getUser());
  session.setMaxInactiveInterval(10);
%>
<textarea rows = "8" cols = "20">
<%
  Vector vector = list.getList();
  if(vector!= null&&vector.size()> 0){
    for(int i = 0;i < vector.size();i++)
      String account = (String)vector.elementAt(i);
    account = new String(account.getBytes("ISO8859_1"), "GBK");
    //将 account 对象中的数据转换成中文
    out.println(account);
```

```
    }
  %>
  </textarea>
    <a href = "loginOut.jsp">返回</a>
  </body>
</html>
```

(4) 用户登录页面 login.jsp 的代码如下。

```
<%@ page contentType = "text/html; charset = gb2312" language = "java" import =
      "java.sql.*" errorPage = "" %>
<html>
<head>
<meta http-equiv = "Content-Type" content = "text/html; charset = gb2312">
<title>使用监听器查看在线用户</title>
</head>
<body>
    <form name = "form" method = "post" action = "showUser.jsp">
    <input type = "text" name = "user"><br><br>
    <input type = "submit" name = "Submit" value = "登录">
  </form>
</body>
</html>
```

(5) 用户退出页面 loginOut.jsp 的代码如下。

```
<%@ page contentType = "text/html; charset = gb2312" language = "java" errorPage = "" %>
<%
 session.invalidate();
 out.println("<script>parent.location.href = 'index.jsp';</script>");
 %>
```

(6) 配置 web.xml。

```
<?xml version = "1.0" encoding = "UTF-8"?>
<web-app xmlns = "http://java.sun.com/xml/ns/j2ee" xmlns:xsi = "http://www.w3.org/2001/
XMLSchema-instance" xsi:schemaLocation = "http://java.sun.com/xml/ns/j2ee http://java
.sun.com/xml/ns/j2ee/web-app_2_4.xsd" version = "2.4">
<display-name>web</display-name>
<servlet>
    <description>Added by JBuilder to compile JSPs with debug info</description>
    <servlet-name>debugjsp</servlet-name>
    <servlet-class>org.apache.jasper.servlet.JspServlet</servlet-class>
    <init-param>
      <param-name>classdebuginfo</param-name>
      <param-value>true</param-value>
    </init-param>
    <load-on-startup>3</load-on-startup>
</servlet>
<servlet-mapping>
```

```
    <servlet-name>debugjsp</servlet-name>
    <url-pattern>*.jsp</url-pattern>
  </servlet-mapping>
</web-app>
```

项目 6　用户管理系统的控制层设计

1. 项目需求

该项目实现对用户信息的管理,包括用户注册、用户登录、修改密码、修改用户信息、删除用户、查找指定用户等。本项目中,主要完成控制层的设计。

2. 项目实施

(1) 数据访问层的代码参照项目 5。

(2) 编写一个监听器,当系统启动时,加载公司信息,用于将公司信息显示在页面的页脚上。公司信息 JavaBean 的定义如下。

```
package cn.edu.mypt.bean;
public class Company {
  //getter setter 略
  public Company(){}
  public Company(String name,String tel,String address){
    this.name = name;
    this.tel = tel;
    this.address = address;
  }
  private String name;
  private String tel;
  private String address;
}
```

监听器 CompanyListener 的定义如下。

```
package cn.edu.mypt.listener;
import javax.servlet.ServletContextEvent;
import javax.servlet.ServletContextListener;
import cn.edu.mypt.bean.Company;
public class CompanyListener implements ServletContextListener {
  @Override
  public void contextDestroyed(ServletContextEvent arg0) {
    //TODO Auto-generated method stub
    arg0.getServletContext().removeAttribute("COMP");
  }
  @Override
  public void contextInitialized(ServletContextEvent arg0) {
```

```
    //TODO Auto - generated method stub
    Company comp = new Company("ABC 公司","028 - 2222091","四川成都");
    arg0.getServletContext().setAttribute("COMP", comp);
  }
```

包含页面 compinfo.jsp 用于显示公司信息，代码如下。

```
<%@ page language = "java" import = "java.util. * " pageEncoding = "utf - 8" %>
<jsp:useBean id = "company" class = "cn.edu.mypt.bean.Company" scope = "application"/>
<hr>
  公司名:<% = company.getName() %>电话:<% = company.getTel() %>
  地址: <% = company.getAddress() %>
<hr>
```

(3) 设计用户注册页面 reg.jsp，代码如下。

```
<%@ page language = "java" import = "java.util. * " pageEncoding = "utf - 8" %>
<!DOCTYPE HTML PUBLIC " - //W3C//DTD HTML 4.01 Transitional//EN">
<html>
  <head><title>用户注册</title></head>
  <body>
  <%
    String uID,uPwd,msg,uName;
    msg = (String)request.getAttribute("message");
    uID = (String)request.getAttribute("userID");
    uPwd = (String)request.getAttribute("userPwd");
    uName = (String)request.getAttribute("userName");
    uID = (uID == null)?"":uID;
    uPwd = (uPwd == null)?"":uPwd;
    msg = (msg == null)?"":msg;
  %>
    <% = msg %>
      <form action = "regServet" method = "post">
          用户 ID: <input type = "text" name = "userID" value = "<% = uID %>"/><br>
          用户名: <input type = "text" name = "userName" value = "<% = uName %>"/><br>
          用户密码: <input type = "text" name = "userPwd" value = "<% = uPwd %>"/><br>
        <input type = "submit" value = "用户注册"/>
    </form>
    <jsp:include page = "compinfo.jsp"/>
  </body>
</html>
```

(4) 编写控制层代码，采用 Servlet 实现对客户端请求的控制转移。用户注册时，用户 ID 不能重复，若用户 ID 重复提示该用户已经存在，否则跳转到 login.jsp 页面。

```
package cn.edu.mypt.servlet;
//导入包(略)
import cn.edu.mypt.bean.UserBean; //参考项目五
import cn.edu.mypt.model.UserModel;
import cn.edu.mypt.model.impl.UserModelImple;
public class RegisterServlet extends HttpServlet {
  //略
```

```
public void doGet(HttpServletRequest request, HttpServletResponse response)
throws ServletException, IOException {
  doPost(request,response);
}
public void doPost(HttpServletRequest request, HttpServletResponse response)
throws ServletException, IOException {
  request.setCharacterEncoding("GBK");
  String uId = request.getParameter("userID");
  String uPwd = request.getParameter("userPwd");
  String uName = request.getParameter("userName");
  UserModel uModel = new UserModelImple();
  UserBean user = uModel.findById(uId);
  String result = "login.jsp";
  if (user!= null) {
    request.setAttribute("message", "用户 ID 已经存在");
    result = "reg.jsp";
  }
  lse {
    user = new UserBean();
    user.setUserID(uId);
    user.setUserName(uName);
    user.setUserPwd(uPwd);
    if (!uModel.userReg(user) ) {
      request.setAttribute("message","用户注册失败");
      result = "reg.jsp";
    }
  }
  request.getRequestDispatcher(result).forward(request, response);
}
public void init() throws ServletException {
}
}
```

(5) 其他代码参见本书素材包。

(6) 配置 web.xml,关键代码如下。

```
<servlet>
  <servlet-name>RegisterServlet</servlet-name>
  <servlet-class>cn.edu.mypt.servlet.RegisterServlet</servlet-class>
</servlet>
<servlet-mapping>
  <servlet-name>RegisterServlet</servlet-name>
  <url-pattern>/RegisterServlet</url-pattern>
</servlet-mapping>
<listener>
  <listener-class>cn.edu.mypt.listener.CompanyListener</listener-class>
</listener>
```

(7) 启动服务器,部署项目并运行。

习　题　6

1. 选择题

(1)（　　）不是 Servlet 接口的方法。

A. doGet()　　B. doPost()　　C. init()　　D. forward()

(2) 在编写 Servlet 时需要（　　）。

A. 继承 Servlet　　B. 实现 HttpRequestServlet

C. 继承 HttpServlet　　D. 实现 HttpRequest

(3) 在 Servlet 中，HttpServletResponse 的（　　）方法用来把 HTTP 请求重定向到其他 URL。

A. sendURL()　　B. redirectURL()

C. sendRedirect()　　D. redirectResponse()

(4) Servlet 程序的入口点是（　　）。

A. init()　　B. main()　　C. service()　　D. doGet()

(5) 编写一个过滤器，需要（　　）。

A. 继承 Filter 类　　B. 实现 Filter 接口

C. 继承 HttpFilter 类　　D. 实现 HttpFilter 接口

(6) 有关 Servlet 的生命周期说法错误的有（　　）。

A. Servlet 的生命周期由 Servlet 实例控制

B. init()方法在创建完 Servlet 实例后对其进行初始化，传递的参数为实现 ServletContext 接口的对象

C. service()方法响应客户端发出的请求

D. destroy()方法释放 Servlet 实例

(7) 在 Servlet 中，response. getWriter()返回的是（　　）。

A. JspWriter 对象　　B. PrintWriter 对象

C. out 对象　　D. ResponseWriter 对象

(8) 在 web. xml 中使用（　　）标签配置过滤器。

A. <filter>和<filter-mapping>　　B. <filter-name>和<filter-class>

C. <filter>和<filter-class>　　D. <filter-pattern>和<filter>

(9) 在访问 Servlet 时，在浏览器地址栏中输入的路径是用（　　）标签配置的。

A. <servlet-name>　　B. <servlet-mapping>

C. <uri-pattern>　　D. <url-pattern>

(10) 在编写过滤器时，必须完成的方法是（　　）。

A. doFilter()　　B. doChain()　　C. doPost()　　D. doDelete()

(11) 过滤器使用（　　）方法才能继续传递到下一个过滤器。

A. request. getRequestDispatcher. forward(request，response)；

B. doFilter()

C. doPut()

D. doChain()

2. 简答题

(1) 简述 Servlet 的生命周期。

(2) 编写一个监听器常用的接口有哪些?

(3) 简述过滤器需要的接口以及过滤器的生命周期。

3. 编程题

(1) 编写一个 Servlet ,模拟验证码的功能,随机生成 4 位验证码,用户输入验证码后,进行校验,输入错误,要求重新输入,输入正确,跳转到正常页面。

(2) 编写一个登录验证过滤器,根据用户的类型和权限,确定用户跳转的页面。

(3) 编写一个监听器,当用户退出时,将用户从用户列表中移除,用户列表用于显示当前在线的用户。

第7章 EL 和 JSTL

通过前面的学习，我们可以开发出具有三层架构的 Web 应用程序，但是如果表示层需要和业务逻辑层交互，仍然需要在 JSP 页面嵌入很多 Java 代码，不利于对表示层的维护和更新。本章讨论使用 JSTL(JSP 标准标签库)和 EL(表达式语言)实现无 Java 代码嵌入的 JSP 页面开发，能很好地实现 Java 代码 与 HTML 的分离。

【技能目标】 能够使用 EL 和 JSTL 简化页面开发。

【知识目标】 EL 表达式；常用的 JSTL 标签。

【关键词】 表达式(expression)　标准(standard)　前缀(prefix)　库(library)　标签(tag)

7.1 EL 表达式

7.1.1 表达式语言简介

1. 为什么要使用表达式语言

在早期的 JSP 中，为了实现与用户的动态交互，或者控制页面输出，需要在 JSP 页面中嵌入大量的 Java 代码。另外，在 JSP 中如果用嵌入 Java 代码的方式访问一个 JavaBean 的属性，需要调用该属性的 getter 方法。如果访问的属性是 String 类型或者其他的基本数据类型，可以比较方便地达到目的。但是如果该属性属于另外一个 JavaBean 对象，就需要多次调用 getter 方法，有时还需要做强制类型转换。不难发现，在 JSP 中嵌入 Java 代码不仅看起来结构混乱，而且导致程序可读性差，不易维护，JSP 2.0 引入了 EL 表达式。使用 EL 表达式，页面代码可以大大简化。

EL 语法很简单，它最大特点就是使用方便。它提供了一些标识符、存取器和运算符，用来检索和操作驻留在 JSP 容器中的数据。EL 在某种程度上以 EcmaScript 和 XML 路径语言(XML path language，XPath)为基础，因此页面设计人员和程序员都应该熟悉它的语法。它不是编程语言，甚至不是脚本编制语言，与 JSTL 标签一起使用时，能使用简单而又方便的符号来表示复杂的行为。

2. EL 表达式语法

1) 语法结构

EL 表达式的主要语法结构如下。

```
${expression}
```

所有 EL 表达式都是以 ${ 为起始、以 } 为结尾的。例如：

```
User user = (User)session.getAttribute("user");
String sex = user.getSex();
```

相当于

```
${sessionScope.user.sex}
```

上述 EL 示例的意思是：从 Session 的范围内取得用户的性别。

两者相比之下，可以发现 EL 的语法更为方便、简洁。

【例 7-1】 使用 EL 表达式。

```
<%@ taglib uri = "http://java.sun.com/jstl/core_rt" prefix = "c" %>
<html>
    <head><title>EL 表达式示例</title></head>
    <% pageContext.setAttribute("color","#FFFFCC"); %>
    <body bgcolor = '${pageScope.color}'>
        <h1>变化的背景色</h1>
    </body>
</html>
```

2) [ ]与. 运算符

EL 提供. 和[ ]两种运算符来存取数据。${sessionScope.user.sex}等价于${sessionScope.user["sex"]}。

. 和[ ]也可以同时混合使用，如${sessionScope.shoppingCart[0].price}。

返回结果为 shoppingCart 中第一项物品的价格。

但在以下两种情况下，两者会有差异。

(1) 当要存取的属性名称中包含一些特殊字符，如. 或-等并非字母或数字的符号，就一定要使用[]，例如，${user.My-Name}应当改为${user["My-Name"]}

(2) 动态取值采用[]运算符。例如，${sessionScope.user[data]}中，data 是一个变量，假若 data 的值为"sex"，那它等价于${sessionScope.user.sex}；假若 data 的值为"name"，它就等价于${sessionScope.user.name}。

【例 7-2】 使用 EL 表达式获取集合元素。

```
<%@ page language = "java" import = "java.util.*" pageEncoding = "utf-8" %>
<!DOCTYPE HTML PUBLIC "-//W3C//DTD HTML 4.01 Transitional//EN">
<html>
  <head><title>集合元素</title></head>
    <body>
      <%
          Map names = new HashMap();
          names.put("one","LiYang");
          names.put("two","WangHua");
          request.setAttribute("names",names);
          List all = new ArrayList() ;                //实例化 List 接口
          all.add("zhangsn") ;                        //向集合中增加内容
          all.add("lisi");                            //向集合中增加内容
          request.setAttribute("allinfo",all);        //向 request 集合中保存
```

```
    %>
    <h3>第一个 List 元素: ${allinfo[0]}</h3>
    <h3>第二个 List 元素: ${allinfo[1]}</h3>
    <h3>Map 元素 one: ${names.one}</h3>
    <h3>Map 元素 Two: ${names["two"] }</h3>
</body>
</html>
```

运行结果如图 7-1 所示。

图 7-1 例 7-2 运行结果

3）变量

EL 存取变量数据的方法很简单,例如,${username}的意思是取出某一范围内名称为 username 的变量。

【例 7-3】 利用 EL 表达式显示用户信息。

```
<jsp:useBean id="person" class="cn,.edu.mypt.bean.UserBean"/>
  <body>
    欢迎您,${uName}.
    <!-- 相当于 request . getAttribute ("uName")或 session.getAttribute ("uName") -->
     ${user.userName }, <!-- 相当于 user.getUsername() -->
    <!-- 也相当于<jsp:getProperty name="user" property "userName" /> -->
  </body>
```

4）自动转变类型

EL 除了提供方便存取变量的语法之外,还能自动转变类型,例如:

```
${param.count + 20}
```

假若窗体传来 count 的值为 10,那么上面的结果为 30。

3. 在 JSP 中禁用 EL 表达式的执行

在 web.xml 的<jsp-property-group>中可以控制一组 JSP 页面是否使用 EL,在每个 JSP 页面中也可以指定是否使用 EL。page 命令的 isELIgnored 属性用来指定是否忽略,格式如下。

```
<%@ page isELIgnored="true|false" %>
```

如果设定为真,那么 JSP 中的表达式被当成字符串处理。比如表达式<p>${2000 % 20}</p>在 isELIgnored="true"时输出为${2000 % 20},而 isELIgnored="false"时输

出为 100。Web 容器默认 isELIgnored="false"。

【例 7-4】 在 JSP 中禁用 EL 表达式。

```
<%@ page language="java" import="java.util.*" pageEncoding="utf-8"%>
<%@page isELIgnored="flase" %>
<!DOCTYPE HTML PUBLIC "-//W3C//DTD HTML 4.01 Transitional//EN">
<html>
  <head><title>EL 启用</title></head>
      <body>
        <h1>启用表达式语言</h1>
        <form method="post" action="expressionexample.jsp">
            <b>${'First Name: '}<input type="text"
            value="${'请输入您的名字'}" /></b>
            <input type="submit" name="Submit" value="${'提交表单'}">
        </form>
    </body>
</html>
```

7.1.2　表达式与内置对象

JSP 有 9 个隐式对象,而 EL 也有自己的隐式对象。EL 隐式对象总共有 11 个,如图 7-2 所示。

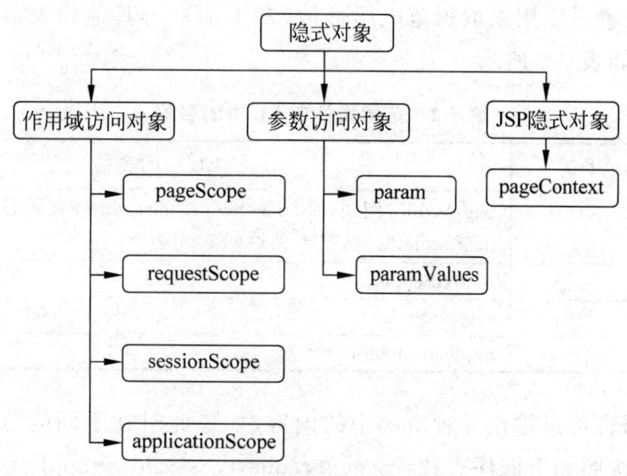

图 7-2　隐式对象

EL 表达式不仅可以读取 request、session 对象中的属性,还可以读取其他 JSP 隐式对象的属性。隐式对象及其说明如表 7-1 所示。

表 7-1　隐式对象及其说明

隐式对象	类　型	说　明
pageContext	javax.servlet.ServletContext	表示此 JSP 页面的上下文
pageScope	java.util.Map	取得 Page 范围的属性名称所对应的值
requestScope	java.util.Map	取得 Request 范围的属性名称所对应的值

续表

隐式对象	类 型	说 明
sessionScope	java.util.Map	取得 session 范围的属性名称所对应的值
applicationScope	java.util.Map	取得 application 范围的属性名称所对应的值
param	java.util.Map	同 ServletRequest.getParameter(String name)。回传 String 类型的值
paramValues	java.util.Map	同 ServletRequest.getParameterValues(String name)。回传 String[]类型的值
header	java.util.Map	同 ServletRequest.getHeader(String name)。回传 String 类型的值
headerValues	java.util.Map	同 ServletRequest.getHeaders(String name)。回传 String[]类型的值
cookie	java.util.Map	如同 HttpServletRequest.getCookies()
initParam	java.util.Map	同 ServletContext.getInitParameter(String name)。回传 String 类型的值

1. 范围属性

与范围有关的 EL 隐式对象包含以下四个：pageScope、requestScope、sessionScope 和 applicationScope，它们基本上和 JSP 的 pageContext、request、session 和 application 一样。这 4 个隐式对象只能用来取得范围属性值，却不能取得其他相关信息。范围属性在 EL 中表示的名称如表 7-2 所示。

表 7-2 范围属性在 EL 中的名称

范围(JSTL 中的名称)	EL 中的名称
page	pageScope，例如，${pageScope.username}表示在 page 范围内查找 username 变量，找不到返回 Null
request	requstScope
session	sessionScope
application	applicationScope

使用表达式语言可以输出 4 种范围中的内容，如果此时在不同的范围中设置了同一个属性名称，则将按照如下顺序查找：page→request→session→application。

【例 7-5】 设置同名属性。

```
<%@ page language="java" import="java.util.*" pageEncoding="utf-8"%>
<!DOCTYPE HTML PUBLIC "-//W3C//DTD HTML 4.01 Transitional//EN">
<html>
  <head><title>设置同名属性</title></head>
    <body>
    <%
      pageContext.setAttribute("info","page 范围");          //设置一个 page 属性
      request.setAttribute("info","request 范围") ;          //设置一个 request 属性
      session.setAttribute("info","session 范围");           //设置一个 session 属性
```

```
    application.setAttribute("info","application 范围");  //设置一个 application 属性
  %>
  <h3>${info}</h3><!-- 表达式输出 -->
 </body>
</html>
```

程序的运行结果如图 7-3 所示。

http://localhost:8080/mytest/setAttribute.jsp

page范围

图 7-3 例 7-5 运行结果

因为没有指定哪一个范围的 info，所以它会依序从 page、request、session、application 范围查找。假如中途找到 info，就直接返回，不再继续找下去，但是假如全部范围都没有找到时，就返回 null。

可以指定一个要取出属性的范围，示例如表 7-3 所示。

表 7-3 取出属性的范围示例

示　　例	说　　明
${pageScope.username}	取出 page 范围的 username 变量
${requestScope.username}	取出 request 范围的 username 变量
${sessionScope.username}	取出 session 范围的 username 变量
${applicationScope.username}	取出 application 范围的 username 变量

其中，pageScope、requestScope、sessionScope 和 applicationScope 都是 EL 的隐式对象，由它们的名称可以很容易猜出它们所代表的意思，例如，${sessionScope.username} 可以取出会话范围的 username 变量。这种写法比之前 JSP 的写法：

```
String username = (String) session.getAttribute("username");
```

要容易、简洁许多。

【例 7-6】 指定取出范围的属性。

```
<%@ page language="java" import="java.util.*" pageEncoding="utf-8"%>
<!DOCTYPE HTML PUBLIC "-//W3C//DTD HTML 4.01 Transitional//EN">
<html>
  <head><title>指定取出范围的属性</title></head>
   <body>
  <%
    pageContext.setAttribute("info","page 范围"); //设置一个 page 属性
    request.setAttribute("info","request 范围"); //设置一个 request 属性
    session.setAttribute("info","session 范围"); //设置一个 session 属性
    application.setAttribute("info" ,"aplication 范围");//设置一个 application 属性
  %>
    <h3>PAGE 属性的范围: ${pageScope.info}</h3>
    <h3>REQUEST 属性的范围: ${requestScope.info}</h3>
    <h3>SESSION 属性的范围: ${sessionScope.info}</h3>
    <h3>APPLICATION 属性的范围: ${applicationScope.info}</h3>
  </body>
</html>
```

程序的运行结果如图 7-4 所示。

此时，由于已经指定了范围，所以可以取出不同范围内的同名属性。

request 对象除可以存取属性之外，还可以取得用户的请求参数或表头信息等。但是在 EL 中，它就只能单纯地用来取得对应范围的属性。

http://localhost:8080/mytest/getAttribute.jsp

PAGE属性的范围：page范围

REQUEST属性的范围：request范围

SESSION属性的范围：session范围

APPLICATION属性的范围：aplication范围

图 7-4　例 7-6 运行结果

2. 接收请求参数

使用表达式语言还可以显示接收的请求参数，功能与 request. getParameter()类似，语法如下。

```
${param.参数名称}
```

【例 7-7】 接收参数。

```
<%@ page language="java" import="java.util.*" pageEncoding="utf-8"%>
<!DOCTYPE HTML PUBLIC "-//W3C//DTD HTML 4.01 Transitional//EN">
<html>
  <head><title>接收参数</title></head>
  <body>
    <h3>通过内置对象接收输入参数: <%=request.getParameter("name")%></h3>
    <h3>通过表达式语言接收输入参数: ${param.name}</h3>
  </body>
</html>
```

本程序同时使用了 request 对象和表达式语言两种方式显示传递的参数。运行时，输入 http:/localhost: 8080/mytest/getparam. jsp? name = zhangsan，则运行结果如图 7-5 所示。

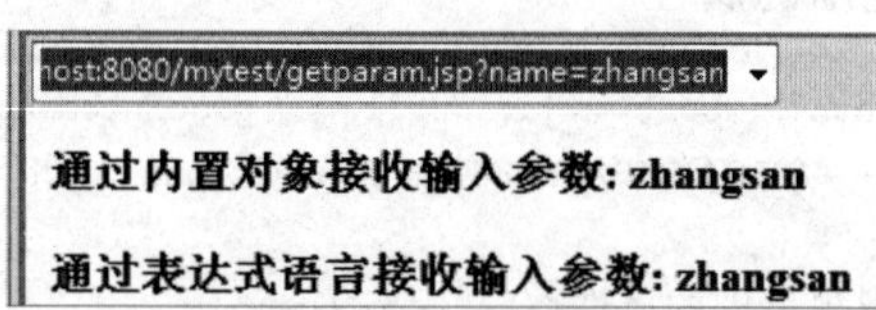

图 7-5　例 7-7 运行结果

以上传递的是一个单独的参数，如果现在传递的是一组参数，则格式如下。

```
${paramValues.参数名称}
```

需要注意的是，接收一组参数时，如果想要取出，则需要分别指定下标。

【例 7-8】 获取多个参数。

form. jsp 的代码如下。

```
<%@ page language="java" import="java.util.*" pageEncoding="utf-8"%>
<!DOCTYPE HTML PUBLIC "-//W3C//DTD HTML 4.01 Transitional//EN">
<html>
```

```
  <head><title>获取多个参数</title></head>
    <body>
      <form action="getparam_value.jsp" method="post">
        兴趣:
        <input type="checkbox" name="inst" value="唱歌">唱歌
        <input type="checkbox" name="inst" value="游泳">游泳
        <input type="checkbox" name="inst" value="旅游">旅游
        <input type="submit" value="提交">
      </form>
    </body>
</html>
```

getparam_value.jsp 的代码如下。

```
<%@ page language="java" import="java.util.*" pageEncoding="utf-8"%>
<!DOCTYPE HTML PUBLIC "-//W3C//DTD HTML 4.01 Transitional//EN">
<html>
  <head><title>获取多个参数</title></head>
    <body>
      <% request.setCharacterEncoding("utf-8"); %>
      <h3>第 1 个参数: ${paramValues.inst[0]}</h3>
      <h3>第 2 个参数: ${paramValues.inst[1]}</h3>
      <h3>第 3 个参数: ${paramValues.inst[2]}</h3>
      </body>
</html>
```

程序的运行结果如图 7-6 所示。

图 7-6 例 7-8 运行结果

3. Cookie

Cookie 通常存在于浏览器的暂存区内。JSTL 并没有提供设置 Cookie 的动作,因为这个动作通常是后端开发者必须去做的事情,而不是交给前端的开发者。假若在 Cookie 中设置一个名称为 userCountry 的值,那么可以使用 ${cookie.userCountry}来取得它。

4. header 和 headerValues

header 储存用户浏览器和服务器用来交流的数据。当用户请求服务器端的网页时会送出一个包含请求信息的文件,这些信息包括用户浏览器的版本、用户计算机所设定的区域等。假若要取得用户浏览器的版本,可使用表达式 ${header["User-Agent"]}。另外,如果同一 header 名称拥有不同的值,此时必须使用 headerValues 来取得这些值。

注意:因为 User-Agent 中包含"-"这个特殊字符,所以必须使用[],而不能写成 $(header.User-Agent)。

5. initParam

就像其他属性一样,可以自行设定 Web 站点的环境参数,例如:

```
<?xml version = "1.0" encoding = "ISO - 8859 - 1"?>
< web - app xmlns = "http://java.sun.com/xml/ns/j2ee"
xmlns:xsi = "http://www.w3.org/2001/XMLSchema - instance"
xsi:schemaLocation = "http://java.sun.com/xml/ns/j2ee/web - app_2_4.xsd"
version = "2.4">
< context - param >
    < param - name > userid </param - name >
    < param - value > mike </param - value >
</context - param >
</web - app >
```

可以直接使用 ${initParam. userid}取得 userid 参数的值,即 mike。下面是之前的方法。

```
String userid = (String)application.getInitParameter("userid");
```

在使用 param 和 paramValues 取得用户参数时通常使用以下方法。

```
request.getParameter(String name)
request.getParameterValues(String name)
```

在 EL 中可以使用 param 和 paramValues 两者来取得数据,格式如下。

```
${param.name}
${paramValues.name}
```

其中,param 的功能和 request. getParameter(String name)相同,而 paramValues 和 request. getParameterValues(String name)相同。

6. pageContext

可以使用${pageContext}来取得其他有关用户请求或页面的详细信息。表 7-4 列出了几个比较常用的表达式。

表 7-4 pageContext 常用表达式

表 达 式	说 明
${pageContext. request. queryString}	取得请求的参数字符串
${pageContext. request. requestURL}	取得请求的 URL,但不包括请求的参数字符串,即 Servlet 的 HTTP 地址
${pageContext. request. contextPath}	服务的 Web 应用的名称
${pageContext. request. method}	取得 HTTP 的方法(Get、Post)
${pageContext. request. protocol}	取得使用的协议(HTTP/1.1、HTTP/1.0)
${pageContext. request. remoteUser}	取得用户名称
${pageContext. request. remoteAddr}	取得用户的 IP 地址

续表

表 达 式	说 明
${pageContext.session.new}	判断 session 是否为新的,所谓新的 session,表示刚由 Server 创建客户端尚未使用
${pageContext.session.id}	取得 session 的 ID
${pageContext.servletContext.serverInfo}	取得主机端的服务信息

7.1.3 EL 表达式运算

为了方便用户操作,EL 定义了许多算术运算符、关系运算符、逻辑运算符等,使用这些运算符可使 JSP 页面更加简洁。但是对于过于复杂的操作还是应该在 Servlet 或 JavaBean 中完成。在使用这些运算符时,所有的操作内容都可以直接使用设置的属性,而不用考虑类型转换的问题。

如果需要在支持表达式语言的页面中正常输出 $ 符号,可在 $ 符号前加转义字符\,因为系统默认 $ 是表达式语言的特殊标记。

JSP 表达式语言提供以下操作符,其中大部分是 Java 中常用的运算符,如表 7-5 所示。

表 7-5 表达式语言的运算符

运 算 符	定 义
算术运算符	+、-(二元)、*、/、div、%、mod、-(一元)
逻辑运算符	and、&&、or、||、!、not
关系运算符	==、eq、!=、ne、>、gt、<=、le、<=、ge。可以与其他值进行比较,或与布尔型、字符串型、整型或浮点型数据进行比较
empty	空操作符是前缀,可用于确定值是否为空
?:运算符	格式为 A ? B :C。根据 A 的结果选择 B 或 C

1. 算术运算符

EL 的算术运算符如表 7-6 所示。

表 7-6 算术运算符

算术运算符	说明	范例	结果
+	加法	${3+2}	5
-	减法	${3-2}	1
*	乘法	${3*2}	6
/或 div	除法	${3/2}	1.5
%或 mod	取模	${3%2}	1

【例 7-9】 算术运算符。

```
<%@ page language = "java" import = "java.util.*" pageEncoding = "utf-8" %>
<!DOCTYPE HTML PUBLIC " - //W3C//DTD HTML 4.01 Transitional//EN">
```

```
<html>
  <head><title>My JSP 'el.jsp' starting page</title></head>
    <body>
      <% //设置 page 范围的属性,基本数据类型自动变为包类
        pageContext.setAttribute("num1",3) ;
        pageContext.setAttribute("num2", 3) ;
      %>
      <h3>加法操作: ${num1 + num2}</h3>
      <h3>减法操作: ${num1 - num2}</h3>
      <h3>乘法操作: ${num1 * num2}</h3>
      <h3>除法操作: ${num1 / num2}和${num1 div num2}</h3>
      <h3>取模操作: ${num1 % num2}和${num1 mod num2}</h3>
    </body>
</html>
```

程序的运行结果如图 7-7 所示。

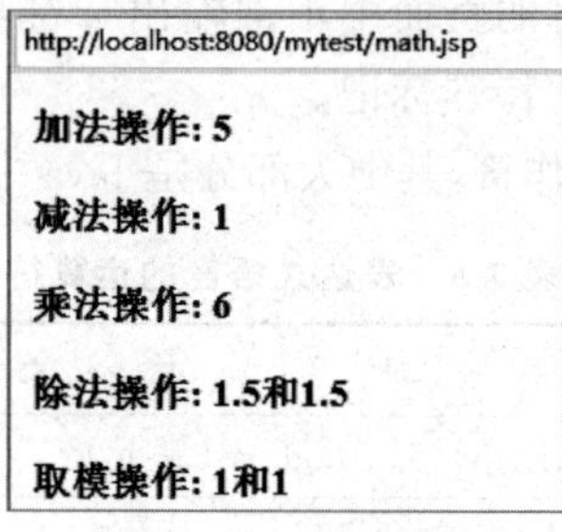

图 7-7 例 7-9 运行结果

2. 关系运算符

EL 的关系运算符如表 7-7 所示。

表 7-7 EL 关系运算符

关系运算符	说明	范 例	结果
==(eq)	等于	${5==5}或${5eq5}	true
!=(ne)	不等于	${5! =5}或${5ne5}	false
<(lt)	小于	${3<5}或${3lt5}	true
>(gt)	大于	${3>5}或{3gt5}	false
<=(le)	小于或等于	${3<=5}或${3le5}	true
>=(ge)	大于或等于	${3>=5}或${3ge5}	false

表达式语言不仅可在数字与数字之间比较,还可在字符与字符之间比较,字符串是根据其对应 Unicode 值来比较大小的。

注意,在使用 EL 关系运算符时,不能写成

```
${param.password1} == ${param.password2}
```

或

```
${ ${param.password1 } == ${ param.password2 } }
```

而应写成

```
${ param.password1 == param.password2 }
```

【例 7-10】 关系运算符。

```
<%@ page language = "java" import = "java.util. * " pageEncoding = "utf - 8" %>
<!DOCTYPE HTML PUBLIC " - //W3C//DTD HTML 4.01 Transitional//EN">
<html>
  <head><title>关系表达式</title></head>
  <body>
    <%
      pageContext.setAttribute("num1", 20) ;
      pageContext.setAttribute("num2", 30);
     %>
    <h3>相等判断 ${num1} == ${num2}: ${num1 == num2}和 ${num1 eq num2}</h3>
    <h3>不等判断 ${num1}!= ${num2}: ${num1!= num2}和 ${num1 ne num2}</h3>
    <h3>小于判断 ${num1}< ${num2}: ${num1 < num2}和 ${num1 gt num2}</h3>
    <h3>大于判断 ${num1}> ${num2}: ${num1 > num2}和 ${num1 gt num2}</h3>
    <h3>小于或等于判断 ${num1}<= ${num2}: ${num1 <= num2}和 ${num1 le num2}</h3>
    <h3>大于或等于判断 ${num1}>= ${num2}: ${num1 >= num2}和 ${num1 ge num2}</h3>
  </body>
</html>
```

程序的运行结果如图 7-8 所示。

图 7-8　例 7-10 运行结果

3. 逻辑运算符

EL 的逻辑运算符如表 7-8 所示。

表 7-8　EL 逻辑运算符

逻辑运算符	说明	范　例	结果
&&(and)	与运算	${A && B}或${A and B}	true/false
||(or)	或运算	${A \|\| B}或${A or B}	true/false
!(not)	非运算	${! A }或${not A}	true/false

【例 7-11】 逻辑运算符。

```
<%@ page language = "java" import = "java.util.*" pageEncoding = "utf-8" %>
<!DOCTYPE HTML PUBLIC "-//W3C//DTD HTML 4.01 Transitional//EN">
<html>
  <head><title>逻辑运算</title></head>
  <body>
    <%
    pageContext.setAttribute("flagA", true) ;
    pageContext.setAttribute("flagB", false);
   %>
    <h3>与操作: ${flagA && flagB}和 ${flagA and flagB}</h3>
    <h3>或操作: ${lagA || flagB}和 ${fagA or flagB}</h3>
    <h3>非操作: ${!flagA}和 ${not flagA}</h3>
  </body>
</html>
```

程序的运行结果如图 7-9 所示。

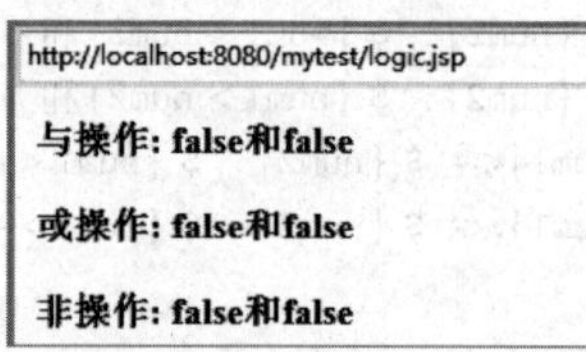

图 7-9 例 7-11 运行结果

4. 其他运算符

除了以上的运算符之外，在表达式语言中还有表 7-9 所示的其他运算符。

表 7-9 其他运算符

运算符	说 明	范 例	结果
empty	判空	${empty name}	true/false
?:	条件运算符	${a>b? a:b}	大值
()	括号运算符	${2*(3+4)}	14

【例 7-12】 其他运算符。

```
<%@ page language = "java" import = "java.util.*" pageEncoding = "utf-8" %>
<!DOCTYPE HTML PUBLIC "-//W3C//DTD HTML 4.01 Transitional//EN">
<html>
  <head><title>其他运算符</title></head>
  <body>
    <%
      pageContext.setAttribute("x", 1) ;
      pageContext.setAttribute("y", 2) ;
      pageContext.setAttribute("z", 3);
    %>
    <h3>empty 操作: ${empty name}</h3>
```

```
    <h3>三目操作: ${x>y?"大于":"小于"}</h3>
    <h3>括号操作: ${x* (y+ z)}</h3>
  </body>
</html>
```

程序的运行结果如图 7-10 所示。

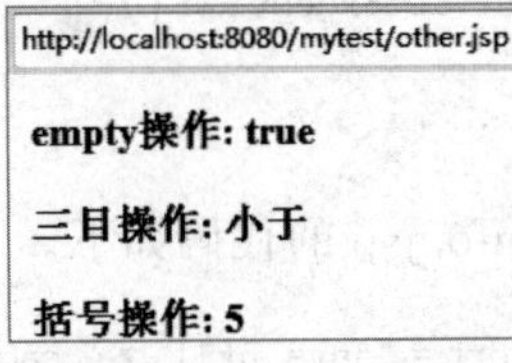

图 7-10　例 7-10 运行结果

7.1.4　任务：查找显示用户信息

1. 任务要求

利用项目 5 的业务逻辑层和数据访问层代码，根据用户 ID 查找用户的基本信息并显示。

2. 任务实施

(1) 编写 Servlet，接收表单参数 userID，根据 userID 通过 UserModelImple 类的 FindByID()方法查找用户，若存在，跳转到用户显示页面，否则跳转到查找页面。

```
package cn.edu.mypt.servlet;
//导入包,略
import cn.edu.mypt.bean.UserBean;
import cn.edu.mypt.model.UserModel;
import cn.edu.mypt.model.impl.UserModelImple;
public class FindServlet extends HttpServlet {
    //略
    public void doGet(HttpServletRequest request, HttpServletResponse response)
    throws ServletException, IOException {
        doPost(request,response);
    }
    public void doPost(HttpServletRequest request, HttpServletResponse response)
    throws ServletException, IOException {
        request.setCharacterEncoding("utf-8");
        String userID = request.getParameter("uID");
        UserModel uM = new UserModelImple(); //参考项目5代码
        UserBean user = uM.findById(userID);
        String result;
        if (user == null){
            result = "findByID.jsp";
            request.setAttribute("Message","用户不存在");
        }
```

```
        else{
            result = "userInfo.jsp";
            request.setAttribute("USER", user);
        }
        request.getRequestDispatcher(result).forward(request, response);
    }
    public void init() throws ServletException {
    }
}
```

（2）显示用户信息页面 userInfo.jsp 的代码如下。

```
<%@ page language="java" import="java.util.*" pageEncoding="utf-8"%>
<!DOCTYPE HTML PUBLIC "-//W3C//DTD HTML 4.01 Transitional//EN">
<html>
  <head><title>用户信息</title></head>
    <body>
      <h1>用户信息</h1>
       用户 ID: ${USER.userID}<br>
       用户名: ${USER.userName}<br>
       用户密码: ${USER.userPwd}<br>
       用户类型: ${USER.userType}
    </body>
</html>
```

（3）查找表单 findByID.jsp 的代码如下。

```
<%@ page language="java" import="java.util.*" pageEncoding="utf-8"%>
<!DOCTYPE HTML PUBLIC "-//W3C//DTD HTML 4.01 Transitional//EN">
<html>
 <head><title>查找用户信息</title> </head>
  <body>
    <h1>查找用户信息</h1>
     ${Message}
    <form action="FindServlet" method="post">
      用户 ID:<input type="text" name="uID" />
      <input type="submit" value="查找"/>
    </form>
  </body>
</html>
```

（4）配置 web.xml。

```
<servlet>
  <servlet-name>FindServlet</servlet-name>
  <servlet-class>cn.edu.mypt.servlet.FindServlet</servlet-class>
</servlet>
<servlet-mapping>
  <servlet-name>FindServlet</servlet-name>
  <url-pattern>/FindServlet</url-pattern>
</servlet-mapping>
```

(5) 启动服务器,部署项目并运行。

7.2 JSTL 标 签

7.2.1 JSTL 简介

JSTL(Java server pages standard tag library,JSP 标准标签库)是 Sun 公司发布的一个针对 JSP 开发的新组件,它允许使用标签开发 JSP 页面。JSTL 是一个实现 Web 应用程序中常见的通用功能的定制标签库集,这些功能包括遍历和条件判断、数据格式化、XML 操作以及数据库访问。

1. 分类

JSTL 分为五大类,如表 7-10 所示。

表 7-10 JSTL 的分类

类 别	前置名称	URI(统一资源标识符)	范 例
核心标签库	c	http://java.sun.com/jsp/jstl/core	<c:out>
格式标签库	fmt	http://java.sun.com/jsp/jstl/fmt	<fmt:formatDate>
SQL 标签库	sql	http://java.sun.com/jsp/jstl/sql	<sql:query>
XML 标签库	xml	http://java.sun.com/jsp/jstl/xml	<x:forEach>
函数标签库	fn	http://java.sun.com/jsp/jstl/functions	<fn:split>

核心标签库提供了定制操作,通过限制作用域的变量管理数据,以及执行页面内容的遍历和条件操作。它还提供了用来生成和操作 URI 的标签。

格式标签库提供了格式化数据(尤其是数字和日期)的操作。它还支持使用本地化资源来进行 JSP 页面的国际化。

XML 标签库用来操作通过 XML 表示的数据。

SQL 标签库提供了用来查询关系数据库的操作。

2. 配置 JSTL

JSTL 现在已经是 Java EE 的一个组成部分,如果采用支持 Java EE 或更高版本的集成开发环境开发 Web 应用程序,就不需要再配置 JSTL 了。

如果用的是 Eclipse 平台,则需要配置 JSTL。配置 JSTL 的步骤如下。

(1) 复制 JSTL 的标准实现。在 Tomcat 的\webapps\examples\WEB-INF\lib 目录下找到 taglibs-standard-impl-1.2.5.jar 和 taglibs-standard-spec-1.2.5.jar 文件,然后复制到 Web 项目的 WEB-INF\lib 目录下。

(2) 使用<taglib>标签定义前缀与 URI 引用。如果要使用核心标签库和函数标签库,需要在 JSP 页面中使用<taglib>标签定义前缀与 URI 引用,代码如下。

```
<%@taglib prefix="c" uri="http://java.sun.com/jsp/jstl/core" %>
<%@taglib prefix="fn" uri="http://java.sun.com/jsp/jstl/functions" %>
```

7.2.2 核心标签库

核心标签库主要有表达式操作、流程控制、遍历操作和 URL 操作，如表 7-11 所示。

表 7-11 核心标签库的分类

类　别	标 签 名 称
表达式操作	out、set、remove、catch
流程控制	if、choose、when、otherwise
迭代操作	foreach、forTokens
URL 操作	import、param、url、redirect

1. 表达式操作

表达式操作包含 4 个标签：<c:out>、<c:set>、<c:remove>和<c:catch>。

1) <c:out>标签

<c:out>标签用于显示输出的内容，与<%=表达式%>或 ${表达式}类似，格式如下。

```
<c:out value = "输出的内容" [default = "defaultValue"]/>
```

或

```
<c:out value = "输出的内容">
    defaultValue
</c:out>
```

其中，value 可以是一个 EL 表达式，也可以是一个字符串；default 可有可无，当 value 不存在时输出 defaultValue。<c:out>标签的属性如表 7-12 所示。

表 7-12 <c:out>标签的属性

属　性	描　　述	EL 类型	必要否	默认值
value	要输出的内容	String	是	无
default	输出的默认值	String	否	body
escapeXml	是否忽略 XML 特殊字符	bool	否	true

【例 7-13】 使用 out 标签输出数据。

```
<%@ page language = "java" contentType = "text/html; charset = UTF - 8" pageEncoding =
"UTF - 8" %>
<%@ taglib uri = "http://java.sun.com/jsp/jstl/core" prefix = "c"  %>
<!DOCTYPE HTML PUBLIC " - //W3C//DTD HTML 4.01 Transitional//EN">
<html>
  <head><title>out 标签</title></head>
    <body>
      Hello<c:out value = "world!"/>
      <!-- 这表示输出一个字符串,结果将打印:Hello world! -->
      <c:out value = "&lt 要显示的数据对象(未使用转义字符)&gt" escapeXml =
          "true" default = "默认值"></c:out><br/>
```

```
        <c:out value="&lt 要显示的数据对象(使用转义字符)&gt" escapeXml=
        "false" default="默认值"></c:out><br/>
        <c:out value="${1+3}"/><br>
        <c:out value="${param.name}" default="欢迎你!"/><br>
        <c:out value="${param.name}">欢迎你</c:out>
    </body>
</html>
```

程序的运行结果如图 7-11 所示。

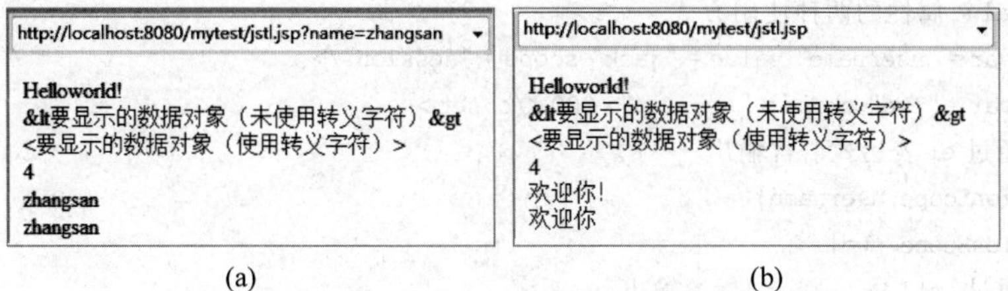

(a) (b)

图 7-11 例 7-13 运行结果

2) <c:set>标签

<c:se>标签主要用来将属性保存在 4 种范围对象中,语法如下。

```
<c:set var="属性名称" value="属性"
  [scope="[page | request |session|application]"]/>
```

或

```
<c:set var="属性名称" [scope="[page | request | session | application]"]>
  属性
</c:set>
```

设置属性到对象的格式如下。

```
<c:set value="属性" target="属性名称 property="属性名称"/>
```

或

```
<c:set target="属性名称" target="属性名称">
  属性
</c:set>
```

<c:set>标签的属性如表 7-13 所示。

表 7-13 <c:set>标签的属性

名 称	说 明	EL 类型	必要否	默认值
value	要存储的值	Object	否	无
var	欲存入的变量名称	String	否	无
scope	var 变量的 JSP 范围	String	否	page
target	要修改的属性所属的对象	String	否	无
property	要修改的属性	Object	否	无

如果指定了 target 属性，那么 property 属性也需要指定。

使用< c:set >时，var 主要用来存放表达式的结果；scope 则是用来设定储存的范围，例如，若 scope="session"，则会把数据储存在 session 对象中。如果< c:set >中没有指定 scope，则它会默认储存在 page 范围里。

【例 7-14】 将变量设置到 JSP 范围对象并输出。

```
<%--将变量定义在 JSP 范围内--%>
<%--value 属性的两种使用方式--%>
<c:set var="username" value="jack" scope="session"/>
<c:set var="pwd" scope="session">000</c:set>
<%--通过 el 表达式语言输出--%>
${sessionScope.username}
${sessionScope.pwd}
<%--通过 jstl 中<c:out>标签输出--%>
<c:out value="${sessionScope.username}"/>
```

【例 7-15】 将变量设置到 JavaBean 对象并输出。

```
<jsp:useBean id="stu" class="net.pcedu.student"/>
<%--通过<c:set>标签给 javaBean 对象的 age 属性设值--%>
<c:set value="16" target="${stu}" property="age"/>
<%--输出 javaBean 对象的属性值--%>
年龄:<c:out value="${stu.age}"/>
```

注意：var 和 scope 这两个属性不能使用表达式来表示，以下是错误的写法。

```
<c:set var="${username}" scope="${ourScope}" value="${1+1}"/>
```

以下是正确的写法。

```
<c:set var="number" scope="session" value="${1+1}"/>
```

【例 7-16】 < c:set >标签的应用。

```
<%@ page language="java" import="java.util.*" pageEncoding="UTF-8"%>
<%@ page import="org.jbit.chp08.bean.User" %>
<%@ taglib uri="http://java.sun.com/jsp/jstl/core" prefix="c"%>
<!DOCTYPE HTML PUBLIC "-//W3C//DTD HTML 4.01 Transitional//EN">
<html>
  <head><title>表达式</title></head>
    <body>
      <% User user = new User();
      request.setAttribute("user", user);
      %>
      <c:set target="${user}" property="name" value="defaultName"></c:set>
      <c:out value="${user.name}" default="noUserName"></c:out>
  </body>
</html>
```

3）<c:remove>标签

<c:remove>标签用于移除一个变量，可以指定这个变量的作用域，若未指定，则默认为变量第一次出现时所在的作用域。这个标签不是特别有用，不过可以用来确保 JSP 完成清理工作。其语法格式如下。

```
<c:remove var = "varName" [scope = "{page|request|session|application}"]/>
```

<c:remove>标签的属性如表 7-14 所示。

表 7-14 <c:remove>标签的属性

名 称	说 明	EL 类型	必要否	默认值
var	欲移除的变量名称	String	否	无
scope	var 变量的 JSP 范围	String	否	page

说明：<c:remove>必须有 var 属性，即要被移除的属性的名称，scope 则可有可无。

例如，将 number 变量从 session 范围中移除。若不设定 scope，则<c:remove>将会从 page、request、session 及 application 中顺序寻找是否存在名称为 number 的数据，若能找到，则将它移除掉，反之则不会做任何事情。

【例 7-17】 <c:remove>标签的应用。

```
<%@ page language = "java" import = "java.util.*" pageEncoding = "UTF-8" %>
<%@ taglib uri = "http://java.sun.com/jsp/jstl/core" prefix = "c" %>
<!DOCTYPE HTML PUBLIC "-//W3C//DTD HTML 4.01 Transitional//EN">
<html>
  <head><title>使用 JSTL 设置变量与移除</title></head>
  <body>
    <!-- 设置之前应该是空值 -->
    设置变量之前的值是:msg = <c:out value = "${msg}" default = "null"/><br>
    <!-- 设置变量 msg -->
    <c:set var = "msg" value = "Hello!" scope = "page"></c:set>
    <!-- 此时 msg 的值应该是上面设置的"已经不是空值了" -->
    设置新值以后:msg = <c:out value = "${msg}"></c:out><br>
    <!-- 把 msg 变量从 page 范围内移除 -->
    <c:remove var = "msg" scope = "page"/>
    <!-- 此时 msg 的值应该显示 null -->
    移除变量 msg 以后:msg = <c:out value = "${msg}" default = "null"></c:out>
  </body>
</html>
```

程序的运行结果如图 7-12 所示。

设置变量之前的值是：msg=null
设置新值以后：msg=Hello!
移除变量msg以后：msg=null

图 7-12 例 7-17 运行结果

4）<c:catch>标签

<c:catch>主要用来处理产生错误的异常状况，并且将错误信息储存起来，其语法格式如下。

```
<c:catch [var = "varName"]>
  … //欲抓取错误的部分
</c:catch>
```

<c:catch>标签的属性如表 7-15 所示。

表 7-15 <c:catch>标签的属性

名称	说　明	EL 类型	必要否	默认值
var	用来存储错误信息的变量	String	否	无

<c:catch>主要将可能发生错误的部分放在<c:catch>和</c:catch>之间。如果发生了错误,可以将错误信息存储在 varName 变量中。

另外,当错误发生在<c:catch>和</c:catch>之间时,则只有<c:catch>和</c:catch>之间的程序会被中止,但整个网页不会被中止。

【例 7-18】 处理异常。

```
<%@ page language="java" import="java.util.*" pageEncoding="utf-8"%>
<%@ taglib uri="http://java.sun.com/jsp/jstl/core" prefix="c" %>
<!DOCTYPE HTML PUBLIC "-//W3C//DTD HTML 4.01 Transitional//EN">
<html>
  <head><title>catch</title></head>
  <body>
    <%--捕获异常,并将异常信息存储在 var 变量中--%>
    <c:catch var="myexp">
    <%
        int i=0;
        int j=3/0;
     %>
    </c:catch>
<h4>异常</h4>
<c:out value="${myexp}"/><%--输出异常--%>
<h4>异常信息</h4>
<c:out value="${myexp.message}"/><%--获取异常信息--%>
<h4>引起原因</h4>
<c:out value="${myexp.cause}"/><%--获取引起异常的原因--%>
</body>
</html>
```

程序的运行结果如图 7-13 所示。

异常
java.lang.ArithmeticException: / by zero
异常信息
/ by zero
引起原因

图 7-13 例 7-18 运行结果

2. 流程控制

流程控制分类中包含 4 个标签:<c:if>、<c:choose>、<c:when>、<c:otherwise>。

1) <c:if>标签

<c:if>的用途和一般在程序中用的 if 一样。其语法格式如下。

```
<c:if test="testCondition"[var="varName"]
                              [scope="{page|request|session|application}"]>
  标签体
<c:if>
```

或

```
<c:if test = "testCondition"[var = "varName"]
                          [scope = "{page|request|session|application}"]/>
```

<c:if>标签的属性如表 7-16 所示。

表 7-16 <c:if>标签的属性

名称	说　明	EL 类型	必要否	默认值
test	如果表达式的结果为 true,则执行标签体,false 则相反	booleab	否	无
var	用来存储 test 运算后的结果,即 true 或 false	String	否	无
scope	var 变量的 JSP 范围	String	否	page

说明:<c:if>标签必须有 test 属性,当 test 中的表达式结果为 true 时,则会执行标签体;如果为 false,则不会执行。<c:if>的标签体除纯文本外,还可以是任何 JSP 程序代码(Scriptlet)、JSP 标签或者 HTML 代码。

在代码 ${param.username=="admin"}中,如果 param.username 等于 admin,则会显示"ADMIN 您好!";若它的内容不等于 admin,则不会执行<c:if>的 body 部分,所以不会显示"ADMIN 您好!"。

```
<c:if test = "${param.username == 'admin'}">
    ADMIN 您好!
</c:if>
```

<c:if>还有另外两个属性 var 和 scope。当执行<c:if>的时候,可以将这次后判断的结果存放到属性 var 里;scope 用于设定 var 的范围。

当表达式过长时,可能希望拆开处理,或是之后还须使用此结果时,也可以用它先将结果暂时保留,以便使用。

```
<c:if test = "${param.username == 'admin'}" var = "condition" scope = "page">
  你好 admin 先生
<c:if/>
${condition }
```

【例 7-19】 登录页面的设计,如果用户已经登录,则不再重复登录。当登录成功后,将用户 ID 保存到 session 对象中。

```
<%@ page language = "java" import = "java.util.*" pageEncoding = "UTF-8"%>
<%@ taglib uri = "http://java.sun.com/jsp/jstl/core" prefix = "c" %>
<!DOCTYPE HTML PUBLIC "-//W3C//DTD HTML 4.01 Transitional//EN">
<html>
    <head><title>登录页面</title></head>
  <body>
    <c:set var = "loggedIn" value = "${not empty sessionScope.userId}"/>
    <c:if test = "${not loggedIn}">
      <form id = "login" method = "post" action = "loginServlet">
```

```
      用户名:<input id="userName" name="userName" type="text"><br>
      密码:<input id="passWord" name="passWord" type="password"><br>
      <input type="submit" value="登录">
      </form>
    </c:if>
    <c:if test="${loggedIn}">
      已经登录!
    </c:if>
  </body>
</html>
```

【例 7-20】 利用<c:set>标签设置 JavaBean 的属性 age,在 JSP 页面中获取 age,如果 age<18,输出相应信息。

```
<%@ page contentType="text/html; charset=gb2312" language="java" %>
<%@ taglib uri="http://java.sun.com/jsp/jstl/core" prefix="c" %>
<%--JSP 页面默认是 true,EL 表达式被忽略--%>
<%@ page isELIgnored="false"%>
<%--定义一个 javaBean 对象--%>
<jsp:useBean id="stu" class="cn.edu.mypt.student"/>
<%--通过<c:set>标签给 JavaBean 对象的 age 属性设值--%>
<c:set value="16" target="${stu}" property="age"/>
<%--输出 JavaBean 对象的属性值--%>
年龄:<c:out value="${stu.age}"/>
<%--当 if 判断为 true 时,输出标签体的内容--%>
<c:if test="${stu.age<18}" var="young" scope="session">对不起,未成年,不能访问这个
网站...</c:if>
<%--输出 if 语句的判断结果--%>
判断结果:<c:out value="${sessionScope.young}"/>
```

2) <c:choose>标签

<c:choose>本身只当作<c:when>和<c:otherwise>的父标签。其语法格式如下。

```
<c:choose>
  标签体(<when>和<otherwise>)
</c:choose>
```

本标签没有属性。

<c:choose>的本体内容只能有:①空白;②1 或多个<c:when>,0 或多个<c:otherwise>。

若使用<c:when>和<c:otherwise>来做流程控制时,两者都必须为<c:choose>的子标签。

3) <c:when>标签

<c:when>的用途和一般程序中用的 when 一样。其语法格式如下。

```
<c:when test="testCondition">
  标签体
</c:when>
```

<c:when>标签的属性如表 7-17 所示。

表 7-17 <c:when>标签的属性

名称	说　明	EL 类型	必要否	默认值
test	如果表达式的结果为 true,则执行标签体,false 则相反	booleab	否	无

<c:when>必须有 test 属性,当 test 表达式的结果为 true 时,则会执行标签体,如果为 false 时,则不会执行。要注意的是:①<c:when>必须在<c:whoose>和</c:choose>之间;②在同一个<c:choose>中时,<c:when>必须在<c:otherwise>之前。

4) <c:otherwise>标签

在同一个<c:choose>中,当所有<c:when>的条件都没有成立时则执行<c:otherwise>的标签体。其语法格式如下。

```
<c:otherwise>
  标签体
</c:otherwise>
```

本标签没有属性。要注意:①<c:otherwise>必须在<c:choose>和</c:choose>之间;②在同一个<c:choose>中,<c:otherwise>必须是最后一个标签;③在同一个<c:choose>中,假若所有<c:when>的 test 属性都不为 true 时,则执行<c:otherwise>的标签体。

【例 7-21】 从 JavaBean 中获取 user 属性,并根据不同的属性值显示用户类型。

```
<%@ page language="java" import="java.util.*" pageEncoding="utf-8"%>
<%@ taglib uri="http://java.sun.com/jsp/jstl/core" prefix="c" %>
<%-- 定义一个 user 对象 --%>
<jsp:useBean id="user" class="cn.edu.mypt.bean.UserBean"/>
<%-- 为 user 对象设置属性 --%>
<c:set value="admin" target="${user}" property="userType"/>
<%-- 获取 user 对象的属性值,并根据不同的属性值显示用户类型 --%>
<c:choose>
  <c:when test="${user.userType eq 'admin'}">
    <h4>该用户是:<c:out value="${user.userType}"/></h4>
  </c:when>
  <c:when test="${user.userType eq 'comm'}">
    <h4>该用户是:<c:out value="${user.userType}"/></h4>
  </c:when>
  <c:otherwise>
    <h4>该用户是类型没有定义</h4>
  </c:otherwise>
</c:choose>
```

程序的运行结果如图 7-14 所示。

该用户是:admin

图 7-14　例 7-21 运行结果

3. 遍历操作

遍历操作类中作主要包含两个标签：<c:forEach>和<c:forTokens>。

1) <c:forEach>标签

<c:forEach>为循环控制，它对集合中的成员循序浏览一遍。运作方式为条件符合时，就会持续重复执行<c:forEach>的循环体。其语法格式如下。

(1) 遍历集合对象的所有成员。

```
<c:forEach[ var = "varName"] items = "collection" [varStatus = "varStatusName"]>
    循环体
</c:forEach>
```

(2) 循环指定的次数。

```
<c:forEach [var = "varName"] [varStatus = "varStatusName"]
    begin = "begin" end = "end" [step = "step"]>
  循环体
</c:forEach>
```

<c:forEach>标签的属性如表 7-18 所示。

表 7-18 <c:forEach>标签的属性

名称	说　明	EL 类型	必要否	默认值
var	用来存放当前成员	String	否	无
items	被遍历的集合对象	Arrays Collection Iterator Enumeration Maop String	否	无
varStatus	用来存放当前对象的信息	String	否	无
begin	开始位置	int	否	0
end	结束位置	int	否	最后一个成员
step	每次循环的间隔数	int	否	1

说明：当有 begin 属性时，begin 必须大于或等于 0；当有 end 属性时，end 必须大于 begin；当有 step 属性时，step 必须大于或等于 0；当 items 为 null 时，则表示集合对象为空；当 begin 大于或等于 items 时，循环不进行；如果要遍历一个集合对象并将它的内容显示出来，就必须有 items 属性。

【例 7-22】 遍历并输出集合对象的内容。

```
<%@ page language = "java" import = "java.util.*" pageEncoding = "UTF-8" %>
<%@ taglib uri = "http://java.sun.com/jsp/jstl/core" prefix = "c" %>
<!DOCTYPE HTML PUBLIC "-//W3C//DTD HTML 4.01 Transitional//EN">
<html>
  <head><title>Hello,world!</title></head>
```

```
<body>
  <% -- 将 JavaBean 对象存放到集合中 -- %>
  <%
    List users = new ArrayList();
    for(int i = 0;i < 3;i++)
    {
      cn.edu.mypt.bean.UserBean u = new cn.edu.mypt.bean.UserBean();
      u.setUserName("zhangsan");
      u.setUserID("Stu" + i);
      users.add(u);
      session.setAttribute("users",users);
    }
  %>
<% -- 注意: 只可通过 11 个隐式对象来输出表达式中的内容(因此不能直接将 List 对象添加到
EL 表达式中) -- %>
<% -- 通过<c:forEach>遍历集合中的信息> -- %>
<h3>用户信息</h3>
<table>
  <tr>
    <th>用户名 ID</th><th>用户名</th><th>当前行的索引</th><th>已遍历的行数</th>
    <th>是否第一行</th>
     <th>是否最后一行</th>
  </tr>
  <c:forEach var = "user" items = "${users}" varStatus = "status">
  <% -- 加上 begin = "1" end = "3" step = "1"属性,将只显示前三条记录 -- %>
  <tr>
    <td><c:out value = "${user.userID}"/></td>
    <td><c:out value = "${user.userName}"/></td>
    <td><c:out value = "${status.index}"/></td><% -- 输出当前行的序号 -- %>
    <td><c:out value = "${status.count}"/></td><% -- 输出已遍历的行数 -- %>
    <td><c:out value = "${status.first}"/></td><% -- 输出当前行是否是第一行 -- %>
    <td><c:out value = "${status.last}"/></td><% -- 输出当前行是否是最后一行 -- %>
  </tr>
  </c:forEach>
</table>
<% -- 通过<c:forEach>输出 1～10 的数据 -- %>
<h3>-- 输出 1～10 的数据 --</h3>
<c:forEach var = "num" begin = "1" end = "10" step = "2">
<c:out value = "${num},"></c:out>
</c:forEach>
<% -- 通过<c:forEach>遍历数组,枚举,集合等 -- %>
<%
    int[]intarr = new int[]{10,20,30};
    String[]strarr = new String[]{"a1","a2","a3"};
    Vector v = new Vector();
    v.add("v1");
    v.add("v2");
    v.add("v3");
    Enumeration e = v.elements();
```

```
        HashMap h = new HashMap();
        h.put("k1","v1");
        h.put("k2","v2");
        h.put("k3","v3");
        request.setAttribute("intarr",intarr);
        request.setAttribute("strarr",strarr);
        request.setAttribute("e",e);
        request.setAttribute("h",h);
%>
    <h3>--遍历整型数组--</h3>
    <c:forEach var="i" items="${intarr}">
        <c:out value="${i}"/>,
    </c:forEach>
    <h3>--遍历字符串数组--</h3>
    <c:forEach var="s" items="${strarr}">
        <c:out value="${s}"/>,
    </c:forEach>
    <h3>--遍历枚举--</h3>
    <c:forEach var="ee" items="${e}">
        <c:out value="${ee}"/>,
    </c:forEach>
    <h3>--遍历 HashMap--</h3>
    <c:forEach var="hh" items="${h}">
        <c:out value="${hh.key}"/>--><c:out value="${hh.value}"/>|
    </c:forEach>
</body>
</html>
```

程序的运行结果如图 7-15 所示。

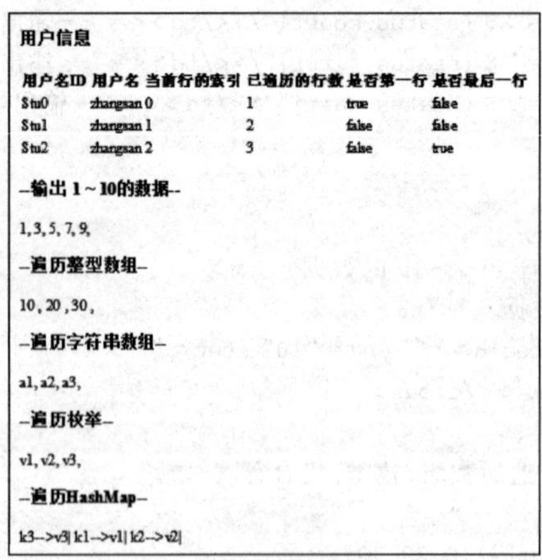

图 7-15　例 7-22 运行结果

2）<c:forTokens>标签

<c:forTokens>标签用来浏览一串字符串中所有的成员，其成员是由定义符号分隔的。其语法格式如下。

```
<c:forTokens items = "StringOfTokens" delims = "delimiters" [var = "varName"]
[varStatus = "varStatusName"] [begin = "begin"] [end = "end"] [step = "step"]>
  标签体
</c:forTokens>
```

<c:forTokens>标签的属性如表 7-19 所示。

表 7-19　<c:forTokens>标签的属性

名称	说　明	EL 类型	必要否	默认值
var	用来存放当前成员	String	否	无
items	被遍历的字符串	String	是	无
delims	定义用来分隔字符串的字符	String	是	无
varStatus	用来存放现在指到的相关成员信息	String	否	无
begin	开始的位置	int	否	0
end	结束的位置	int	否	最后一个成员
step	每次遍历的间隔数	int	否	1

说明：当有 begin 属性时，begin 必须大于或等于 0；当有 end 属性时，end 必须大于 begin；当有 step 属性时，step 必须大于或等于 0；当 items 为 null 时，则表示集合对象为空；当 begin 大于或等于 items 时，遍历不进行。

<c:forTokens>的 begin、end、step、var 和 varStatus 的用法都和<c:forEach>一样，因此这里就只介绍 items 和 delims 两个属性：items 的内容必须为字符串；而 delims 是用来分隔 items 中定义的字符串的字符。

【例 7-23】　在网页中输出 abcde，它会把符号“,”当作分隔标记，因此循环 5 次，但是并没有将“a,b,c,d,e”中的“,”显示出来。

```
<c:forTokens items = "a,b,c,d,e" delims = "," var = "item">
    ${item}
</c:forTokens>
```

输出结果为：abcde。

【例 7-24】　在网页中输出 123456789，也就是把 123-456-789 中的-当作分隔标记，将字符串拆为 3 份，每执行一次循环就将当前部分放到 item 属性中。

```
<%
  String phoneNumber = "123 - 456 - 789";
  Request.setAttribute("userPhone",phoneNumber);
%>
<c:forTokens items = "${userPhone}" delims = "-" var = "item">
  ${item}
</c:forTokens>
```

输出结果为：123 456 789。

【例 7-25】　用 delims 一次设定多个分隔字符串用的字符。

```
<c:forTokens items = "a,b;c - d,e" delims = ",; - " var = "item">
```

```
    ${item}
</c:forTokens>
```

输出结果为：abcde。

<c:forEach>没有 delims 属性，因此<c:forEach>无法设定分隔字符串用到的字符，而<c:forEach>分隔字符串用的字符只有“,”，这和使用<c:forTokens>将 delims 属性设为“,”的结果相同。所以如果使用<c:forTokens>来分隔字符串，功能和弹性会比使用<c:forEach>大。

```
<c:forEach items = "a,b,c,d,e" var = "item">
    ${item}
</c: forEach>
```

【例 7-26】 将字符串划分为数组。

```
<%@ taglib prefix = "c" uri = "http://java.sun.com/jstl/core_rt" %>
<%@ page contentType = "text/html; charset = gb2312" language = "java" %>
<%-- 通过分隔符将字符串划分为数组,并输出 --%>
<c:forTokens var = "ele" items = "blue,red,green|yellow|pink,black|white" delims = "|">
<c:out value = "${ele}"/>||
</c:forTokens>
<br>
<%-- 通过多个分隔符将字符串划分数组,并输出 --%>
<c:forTokens var = "ele" items = "blue,red!green|yellow;pink;black|white" delims = "|;,!">
<c:out value = "${ele}"/>||
</c:forTokens>
```

输出结果如下。

```
blue,red,green|| yellow|| pink,black|| white||
blue|| red|| green|| yellow|| pink|| black|| white||
```

4. URL 操作

JSTL 包含 3 个与 URL 操作有关的标签，它们分别为<c:import>、<c:url>和<c:redirect>。

1）<c:import>标签

<c:import>可以把其他静态或动态文件包含至本身 JSP 网页。它和 JSP 动作<jsp:include>最大的区别在于：<jsp:include>只能包含和自己在同一个 Web 应用下的文件；而<c:import>除了能包含和自己在同一个 Web 应用的文件外，也可以包含不同 Web 应用或者是其他网站的文件。其语法格式如下。

```
<c:import url = "包含地址的 URL" [context = "上下文路径"] [var = "保存内容的属性名称"]
[scope = "{page|request|session|application}"][charEncoding = "charEncoding"]
[varReader = "以 Reader 方式读取内容"]>
标签体
  [<c:param name = "参数名称" value = "参数内容"/>]
</c:import>
```

<c:import>标签的属性如表 7-20 所示。

表 7-20 <c:import>标签的属性

名 称	说 明	EL 类型	必要否	默认值
url	指定被包含的地址	String	否	无
context	如果访问其他 Web 站点必须以"/"开头	String	是	无
var	存储包含的文件的内容(以 String 类型存入)	String	否	无
scope	var 变量的 JSP 范围	String	否	Page
charEncoding	被包含文件的内容的编码格式	String	否	无
varReader	存储被包含的文件的内容(Reader 类型存入)	String	否	无

<c:import>中必须有 url 属性,它用来设定被包含网页的地址。它可以是绝对地址或是相对地址,url 为 null 或空时,会抛出 JspException 异常。

使用绝对地址的写法如下,利用此写法就会把 http://java.sun.com 的内容加到网页中。

```
<c:import url = "http://java.sun.com"/>
```

如果是使用相对地址,假设存在一个文件名为 Hello.jsp,它和使用<c:import>的网页存在于同一个 webapps 目录时,<c:import>的写法如下。

```
<c:import url = "Hello.jsp"/>
```

如果以"/"开头,那么就表示跳到 Web 站点的根目录下,以 Tomcat 为例,即 webapps 目录。假设一个文件为 hello.txt,存放在 webapps/examples/images 目录下,而 context 为 exampes,可以写成以下方式将 hello.txt 文件包含进我们的 JSP 页面中。

```
<c:import url = "/images/hello.txt"/>
```

<c:import>也提供 var 和 scope 属性。当 var 属性存在时,虽然同样会把其他文件的内容包含进来,但是它并不会输出至网页上,而是以 String 的类型存储在 varName 中。Scope 是设 varName 的范围。存储数据后,在需要用时可以将它取出来,代码如下。

```
<c:import url = "/images/hello.txt" var = "s" scope = "session"/>
```

假设有以下 jstltestb.jsp 页面。

```
<%@ page language = "java" import = "java.util.*" pageEncoding = "utf-8" %>
<!DOCTYPE HTML PUBLIC "-//W3C//DTD HTML 4.01 Transitional//EN">
<html>
  <head><title>Hello,world!</title></head>
    <body>
      <h1>欢迎你! ${param.name}</h1>
</body>
</html>
```

【例 7-27】 导入页面。

```
<%@ page language="java" import="java.util.*" pageEncoding="utf-8"%>
<%@ taglib uri="http://java.sun.com/jsp/jstl/core" prefix="c" %>
<!DOCTYPE HTML PUBLIC "-//W3C//DTD HTML 4.01 Transitional//EN">
<html>
  <head><title>Hello,world!</title></head>
<body>
  <c:import url="jstl3_b.jsp">
     <c:param name="name" value="jack"/>
  </c:import>
  </body>
</html>
```

程序的运行结果如图 7-16 所示。

【例 7-28】 包含同一个 Web 应用程序的文件和不同 Web 应用程序的文件。

(1) jstltest1.jsp 页面的代码如下。

```
<%@ page language="java" import="java.util.*" pageEncoding="utf-8"%>
<%@ taglib uri="http://java.sun.com/jsp/jstl/core" prefix="c" %>
<!DOCTYPE HTML PUBLIC "-//W3C//DTD HTML 4.01 Transitional//EN">
<html>
  <head><title>Hello,world!</title></head>
  <body>
    <h3>引入绝对路径的文件</h3>
    <c:import url="http://127.0.0.1:8080/mytest/jstltestb.jsp" var="file" charEncoding=
"utf-8"/>
    <pre><c:out value="${file}"/></pre>
</body>
</html>
```

其运行结果如图 7-17 所示。

欢迎你！jack

图 7-16　例 7-27 运行结果

引入绝对路径的文件

```
<!DOCTYPE HTML PUBLIC "-//W3C//DTD HTML 4.01 Transitional//EN">
<html>
  <head><title>Hello,world!</title></head>
    <body>
      <h1>欢迎你！</h1>
</body>
</html>
```

图 7-17　例 7-28 的运行结果一

(2) jstltest2.jsp 页面的代码如下。

```
<%@ page language="java" import="java.util.*" pageEncoding="utf-8"%>
<%@ taglib uri="http://java.sun.com/jsp/jstl/core" prefix="c" %>
<!DOCTYPE HTML PUBLIC "-//W3C//DTD HTML 4.01 Transitional//EN">
<html>
  <head><title>Hello,world!</title></head>
  <body>
    <h3>引入相对路径的文件</h3>
      <c:import url="jstltestb.jsp" var="f"/>
      <pre><c:out value="${f}"/></pre>
```

```
    </body>
</html>
```

其运行结果如图 7-18 所示。

(3) jstltest3.jsp 页面的代码如下。

```
<%@ page language = "java" import = "java.util.*" pageEncoding = "utf-8" %>
<%@ taglib uri = "http://java.sun.com/jsp/jstl/core" prefix = "c" %>
<!DOCTYPE HTML PUBLIC "-//W3C//DTD HTML 4.01 Transitional//EN">
<html>
  <head><title>Hello,world!</title></head>
<body>
  <h3>传递参数到被引入文件</h3>
    <c:import url = "jstltestb.jsp" var = "ff">
      <c:param name = "name" value = "jack"/>
    </c:import>
  <pre><c:out value = "${ff}"/></pre>
</body>
</html>
```

其运行结果如图 7-19 所示。

```
引入相对路径的文件
<!DOCTYPE HTML PUBLIC "-//W3C//DTD HTML 4.01 Transitional//EN">
<html>
  <head><title>Hello,world!</title></head>
   <body>
    <h1>欢迎你! </h1>
</body>
</html>
```

图 7-18 例 7-28 的运行结果二

```
传递参数到被引入文件
<!DOCTYPE HTML PUBLIC "-//W3C//DTD HTML 4.01 Transitional//EN">
<html>
  <head><title>Hello,world!</title></head>
   <body>
    <h1>欢迎你! jack</h1>
</body>
</html>
```

图 7-19 例 7-28 运行结果三

2) <c:url>标签

<c:url>主要用来生成一个 URL,其语法格式如下。

```
<c:url value = "操作的 url" [context = "上下文路径"] [var = "保存的属性名称"]
[scope = {page|request|session|application}]/>
```

或

```
<c:url value = "操作的 url" [context = "上下文路径"] [var = "保存的属性名称"]
[scope = {page|request|session|application}]>
  <c:param name = "参数名称" value = "参数内容"/>
</c:url>
```

url 属性如表 7-21 所示。

表 7-21 url 属性

名称	说　明	EL 类型	必要否	默认值
value	操作的 URL	String	是	无
context	如果访问其他 Web 站点必须以"/"开头	String	否	无
var	存储包含的文件的内容(以 String 类型存入)	String	否	无
scope	var 变量的 JSP 范围	String	否	page

<c:url>有三个属性,分别为 context、var、scope。context 属性和<c:import>的同名属性相同,可以用来生成一个其他 Web 站点的网址。如果<c:url>有 var 属性,则网址会被存放到 varName 中,而不会直接输出网址。

当需要动态产生网址时,有可能传递的参数不固定,或者是需要一个网址能连至同服务器的其他 Web 站点的文件,这时采用<c:url>标签可以将生成的网址存储起来重复使用。

<c:url>也可以搭配<c:param>使用,下面代码的执行结果将会生成一个网址 http://www.javafan.net? param=value。

```
<c:url value = "http://www.java.net">
    <c:param name = "param" value = "value">
</c:url>
```

上面的代码可以搭配 HTML 的<a>标签使用,例如:

```
<a href = "<c:url value = "http://www.java.net"> <c:param name = "param" value = "value"/>
</c:url>"> Java </a>
```

利用<c:url>从 Web 站点的角度来设定需要的文件,自动生成 image 目录下的 code.gif 文件的地址,并且当域名改变时,也不用地址。

```
<img src = "<c:url value = '/images/code.gif'>"/>
```

【例 7-29】 将一个 URL 存放到一个变量中,并输出。

```
<%@ page language = "java" contentType = "text/html; charset = UTF - 8" pageEncoding =
        "UTF - 8" %>
<%@ taglib uri = "http://java.sun.com/jsp/jstl/core" prefix = "c" %>
<% -- 将一个 URL 存放到一个变量中,并输出 -- %>
<c:url var = "myurl" value = "hello.jsp" scope = "session">
<c:param name = "name" value = "zhangsan"/>
</c:url>
<c:out value = " ${myurl}"/>
```

程序的运行结果如图 7-20 所示。

hello.jsp?name=zhangsan

图 7-20 例 7-29 运行结果

3) <c:redirect> 标签

<c:redirect>可以将客户端的请求从一个 JSP 网页导向其他文件,其语法格式如下。

```
<c:redirect url = "导向的目标地址" context = "上下文路径"/>
```

或

```
<c:redirect url = "导向的目标地址" context = "上下文路径">
<c:param name = "参数名称" value = "参数内容"/>
</c:redirect>
```

<c:redirect>标签的属性如表 7-22 所示。

表 7-22 <c:redirect>标签的属性

名 称	说 明	EL 类型	必要否	默认值
url	导向的目标地址	String	是	无
context	其他 Web 站点必须以/开头	String	否	无

url 可以是相对地址或绝对地址，例如，将网页自动导向到 http://www.sina.net，可编写如下代码。

```
<c:redirect url = "http://www.sina.net"/>
```

若加上 context 属性，可导向至其他 Web 站点中的文件。例如，导向至/others 下的/jsp/index.html 文件，可采用如下代码。

```
<c:redirect url = "/jsp/index.html" context = "/others"/>
```

<c:redirect>的功能不止可以导向网页，同样它还可以传递参数给目标文件。在这里同样使用<c:param>来设定参数。这样就可以通过 getParameter("param")得到传递的值。

```
<c:redirect url = "http://www.javafan.net">
<c:param name = "param" value = "value">
</c:redirect>
```

【例 7-30】 获取 URL 实现重定向。

hello.jsp 页面的代码如下。

```
<%@ page language = "java" import = "java.util.*" pageEncoding = "utf-8" %>
<!DOCTYPE HTML PUBLIC "-//W3C//DTD HTML 4.01 Transitional//EN">
<html>
  <head><title>Hello,world!</title></head>
    <body>
      <h1>欢迎你! ${param.name}</h1>
</body>
</html>
```

Urltest01.jsp 页面的代码如下。

```
<%@ page language = "java" import = "java.util.*" pageEncoding = "utf-8" %>
<%@ taglib uri = "http://java.sun.com/jsp/jstl/core" prefix = "c" %>
<% -- 通过<c:url>获得 URL -- %>
<c:url value = "hello.jsp" var = "test">
    <c:param name = "name" value = "zhangsn"/>
</c:url>
<% -- 通过<c:redirect>重定向到获得的 url 上 -- %>
<c:redirect url = "${test}"/>
```

Urltest02.jsp 页面的代码如下。

```
<%@ page language = "java" import = "java.util.*" pageEncoding = "utf-8" %>
<%@ taglib uri = "http://java.sun.com/jsp/jstl/core" prefix = "c" %>
<% -- 通过<c:url>获得 URL -- %>
```

```
<c:url value = "hello.jsp" var = "test"></c:url>
<% -- 通过<c:redirect>重定向到获得的 url 上(在<c:redirect>内部传参) -- %>
<c:redirect url = " ${test}">
    <c:param name = "name" value = "admin"/>
</c:redirect>
```

7.2.3 SQL 标签库

JSTL 提供了与数据库操作有关的标签,可以使用这些标签进行数据库的更新及查询操作。SQL 标签库主要包含如表 7-23 所示的标签。

表 7-23 SQL 标签库包含的标签

标签名称	描　述
<sql:setDataSource>	设置要使用的数据源名称
<sql:query>	执行查询操作
<sql:update>	执行更新操作
<sq:ransaction>	执行事务的处理操作,并设置操作的安全级别

1. <sql:setDataSource>标签

在进行 SQL 操作前,可以通过<sql:setDateSource>标签来设置数据源,其语法格式如下。

(1) 设置数据源。

```
<sql:setDataSource dataSource = "数据源名称" [var = "保存的属性名称"]
[scope = "[page I request I session I applicatin]"]/>
```

(2) 设置 JDBC。

```
<sql:setDataSource driver = "数据库驱动程序" url = "数据库链接地址"
  user = "用户名" password = "密码"
  [var = "保存的属性名称"][scope = "[page | request | session| application]"/>
```

本标签的属性如表 7-24 所示。

表 7-24 <sql:setDataSource>标签的属性

属性名称	说　明	EL 类型	必要否	默认值
dataSource	数据源	String	否	无
driver	JDBC 数据库驱动程序	String	否	无
url	数据库的 URL 地址	String	否	无
user	数据库的用户名	String	否	无
password	数据库的密码	String	否	无
var	存储数据库连接的属性名称	String	否	默认设置
scope	var 属性的保存范围,默认为 page 范围	String	否	page

【例 7-31】　使用 SQL 标签定义数据源。

```
<sql:setDataSource dataSource="jdbc/test" var="ds"/>
```

2. 数据库操作标签

数据库的主要操作就是查询、更新及事务处理，JSTL 主要提供了< sql:query >、< sql:update >、< sql:param >、< sql: dateParam >操作标签。

1）< sql:query >标签

< sql:query >标签用来执行 SQL 查询语句，并将结果存储在作用域变量中，其语法格式如下。

```
<sql:query sql="SQL 查询语句" var="保存查询结果的属性名称"
[scope="[page|request|session|application]"]
[dataSource="数据源的名称] maxRows="最多显示的记录数" startRow="记录的开始行数"/>
```

或

```
<sql:query var="保存查询结果的属性名称" [scope="[page|request|session|aplication"]
[dataSource="数据源的名称"] maxRows="最多显示的记录数 startRow="记录的开始行数">
    SQL 查询语句
</sql:query>
```

< sql:query >标签的属性如表 7-25 所示。

表 7-25　< sql:query >标签的属性

属性名称	说　明	EL 类型	必要否	默认值
sql	需要执行的 SQL 命令(返回一个 ResultSet 对象)	String	否	bady
dataSource	所使用的数据库源	String	否	默认数据库
maxRows	存储在变量中的最大结果数	String	否	无穷大
startRow	开始记录的结果的行数	String	否	0
var	代表数据库的变量	String	否	默认设置
scope	var 属性的作用域	String	否	page

SQL 语句的查询结果保存在 result 对象中，如果要想取出数据表的具体信息，可以通过其属性来实现，如表 7-26 所示。

表 7-26　result 对象的属性

属　性	描　述
rows	根据字段名称取出列的内容
rowsByIndex	根据字段索引取出列的内容
columnNames	取得字段的名称
rowCount	取得全部的记录数
limitedByMaxRows	取出最大的数据长度

【例 7-32】 查询 userable 表的数据。

```
<%@ page language="java" contentType="text/html; charset=UTF-8"pageEncoding="UTF-8" %>
<%@ taglib uri="http://java.sun.com/jsp/jstl/core" prefix="c" %>
<%@ taglib uri="http://java.sun.com/jsp/jstl/sql" prefix="sql" %>
<html>
  <head><title>用户信息</title></head>
  <body>
  <sql:setDataSource var="testDb"
        driver="com.microsoft.sqlserver.jdbc.SQLServerDriver"
        url="jdbc:sqlserver://127.0.0.1:1433;DatabaseName=test"
        user="sa"
        password="123456"/>
  <sql:query dataSource="${testDb}" sql="select * from userTable" var="result" />
  <table>
    <tr><th>用户 ID</th><th>用户名</th></tr>
    <c:forEach var="row" items="${result.rows}">
    <tr>
      <td><c:out value="${row.userID}"/></td>
      <td><c:out value="${row.userName}"/></td>
    </tr>
    </c:forEach>
  </table>
  <h3>共有${result.rowCount}条记录! </h3>
  </body>
</html>
```

程序的运行结果如图 7-21 所示。

用户ID	用户名
1	zhangsan

共有1条记录!

图 7-21 例 7-32 运行结果

查询结果保存在 javax.servlet.jsp.jstl.sql.Result 类的实例 resalt 中。要取得结果集中的数据可以使用<c:forEach>循环来进行。

rows 是 result 对象的属性之一,用来表示数据库表中的“列”集合,循环时,通过${rows.×××}表达式可以取得每一列的数据,×××是表中的列名。

对上面的程序做如下修改。

```
<%@ page language="java" contentType="text/html; charset=UTF-8"
      pageEncoding="UTF-8" %>
<%@ taglib uri="http://java.sun.com/jsp/jstl/core" prefix="c" %>
<%@ taglib uri="http://java.sun.com/jsp/jstl/sql" prefix="sql" %>
<html>
  <head><title>用户信息</title></head>
  <body>
  //略(参考例 7-24)
  <table>
    <tr>
      <c:forEach var="columnName" items="${result.columnNames}">
        <td><c:out value="${columnName}"/></td>
      </c:forEach>
```

```
  </tr>
    <c:forEach var="row" items="${result.rowsByIndex}">
    <tr>
      <c:forEach var="column" items="${row}">
        <td><c:out value="${column}"/></td>
      </c:forEach>
    </tr>
  </c:forEach>
</table>
```

使用<sql:query>标签最大的好处是它已经提供好了分页显示的功能，所以直接使用 maxRows 和 startRow 这两个属性即可完成。例如：

```
//查询总数
<sql:query var="result" dataSource="${dataSrc}">
  Select count(*) from usertable
</sql:query>
<c:forEach var="row" items="${result.rows}">
//设置 pagination 属性
<jsp: setProperty name- "pagination" property= "recordCount" value= "${ row. Count}"/>
//分页显示
<sql:query var = "result" dataSource = "${dataSrc}" maxRows = "${pagination. pageSize
" statrtRow = "${ pagination. firstResult}">
  Select userID,userName from user
</sql:query>
...(略)
```

2）<sql:update>标签

<sql: update>标签用于执行 SQL 更新语句，其语法格式如下。

```
<sql:update sql = "SQL 语句" var = " 保存更新的记录数"
  [scope = "[page|request|session|application]"][dataSource = "数据源的名称"/>
```

或

```
<sql: update var = "保存更新的记录数[scope = "[page|request|session|application'"]
[dataSource - "数据源的名称"]>
SQL 更新语句
</sql:update>
```

<sql:update>标签的属性如表 7-27 所示。

表 7-27 <sql: update>标签的属性

属性	描　　述	EL 类型	必要否	默认值
sql	需要执行的 SQL 命令(不返回 ResultSet 对象)	String	否	body
dataSource	数据源	String	否	默认数据库
var	影响的行数	String	否	无
scope	var 属性的作用域	String	否	page

在< sql:update >语句中所有的更新记录数都保存在 var 变量中。

3）< sql: param >和< sql:dateParam >标签

< sql:param >和< sql:query >标签和< sql: update >标签嵌套使用，用来提供一个值占位符。其语法格式如下。

```
< sql:param value = "参数内容"/>
```

< sql:param >标签的属性如表 7-28 所示。

表 7-28　< sql:param >标签的属性

属性	描　述	EL 类型	必要否	默认值
value	需要设置的参数值	String	否	body

< sql:dateParam >标签与< sql:query >标签和< sql:update >标签嵌套使用，用来提供日期和时间的占位符。其语法格式如下。

```
< sql:dateParam type = "date 类型" value = "参数内容"/>
```

< sql:dateParam >标签的属性如表 7-29 所示。

表 7-29　< sql:dateParam >标签的属性

属性	描　述	EL 类型	必要否	默认值
value	需要设置的日期参数(java. util. Date)	String	否	body
type	date（只有日期，ime（只有时间），timestamp(日期和时间)	String	否	timestamp

【例 7-33】 实现对数据库的常见操作。

```
<%@ page language = "java" contentType = "text/html; charset = UTF - 8"
    pageEncoding = "UTF - 8" %>
<%@ taglib uri = "http://java.sun.com/jsp/jstl/core" prefix = "c" %>
<%@ taglib uri = "http://java.sun.com/jsp/jstl/sql" prefix = "sql" %>
< html >
  < head >< title >实现对数据库的常见操作</title ></head >
  < body >
    < sql:setDataSource var = "ds"
      driver = "com.microsoft.sqlserver.jdbc.SQLServerDriver"
      url = "jdbc:sqlserver://127.0.0.1:1433;DatabaseName = test"
      user = "sa"
      password = "123456"/>
        < sql:update var = "up"
      sql = "insert into userTable(userId,username,userpwd) values(1002,'yang','123456')"
      dataSource = " ${ds}"/>
    <% -- 使用数据源进行更新 -- %>
    < sql:update var = "up" dataSource = " ${ds}">
      update usertable set userpwd = '123' where userID = '1'
    </sql:update >
```

```
<%--含参数的更新--%>
<sql:update var="up"
    sql="update usertable set userPwd=? where userid=?"
    dataSource="${ds}">
    <sql:param value="234"/>
    <sql:param value="1"/>
  </sql:update>

<%--删除记录--%>
  <sql:update var="up" sql="delete from usertable where userID='1'" dataSource=
"${ds}"/>

<%--创建表--%>
<sql:update var="up" sql="create table student(name varchar(20))" dataSource=
"${sc}"/>
<%--删除表--%>
<sql:update var="up" sql="drop table student" dataSource="${sc}"/>
</body>
</html>
```

3. 事务处理标签

事务处理可以保证数据库更新操作的完整性，在 JSTL 中通过<sql:transaction>标签控制事务的处理。其语法格式如下。

```
<sql:transaction [dataSource="数据源名称"
  [isolation="read_committed|read_uncomitted|repeatable|serializable]"]>
  <sql:update>|<sql:query>
</sql:transaction>
```

<sql:transaction>标签的属性如表 7-30 所示。

表 7-30　<sql:transaction>标签的属性

属　性	描　述	EL 类型	必要否	默认值
dataSource	数据源	String	否	默认数据库
isolation	事务隔离等级	String	否	数据库默认

该标签对 isolation 定义了 4 种安全级别，如表 7-31 所示。

表 7-31　事务的 4 种安全级别

安全级别	脏读	不可重复读	幻象读
read_committed			
read_uncommitted	√		
repeatable	√	√	
serializable	√	√	√

在使用< sql:transaction >标签时，其中往往嵌套多个< sql:query >或< sql:update >标签。

【例 7-34】 使用< sql:ransaction >标签。

```
<%@ page language="java" contentType="text/html; charset=UTF-8"pageEncoding="UTF-8" %>
<%@ taglib uri="http://java.sun.com/jsp/jstl/core" prefix="c" %>
<%@ taglib uri="http://java.sun.com/jsp/jstl/sql" prefix="sql" %>
<html>
  <head><title>实现对数据库的常见操作</title></head>
  <body>
    <sql:setDataSource dataSource="jdbc/test" var="ds"/>
    <sql:transaction
      isioatios="serializable"
      dataSource="${ds}">
        <sql:update var="result">
          sql="insert into userTable(userId,username,userpwd) values(1002,'yang','123456')"
        </sql:update>
    </sql:transaction>
  </body>
</html>
```

7.2.4 格式化标签

格式化标签用来格式化并输出文本、日期、时间、数字。引用格式化标签库的语法如下。

```
<%@ taglib prefix="fmt" uri="http://java.sun.com/jsp/jstl/fmt" %>
```

格式化标签库中主要包含的标签如表 7-32 所示。

表 7-32 格式化标签库分类

类　别	标签名称	描　述
国际化标签	< fmt:setLocale >	指定地区
	< fmt:requestEncoding >	设置 request 对象的字符编码
资源文件标签	< fmt:bundle >	设置临时的要读取资源文件的名称
	< fmt:message >	显示资源配置文件信息
	< fmt:setBundle >	设置一个全局的要读取资源文件的名称
数字及日期时间格式化	< fmt:formatNumber >	使用指定的格式或精度格式化数字
	< fimt:parseNumber >	解析一个代表数字、货币或百分比的字符串
	< fmt:formatDate >	使用指定的风格或模式格式化日期和时间
	< fmt:parseDate >	解析一个代表着日期或时间的字符串
设置时区	< fmt:setTimeZone >	设置一个全局的时区
	< fmt:timeZone >	设置一个临时的时区

1. 国际化标签

国际化是程序的重要组成部分，一个程序可以根据所在区域进行相应信息的显示，如各个地区的数字、日期时间显示风格都是不同的。

1）设置本地化环境标签<fmt:setLocale>

HTML请求到达服务器时，浏览器提供的HTTP头可以指出用户的首选本地化环境(可能是多个本地化环境的列表)。这个列表放在Accept-Language中，JSP容器会访问这个头信息。如果没有使用标签<fmt:setLocale>明确地指出引用本地化环境，就会使用这个列表中的首选本地化环境。

<fmt:setLocale>标签专门用于设置当前本地化环境，其语法格式如下。

```
<fmt:setLocale value="地区编码"[variant="浏览器"]
  [scope="page|request|session|application"] />
```

<fmt:setLocale>标签的属性如表7-33所示。

表7-33 <fmt:setLocale>标签的属性

属　性	描　　述	EL类型	必要否	默认值
value	设置本地环境名，例如en_us或者zh_cn	String	是	en_us
variant	如果要访问在同一个Web容器下的其他资源时设置，必须以"\"开头	String	否	无
Scope	设置的本地化环境名的有效范围	String	否	page

【例7-35】 设置本地日期。

```
<%@page import="java.util.Date"%>
<%@ page language="java" contentType="text/html; charset=UTF-8" pageEncoding="UTF-
8"%>
<%@ taglib prefix="fmt" uri="http://java.sun.com/jsp/jstl/fmt" %>
<!DOCTYPE HTML PUBLIC "-//W3C//DTD HTML 4.01 Transitional//EN">
<html>
  <head><title>Hello,world!</title></head>
    <body>
    <% //设置一个page范围的属性
      pageContext.setAttribute("date", new Date());
   %>
      <h3>中文日期显示:
      <fmt:setLocale value="zh_CN"/>
      <fmt:formatDate value="${date}"/></h3>
        <h3>英文日期显示:
        <fmt:setLocale value="en_US"/>
        <fmt:formatDate value="${date}"/></h3>
    </body>
</html>
```

程序的运行结果如图7-22所示。

中文日期显示: 2020-4-17
英文日期显示: Apr 17, 2020

图7-22 例7-35运行结果

2）<fmt:requestEncoding>标签

<fimt:requestEncoding>标签的主要功能与

setCharacterEncoding()一样,用来指定返回给 Web 应用程序的表单编码类型。其语法格式如下。

```
<fmt:requestEncoding value="字符集"/>
```

<fmt:requestEncoding>标签的属性如表 7-34 所示。

表 7-34 <fmt:requestEncoding>标签的属性

属性	描　述	EL 类型	必要否	默认值
key	字符编码集的名称,用于解码 request 参数	String	是	无

【例 7-36】 设置统一编码为 GBK。

```
<fmt:requestEncoding value="GBK"/>
```

2. 资源文件标签

JSTL 中提供了 4 个标签用于资源文件的读取和操作,分别是<fmt:bundle>、<fmt:message>、<fmt:param>和<fmt:setBundle>。所有的资源文件都是以. propertes 为扩展名,所有的内容要按照 key="value"的格式进行编写。所有 Web 资源文件要保存在 WEB-INF/classes 目录下。

1) 绑定信息资源标签<fmt:bundle>

一旦已经设置了 Web 引用的本地化环境后,就可以使用<fmt:bundle>标签绑定信息资源,其中可以包括一些调用本地文本的<fmt:message>标签。其语法格式如下。

```
<fmt:bundle basename="资源文件基础名称" [prefix="前缀标志"]>
  <fmt:message key="键名"/>
</fmt:bundle>
```

<fmt:bundle>标签的属性如表 7-35 所示。

表 7-35 <fmt:bundle>标签的属性

属　性	描　述	EL 类型	必要否	默认值
basename	指定被绑定的资源文件的基础名称	String	是	无
prefix	指定<fmt:message>标签 key 属性的前缀	String	否	无

basename: 资源文件的基础名称,例如,某资源文件为 Res_en. property,则基础名称为 Res。

prefix: 指定这个属性,就会为标签体中嵌套的<fmt:message>标签附加一个前缀。

当<fmt:bundle>标签中嵌套<fmt:message>标签时,<fmt:message>标签默认使用<fmt:bundle>标签中的 basename 所指定的资源文件。

2) 获取资源属性值标签<fmt:message>

当通过<fmt:bundlc>标签指定好资源文件名称后,即可使用<imt:message>标签按照 key 读取 value。其语法格式如下。

```
<fmt:message key="资源文件的指定 key" [bundle="资源文件名称"
[var="存储内容的属性名称" [scope="[page|request|session|aplication"/>
```

或

```
<fmt:message key="资源文件的指定 key" [bundle="资源文件名称]
[var="存储内容的属性名称" [scope="[page|request|session|aplication]/>
    <fmt:param value="设置占位符内容"/>
</fmt:message>
```

<fmt:message>标签的属性如表 7-36 所示。

表 7-36 <fmt:message>标签的属性

属　性	描　述	EL 类型	必要否	默认值
key	要检索的消息关键字	String	否	body
bundle	要使用的资源名	String	否	默认资源名
var	存储局部消息的变量名	String	否	显示在页面
scope	var 属性的作用域	String	否	page

3）获取参数值标签<fmt:param>

该标签一般与<fmt:message>标签配套使用，用来在获取的消息中替换一个值。例如，资源文件中的一条消息如下：

```
密码错误="{0}的密码错误"
```

<fmt:message>标签首先使用 key="密码错误"这个关键字找到以上这条消息，然后在<fmt:message>标签中使用<fmt:param>标签赋一个值来替代{0}部分。

<fmt:param>标签的语法格式如下。

```
<fmt:message ...>
    <fmt:param value="value"/>
</fmt:message>
```

其中 value 属性的值即为要替代{0}部分的值。其属性如表 7-37 所示。

表 7-37 <fmt:param>标签的属性

属性	描　述	EL 类型	必要否	默认值
value	要设置的参数内容	String	否	

【例 7-37】 读取资源文件。

```
<%@ page language="java" contentType="text/html; charset=UTF-8" pageEncoding="UTF-8" %>
<%@ taglib prefix="fmt" uri="http://java.sun.com/jsp/jstl/fmt" %>
<!DOCTYPE HTML PUBLIC "-//W3C//DTD HTML 4.01 Transitional//EN">
<html>
  <head><title>Hello,world!</title></head>
    <body>
```

```
        <fmt:bundle basename = "Message">
          <fmt:message key = "name" var = "nameref"/>
        </fmt:bundle>
        <h3>姓名: ${nameref}</h3>
        <fmt:bundle basename = "Message">
          <fmt:message key = "info" var = "inforef">
            <fmt:param value = "${nameref}"/>
          </fmt:message>
        </fmt:bundle>
      <h3>信息: ${inforef}</h3>
    </body>
</html>
```

在 src 目录下创建资源文件 Message. properties,其内容如下。

```
name = zhangsan
info = welcome,{0}\!
```

程序的运行结果如图 7-23 所示。

姓名: zhangsan
信息: welcome,zhangsan!

图 7-23 例 7-37 运行结果

4) 设置资源文件标签<fmt:setBundle>

在进行资源文件读取时,可以通过<fmt:seBundle>标签设置一个默认的资源名称,这样每次在使用<fmt:message>标签进行信息读取时,直接设置要读取资源的属性即可。<fmt:setBundle>标签的语法格式如下。

```
<fmt:setBundle basename = "资源文件名称"[var = "保存资源文件的属性名称"]
[scope = [page|request|session|aplication]"/>
```

<fmt:setBundle>标签的属性如表 7-38 所示。

表 7-38 <fmt:setBundle>标签的属性

属 性	描 述	EL 类型	必要否	默认值
basename	资源文件的基础名称,供作用域变量或配置变量使用	String	是	无
var	存储资源文件的变量名	String	否	
scope	变量的作用域	String	否	page

【例 7-38】 设置默认资源文件。

```
<%@page import = "java.util.Date" %>
<%@ page language = "java" contentType = "text/html; charset = UTF - 8" pageEncoding = "UTF -
8" %>
<%@ taglib prefix = "fmt" uri = "http://java.sun.com/jsp/jstl/fmt" %>
<!DOCTYPE HTML PUBLIC " - //W3C//DTD HTML 4.01 Transitional//EN">
<html>
  <head><title>Hello,world!</title></head>
    <body>
      <fmt:setBundle basename = "Message" var = "Src"/>
      <fmt:message key = "name" var = "nameref" bundle = "${Src}"/>
```

```
        <h3>姓名: ${nameref}</h3>
        <fmt:message key="info" var="inforef" bundle="${Src}">
          <fmt:param value="${nameref}"/>
        </fmt:message>
          <h3>信息: ${inforef}</h3>
    </body>
</html>
```

运行结果同例 7-37。

3. 数字格式化标签

要进行数字格式化操作,可以使用< fmt:formatNumber > 和< fmt:parseNumber > 两个标签来完成。

1) < fmt:formatNumber >标签

< fmt:formatNumber >标签用于格式化数字、百分比、货币。其语法格式如下。

```
<fmt:formatNumber value="数字"
[type="[number|currency|percent]"]
[pattern="格式化格式"]
[currencyCode="货币的 ISO 编码"]
[currencySymbol="货币符号"]
[groupingUsed="[true|false]"]
[maxIntegerDigits="整数位的最大显示长度"]
[minIntegerDigits="整数位的最小显示长度"]
[maxFractionDigits="小数位的最大显示长度"]
[minFractionDigits="小数位的最小显示长度"]
[var="格式化数据的保存属性"]
[scope="[page|requaet|session|application]"]/>
```

该标签可以有标签体,其属性如表 7-39 所示。

表 7-39 < fmt:formatNumber >标签的属性

属 性	描 述	EL 类型	必要否	默认值
value	要显示的数字	String	是	无
type	number、currency 或 percent	String	否	number
pattern	指定一个自定义的格式化模式用于输出	String	否	无
currencyCode	货币码(当 type="currency"时)	String	否	取决于默认区域
currencySymbol	货币符号(当 type="currency"时)	String	否	取决于默认区域
groupingUsed	是否对数字分组(true 或 false)	String	否	true
maxIntegerDigits	整型数最大的位数	String	否	无
minIntegerDigits	整型数最小的位数	String	否	无
maxFractionDigits	小数点后最大的位数	String	否	无
minFractionDigits	小数点后最小的位数	String	否	无
var	存储格式化数字的变量		否	显示在页面
scope	var 属性的作用域		否	page

pattern 的格式符如表 7-40 所示。

表 7-40　pattern 的格式符

符号	含　义	符号	含　义
0	代表一位数字	;	分隔格式
E	使用指数格式	-	使用默认负数前缀
#	代表一位数字,若没有则显示 0,前导 0 和追尾 0 不显示	%	百分数
.	小数点	?	千分数
,	数字分组分隔符	¥	货币符号,使用实际的货币符号代替
X	指定可以作为前缀或后缀的字符	'	在前缀或后缀中引用特殊字符

【例 7-39】　数字的格式化输出。

```
<%@ page language="java" contentType="text/html; charset=UTF-8" pageEncoding="UTF-8"%>
<%@ taglib prefix="fmt" uri="http://java.sun.com/jsp/jstl/fmt" %>
<!DOCTYPE HTML PUBLIC "-//W3C//DTD HTML 4.01 Transitional//EN">
<html>
  <head><title>格式标签</title></head>
   <body>
    <fmt:formatNumber value="123456.123456"
      maxIntegerDigits="7" maxFractionDigits="3"
      groupingUsed ="true" var="num"/>
      <h3>格式化数字: ${num}</h3>
      <fmt:formatNumber value="123456.456789"
      pattern="##.###E0" var="num"/>
      <h3>科学计数法: ${num}</h3>
    </body>
</html>
```

程序的运行结果如图 7-24 所示。

格式化数字: 123,456.123
科学计数法: 12.346E4

图 7-24　例 7-39 运行结果

2) <fmt:parseNumber>标签

<fmt:parseNumber>标签用来解析数字、百分数、货币,其语法格式如下。

```
<fmt:parseNumber value="格式化的数字"
[type="[number|currency|percent]"]
[pattern="格式化格式"]
[parseLocale="区域编码"]
[integerOnly="[true|false]"]
[var="格式化数据的保存属性"]
[scope="[page|requaet|session|application]"]/>
```

该标签可以有标签体,其属性如表 7-41 所示。

表 7-41　<fmt:parseNumber>标签的属性

属 性	描　述	EL 类型	必要否	默认值
value	要解析的数字	String	否	body
type	number、currency 或 percent	String	否	number
parseLocale	解析数字时所用的区域	String	否	默认区域
integerOnly	是否只解析整型数(true)或浮点数(false)	String	否	false
pattern	自定义解析模式	String	否	无
timeZone	要显示的日期的时区	String	否	默认时区
var	存储待解析数字的变量名		否	显示在页面
scope	var 属性的作用域		否	page

pattern 属性与<fmt:formatNumber>标签中的 pattern 属性有相同的作用。在解析时，pattern 属性告诉解析器期望的格式。

【例 7-40】 数字的反格式化处理。

```
<%@ page language = "java" contentType = "text/html; charset = UTF - 8" pageEncoding = "UTF -
8" %>
<%@ taglib prefix = "fmt" uri = "http://java.sun.com/jsp/jstl/fmt" %>
<! DOCTYPE HTML PUBLIC " - //W3C//DTD HTML 4.01 Transitional//EN">
<html>
  <head><title>逆格式化</title></head>
    <body>
      <fmt:parseNumber value = "1,234,567.345"var = "num"/>
      <h3>格式化数字: ${num}</h3>
      <fmt:parseNumber value = "7.891E6" pattern = "##.##E0"var = "num"/>
        <h3>科学计数法: ${num}</h3>
        <fmt:parseNumber value = "1.5%" pattern = "00%" var = "num"/>
        <h3>百分比: ${num}</h3>
    </body>
</html>
```

程序的运行结果如图 7-25 所示。

格式化数字: 1234567.345
科学计数法: 7891000
百分比: 0.015

图 7-25　例 7-40 运行结果

4. 日期时间格式化标签

日期时间格式化标签实现对日期和时间的格式化显示，包括<fmt:formatDate>和<fmt:parseDate>两个标签。

1) <fmt:formatDate>标签

<fmt:formatDate>标签用于使用不同的方式格式化日期，其语法格式如下。

```
<fmt:formatDate
  value = "date"
  [type = "[time|date|both]"]
  [dateStyle = "[default|short|medium|long|full]"]
  [timeStyle = "[default|short|medium|long|full]"]
  [pattern = "格式化样式"]
  [timeZone = "时域"]
```

```
    [var = "格式化数据的保存属性"]
    [scope = "[page|requaet|session|application]"]/>
```

该标签可以有标签体,其属性如表 7-42 所示。

表 7-42　<fmt:formatDate>标签的属性

属 性	描　述	EL 类型	必要否	默认值
value	要显示的日期	String	是	无
type	date、time 或 both	String	否	date
dateStyle	full、long、medium、short 或 default	String	否	default
timeStyle	full、long、medium、short 或 default	String	否	default
pattern	自定义格式模式	String	否	无
timeZone	显示日期的时区	String	否	默认时区
var	存储格式化日期的变量名		否	显示在页面
scope	存储格式化日志变量的范围		否	页面

<fmt:formatDate> 标签的格式模式如下。

G:时代标志。

y:不包含纪元的年份。如果不包含纪元的年份小于 10,则不显示前导零。

M:月份数字。一位数的月份没有前导零。

d:月中的某一天。一位数的日期没有前导零。

h:12 小时制的小时。一位数的小时数没有前导零。

H:24 小时制的小时。一位数的小时数没有前导零。

m:分钟。一位数的分钟数没有前导零。

s:秒。一位数的秒数没有前导零。

S:毫秒。

E:星期几。

D:一年中的第几天。

F:一个月中的第几个星期几。

w:一年中的第几周。

W:一个月中的第几周。

a:a. m. /p. m. 指示符。

k:小时(12 小时制的小时)。

K:小时(24 小时制的小时)。

z:时区。

':转义文本。

'':单引号。

【例 7-41】 日期显示格式。

```
<%@page import = "java.util.Date" %>
<%@ page language = "java" contentType = "text/html; charset = UTF - 8" pageEncoding = "UTF -
```

```
8"%>
<%@ taglib prefix="fmt" uri="http://java.sun.com/jsp/jstl/fmt" %>
<!DOCTYPE HTML PUBLIC "-//W3C//DTD HTML 4.01 Transitional//EN">
<html>
  <head><title>日期格式</title></head>
    <body>
    <%  pageContext.setAttribute("dateRef", new Date()); %>
<fmt:formatDate value="${dateRef}"
      type="both" dateStyle="default" timeStyle="default" var="date"/>
  <h3>default 显示日期时间: ${date}</h3>
    <fmt:formatDate value="${dateRef}"
      type="both" dateStyle="short" timeStyle="short" var="date"/>
<h3>short 显示日期时间: ${date}</h3>
  <fmt:formatDate value="${dateRef}"
      type="both" dateStyle="medium" timeStyle="medium" var="date"/>
<h3>medium 显示日期时间: ${date}</h3>
<fmt:formatDate value="${dateRef}"
      type="both" dateStyle="long" timeStyle="long" var="date"/>
<h3>long 显示日期时间: ${date}</h3>
<fmt:formatDate value="${dateRef}"
      type="both" dateStyle="full" timeStyle="full" var="date"/>
    <h3>full 显示日期时间: ${date}</h3>
    <fmt:formatDate value="${dateRef}"
      type="both" pattern="yyyy年MM月dd日HH时mm分ss秒sss毫秒" var="date"/>
    <h3>自定义格式显示日期时间: ${date}</h3>
  </body>
</html>
```

程序的运行结果如图 7-26 所示。

default显示日期时间: 2020-4-17 19:38:22
short显示日期时间: 20-4-17 下午7:38
medium显示日期时间: 2020-4-17 19:38:22
long显示日期时间: 2020年4月17日 下午07时38分22秒
full显示日期时间: 2020年4月17日 星期五 下午07时38分22秒 CST
自定义格式显示日期时间: 2020年04月17日19时38分22秒022毫秒

图 7-26 例 7-41 运行结果

2) <fmt:parseDate>标签

<fmt:parseDate>标签用于解析日期,其语法格式如下。

```
<fmt:parseDate value="date"
  [type="[time|date|both]"]
  [dateStyle="[default|short|medium|long|full]"]
  [timeStyle="[default|short|medium|long|full]"]
  [pattern="格式化样式"]
  [timeZone="时域"]
  [var="格式化数据的保存属性"]
  [scope="[page|requaet|session|application]"]/>
```

其属性如表 7-43 所示。

表 7-43 <fmt:parseDate>标签的属性

属 性	描 述	EL 类型	必要否	默认值
value	要显示的日期	String	是	无
type	date、time 或 both	String	否	date
dateStyle	full、long、medium、short 或 default	String	否	default
timeStyle	full、long、medium、short 或 default	String	否	default
pattern	自定义格式模式	String	否	无
timeZone	显示日期的时区	String	否	默认时区
var	存储格式化日期的变量名		否	显示在页面
scope	存储格式化日志变量的范围		否	page

【例 7-42】 日期格式解析。

```
<%@ page language = "java" contentType = "text/html; charset = UTF - 8" pageEncoding = "UTF -
8" %>
<%@ taglib prefix = "fmt" uri = "http://java.sun.com/jsp/jstl/fmt" %>
<!DOCTYPE HTML PUBLIC " - //W3C//DTD HTML 4.01 Transitional//EN">
<html>
  <head><title>日期解析!</title></head>
   <body>
    <fmt:parseDate value = "2020 年 4 月 17 日 星期五 下午 08 时 04 分 41 秒 CST"
    type = "both" dateStyle = "full" timeStyle = "full" var = "date"/>
    <h3>字符串变为日期: ${date}</h3>
    <fmt:parseDate value = "2020 年 04 月 17 日 20 时 01 分 21 秒 021 毫秒"
        pattern = "yyyy 年 MM 月 dd 日 HH 时 mm 分 ss 秒 sss 毫秒" var = "date"/>
    <h3>字符串变为日期: ${date}</h3>
</body>
</html>
```

程序的运行结果如图 7-27 所示。

字符串变为日期: Fri Apr 17 20:04:41 CST 2020
字符串变为日期: Fri Apr 17 20:01:21 CST 2020

图 7-27 例 7-42 运行结果

5. 设置时区标签

设置时区有<fmt:setTimeZone>标签和<fmt:timeZone>标签。

1) <fmt:setTimeZone>标签

<fmt:setTimeZone>标签可以设置显示的时区或者将设置的时区存储到一个属性范围中，其语法格式如下。

```
<fmtsetTimeZone value = "设的时区"[var = "存储时区的属性名称"][scope - "page|request|
session|aplication]"/>
```

该标签的属性如表 7-44 所示。

表 7-44　<fmt:setTimeZone>标签的属性

属性	描　　述	EL 类型	必要否	默认值
value	时区	String	是	无
var	存储新时区的变量名		否	Replace default
scope	变量的作用域		否	Page

【例 7-43】 修改时区。

```
<%@page import = "java.util.Date" %>
<%@ page language = "java" contentType = "text/html; charset = UTF - 8" pageEncoding = "UTF -
8" %>
<%@ taglib prefix = "fmt" uri = "http://java.sun.com/jsp/jstl/fmt" %>
<!DOCTYPE HTML PUBLIC " - //W3C//DTD HTML 4.01 Transitional//EN">
<html>
  <head><title>修改时区</title></head>
    <body>
    <% pageContext.setAttribute("now", new Date()); %>
当前时区时间:<fmt:formatDate value = "${now}" type = "both" timeStyle = "long" dateStyle =
"long" /><br>
<p>修改为 GMT - 8 时区:</p>
<fmt:setTimeZone value = "GMT - 8" />
Changed Zone: <fmt:formatDate value = "${now}" type = "both" timeStyle = "long" dateStyle =
"long" />
</body>
</html>
```

程序的运行结果如图 7-28 所示。

当前时区时间:2020年4月17日 下午08时21分06秒
修改为 GMT-8 时区:
Changed Zone: 2020年4月17日 上午04时21分06秒

图 7-28　例 7-43 运行结果

2）<fmt:timeZone>标签

使用<fmt:timeZone>标签可以设定一个暂时的时区,其语法格式如下。

```
<fmt:timeZone value = "设置的时区">
  标签体
</fmt:timeZone>
```

该标签的属性如表 7-45 所示。

表 7-45　<fmt:timeZone>标签的属性

属性	描述	EL 类型	必要否	默认值
value	时区	String	是	无

【例 7-44】 设置临时时区。

```
<%@page import = "java.util.Date" %>
<%@ page language = "java" contentType = "text/html; charset = UTF - 8" pageEncoding = "UTF - 8" %>
<%@ taglib prefix = "fmt" uri = "http://java.sun.com/jsp/jstl/fmt" %>
<!DOCTYPE HTML PUBLIC " - //W3C//DTD HTML 4.01 Transitional//EN">
<html>
```

```
  <head><title>修改时区</title></head>
    <body>
      <% pageContext.setAttribute("now", new Date()); %>
      本地时间: ${now}<br>
    <fmt:timeZone value = "GMT - 8">
    <fmt:formatDate value = "${now}"
      var = "date" type = "both"
      dateStyle = "short"
      timeStyle = "short"
    />
      </fmt:timeZone>
      GMT - 8 时间: ${date}
    </body>
</html>
```

程序的运行结果如图 7-29 所示。

本地时间:Fri Apr 17 20:35:29 CST 2020
GMT-8时间:20-4-17 上午4:35

图 7-29　例 7-44 运行结果

7.2.5　函数标签库

JSTL 包含一系列标准函数，大部分是通用的字符串处理函数。引用 JSTL 函数库的语法如下。

```
<%@ taglib prefix = "fn" uri = "http://java.sun.com/jsp/jstl/functions" %>
```

常用的函数标签如表 7-46 所示。

表 7-46　常用的函数标签

函　　数	描　　述
fn:contains(string,string)	检查输入的字符串是否包含指定的子字符串
fn:containsIgnoreCase(string,string)	检查输入的字符串是否包含指定的子字符串,大小写不敏感
fn:endsWith(string,string)	检查输入的字符串是否以指定的后缀结尾
fn:escapeXml(string)	跳过可以作为 XML 标签的字符
fn:indexOf(string,string)	返回指定字符串在输入字符串中出现的位置
fn:join(string[],string)	将数组中的元素合成一个字符串然后输出
fn:length(string)	返回字符串长度
fn:replace(String,string,string)	将输入字符串中指定的位置替换为指定的字符串然后返回
fn:split(string,string)	将字符串用指定的分隔符分隔然后组成一个子字符串数组并返回
fn:startsWith(string,string)	检查输入字符串是否以指定的前缀开始
fn:substring(string,string)	返回字符串的子集
fn:substringAfter(string,string)	返回字符串在指定子字符串之后的子集
fn:substringBefore(string,string)	返回字符串在指定子字符串之前的子集
fn:toLowerCase(string)	将字符串中的字母转为小写

续表

函数	描述
fn:toUpperCase(string)	将字符串中的字母转为大写
fn:trim(string)	移除首尾的空白符

【例 7-45】 字符串的运算。

```
<%@page import = "java.util.Date" %>
<%@ page language = "java" contentType = "text/html; charset = UTF-8" pageEncoding = "UTF-8" %>
<%@ taglib prefix = "fn" uri = "http://java.sun.com/jsp/jstl/functions" %>
<%@ taglib uri = "http://java.sun.com/jsp/jstl/core" prefix = "c" %>
<!DOCTYPE HTML PUBLIC " - //W3C//DTD HTML 4.01 Transitional//EN">
<html>
  <head><title>字符串</title></head>
    <body>
      <c:set var = "string1" value = "this is a tag"/>
      替换: ${fn:replace(string1,"tag","table")}<br>
      子串: ${fn:substring(string1,0,10)}<br>
      拆分: ${fn:split(string1,"")[0]}<br>
      长度:${fn:length(string1)}<br>
      查找:${fn:indexOf(string1,"is")}<br>
      <c:set var = "string2" value = "${fn:split(string1, ' ')}" />
      <c:set var = "string3" value = "${fn:join(string2, '-')}" />
      连接 : ${string3}<br>
    </body>
</html>
```

7.2.6 任务:用户管理的界面设计

1. 任务要求

利用 SQL 标签实现对用户的增加、注册、删除、修改、查找等功能。

2. 任务实施

(1) 创建 Java Web 项目 UserManger。

(2) 创建一个用户 JavaBean,包括用户名、用户 ID、用户密码、用户类型(普通用户、系统管理员等)等基本信息,参考 4.1 节中的任务。

(3) 编写用户管理的主页面 jstlUserM.jsp。

```
<%@ page language = "java" contentType = "text/html; charset = UTF-8"
    pageEncoding = "UTF-8" %>
<%@ taglib uri = "http://java.sun.com/jsp/jstl/core" prefix = "c" %>
<%@ taglib uri = "http://java.sun.com/jsp/jstl/sql" prefix = "sql" %>
<html>
<head><title>用户信息</title></head>
```

```
  <body>
    <sql:setDataSource var = "ds" scope = "session"
      driver = "com.microsoft.sqlserver.jdbc.SQLServerDriver"
      url = "jdbc:sqlserver://127.0.0.1:1433;DatabaseName = test"
      user = "sa"
      password = "123456"/>
  <sql:query dataSource = "${ds}" sql = "select * from userTable" var = "result" />
  <table>
    <tr><th>用户 ID</th><th>用户名</th><th>修改</th>
      <th>删除</th><th>更多</th></tr>
  <c:forEach var = "row" items = "${result.rows}">
  <tr>
    <td><c:out value = "${row.userID}"/></td>
    <td><c:out value = "${row.userName}"/></td>
      <c:url var = "delurl" value = "jstldel.jsp">
        <c:param name = "uID" value = "${row.userID}"/>
      </c:url>
      <c:url var = "modurl" value = "jstledit.jsp">
        <c:param name = "uID" value = "${row.userID}"/>
      </c:url>
      <c:url var = "geturl" value = "jstlmore.jsp">
        <c:param name = "uID" value = "${row.userID}"/>
      </c:url>
        <td><a href = "${delurl}">删除</a></td>
        <td><a href = "${modurl}">修改</a></td>
        <td><a href = "${geturl}"> more </a></td>
      </tr>
    </c:forEach>
  </table>
  <a href = "jstladdform.jsp">增加</a>
  </body>
</html>
```

(4) 编写删除用户页面 jstldel.jsp。

```
<%@ page language = "java" contentType = "text/html; charset = UTF-8"
        pageEncoding = "UTF-8" %>
<%@ taglib uri = "http://java.sun.com/jsp/jstl/core" prefix = "c" %>
<%@ taglib uri = "http://java.sun.com/jsp/jstl/sql" prefix = "sql" %>
<c:set var = "logdIn" value = "${not empty param.uID}"/>
<c:if test = "${logdIn}">
<sql:update var = "up"
  sql = "delete from usertable where userID = ?" dataSource = "${ds}">
    <sql:param value = "${param.uID}"/>
  </sql:update>
</c:if>
<c:redirect url = "jstlUserM.jsp"/>
```

(5) 编写输入用户信息的页面 jstladdform.jsp。

```
<%@ page language = "java" contentType = "text/html; charset = UTF-8"
        pageEncoding = "UTF-8" %>
```

```
<%@ taglib uri="http://java.sun.com/jsp/jstl/core" prefix="c" %>
<%@ taglib uri="http://java.sun.com/jsp/jstl/sql" prefix="sql" %>
<html>
<head><title>增加用户</title></head>
<body>
      <form action="jstladd.jsp" method="post">
      用户 ID:<input type="text" name="userID" value="${user.userID}"/><br>
      用户名:<input type="text" name="userName" value="${user.userName}"/><br>
      用户密码:<input type="text" name="userName" value="${user.userPwd}"/><br>
      <input type="submit" value="submit"/>
    </form>
  </body>
</html>
```

(6) 编写增加用户信息的页面 jstladd.jsp。

```
<%@ page language="java" contentType="text/html; charset=UTF-8"
    pageEncoding="UTF-8" %>
<%@ taglib uri="http://java.sun.com/jsp/jstl/core" prefix="c" %>
<%@ taglib uri="http://java.sun.com/jsp/jstl/sql" prefix="sql" %>
<c:set var="logdIn" value="${not empty param.userID}"/>
<c:if test="${logdIn}">
  <sql:update var="up"
    sql="insert into userTable(userId,username,userpwd) values(?,?,?)"
    dataSource="${ds}">
      <sql:param value="${param.userID}"/>
      <sql:param value="${param.userName}"/>
      <sql:param value="${param.userPwd}"/>
  </sql:update>
</c:if>
<c:redirect url="jstlUserM.jsp"/>
```

(7) 编写修改用户信息的页面 jstledit.jsp。

```
<%@ page language="java" contentType="text/html; charset=UTF-8"
          pageEncoding="UTF-8" %>
<%@ taglib uri="http://java.sun.com/jsp/jstl/core" prefix="c" %>
<%@ taglib uri="http://java.sun.com/jsp/jstl/sql" prefix="sql" %>
<html>
  <head><title>修改用户信息</title></head>
  <body>
    <c:set var="logdIn" value="${empty param.uID}"/>
    <c:if test="${logdIn}">
      <c:redirect url="jstlUserM.jsp"/>
    </c:if>
    <sql:query dataSource="${ds}"
      sql="select * from userTable where userID=?" var="result">
      <sql:param value="${param.uID}"/>
    </sql:query>
    <jsp:useBean id="user" class="cn.edu.mypt.bean.UserBean"/>
```

```
    <c:forEach var = "row" items = "${result.rows}">
        <c:set property = "userID" target = "${user}" value = "${row.userID}"/>
        <c:set property = "userName" target = "${user}" value = "${row.userName}"/>
    </c:forEach>
    用户 ID:<c:out value = "${user.userID}"/>
    <form action = "jstlmod.jsp" method = "post">
        用户名:<input type = "text" name = "userName" value = "${user.userName}"/>
        用户密码: <input type = "text" name = "userName" value = "${user.userPwd}"/>
        <input type = "hidden" name = "userID" value = "${user.userID}"/>
        <input type = "submit" value = "submit"/>
    </form>
  </body>
</html>
```

(8) 编写修改用户信息的处理程序 jstlmod.jsp。

```
<%@ page language = "java" contentType = "text/html; charset = UTF - 8" pageEncoding = "UTF -
8" %>
<%@ taglib uri = "http://java.sun.com/jsp/jstl/core" prefix = "c" %>
<%@ taglib uri = "http://java.sun.com/jsp/jstl/sql" prefix = "sql" %>
<c:set var = "logdIn" value = "${not empty param.userID}"/>
<c:if test = "${logdIn}">
    <sql:update var = "up" dataSource = "${ds}"
      sql = "update usertable set username = ?,userpwd = ? where userID = ?">
    <sql:param value = "${param.userName}"/>
    <sql:param value = "${param.userPwd}"/>
    <sql:param value = "${param.userID}"/>
  </sql:update>
</c:if>
<c:redirect url = "jstlUserM.jsp"/>
```

(9) 编写显示用户信息页面 jstluser.jsp。

```
<%@ page language = "java" contentType = "text/html; charset = UTF - 8"
         pageEncoding = "UTF - 8" %>
<%@ taglib uri = "http://java.sun.com/jsp/jstl/core" prefix = "c" %>
<%@ taglib uri = "http://java.sun.com/jsp/jstl/sql" prefix = "sql" %>
<html>
<head><title>用户信息</title></head>
<body>
<c:set var = "logdIn" value = "${empty param.uID}"/>
<c:if test = "${logdIn}">
    <c:redirect url = "jstlUserM.jsp"/>
</c:if>
  <sql:query dataSource = "${ds}"
    sql = "select * from userTable where userID = ?"
    var = "result" >
    <sql:param value = "${param.uID}"/>
  </sql:query>
  <jsp:useBean id = "user" class = "cn.edu.mypt.bean.UserBean"/>
```

```
    <c:forEach var = "row" items = "${result.rows}">
        <c:set property = "userID" target = "${user}" value = "${row.userID}"/>
        <c:set property = "userName" target = "${user}" value = "${row.userName}"/>
    </c:forEach>
    用户 ID:<c:out value = "${user.userID}"/><br>
    用户名:<c:out value = "${user.userName}"/><br>
    <a href = "jstlUserM.jsp">back</a>
  </body>
</html>
```

（10）启动服务器，部署项目并运行 http://127.0.0.1:8080.UserManger/jstlUserM.jsp。

7.3 自定义标签和函数

JSTL 只提供了简单的输出等功能，没有实现任何的 HTML 代码封装。利用 JSTL 实现一个表格每次都要编写 table、tr、td 等 HTML 代码。而自定义标签就可以实现 HTML 代码的封装，甚至能做到只要往自定义标签里传入一个集合对象，就能够用表格列出所有的对象及其属性。

JSP 中设计自定义标签的目的是实现 HTML 代码复用。自定义标签可以实现非常复杂的功能，使用起来却很简单。用户可以将常用的表现元素如多样化的表格、复杂的报表等封装成自定义标签，这样可以大大提高开发的效率。

7.3.1 什么是自定义标签

自定义标签是使用 XML 语法格式完成程序操作的一种方法，其使用的形式类似于 JavaBean 的语法<jsp:useBean>。与 JavaBean 一样，可以将大量的复杂操作写在类中完成，而且自定义标签最大的优势是它可以按照 HTML 标签的形式表现，这样可以方便地处理 JSP 页面的数据显示。使用自定义标签具有以下好处。

（1）可以简洁地在 JSP 页面上构建模板。

（2）便于程序员和页面编辑人员分工协作，各自维护代码和页面。

（3）非常好的分离了页面内容和表现形式。

（4）标签文件具有良好的可重用性。

1. 自定义标签的分类

在自定义标签的开发中，主要有以下几种典型的标签类型。

（1）简单标签（空标签），如<mytag：helloworld/>。

（2）带属性标签，如<imytag：max num＝"3" num2＝"5"/>。

（3）带内容的标签，在自定义标签的起始和结束标签之间的部分为标签体（内容），内容可以是 JSP 中的标准标签，也可以是 HTML、脚本语言或其他的自定义标签，如<imytag：greeting>hello</greeting>。

（4）带有内容和属性的标签，如<imytag：greeting name＝"Tom">hello</greeting>。

2. 创建和使用自定义标签的基本步骤

在创建和使用自定义标签时，主要按照下面的步骤操作。

(1) 创建标签的处理类。

在 JSP 2.0 版本以前，标签必须直接或间接地实现 javax.servlet.jsp.tagext.Tag 接口，在 Tag 接口中主要定义的是和标签生命周期相关的方法。

(2) 创建标签库描述文件。

标签库描述文件是一个扩展名为.tld 的 XML 文档，它描述了标签处理程序的属性、信息和位置，JSP 通过这个文件获取调用标签处理类。

(3) 在 web.xml 文件中配置元素。

(4) 在 JSP 文件中引入标签库。

一般情况下，开发 JSP 自定义标签需要引用以下两个包。

```
import javax.servlet.jsp.*;
import javax.servlet.jsp.tagext.*;
```

7.3.2 标签处理程序的接口和类

标签处理程序是一个运行时调用的 Java 类，包含在 JSP 文件中定义的标签所实现的代码。标签处理程序类必须实现或扩展 javax.servlet.jsp.tagext 包中的类和接口。

1. Tagext 包的接口

Tagext 包的接口和类的继承关系如图 7-30 所示。

Tag 接口、IterationTag 接口和 BodyTag 接口以及 TagSupport 类和 BodyTagSupport 类为传统标签。SimpleTag 接口和 SimpleTagSupport 类为自定义标签。

所有标签处理类中，实现了 Tag 接口的标签为传统标签，实现 SimpleTag 接口的标签为简单标签。简单标签可以实现传统标签的一切功能，而且开发起来更容易。开发自定义标签的核心就是编写处理类，需要继承的就是 SimpleTagSupport 类。SimpleTagSupport 类实现了 SimpleTag 接口。

(1) Tag：定义了标签处理程序和 JSP 页面生成的 Servlet 之间的基本协议，定义了所有标签处理程序的基本方法。Tag 接口的定义如下。

```
public Interface Tag extends JspTag{
  public static final int SKIP_BODY;//忽略标签体的内容,将内容转交给 doEndTag()
  public static final int EVAL_BODY_INCLUDE;//正常执行标签体内容,但不做任何处理
  public static final int SKIP_PAGE;//所有在 JSP 上的操作都将停止,会将所有输出的内容立刻
                                    //显示在浏览器上
  public static final int EVAL_PGE;//正常执行 JSP 页面
  public Tag getParent(Tag);
  public int doStartTag()throws JspException;
  public int doEndTag()throws JspException;
  public void release();
}
```

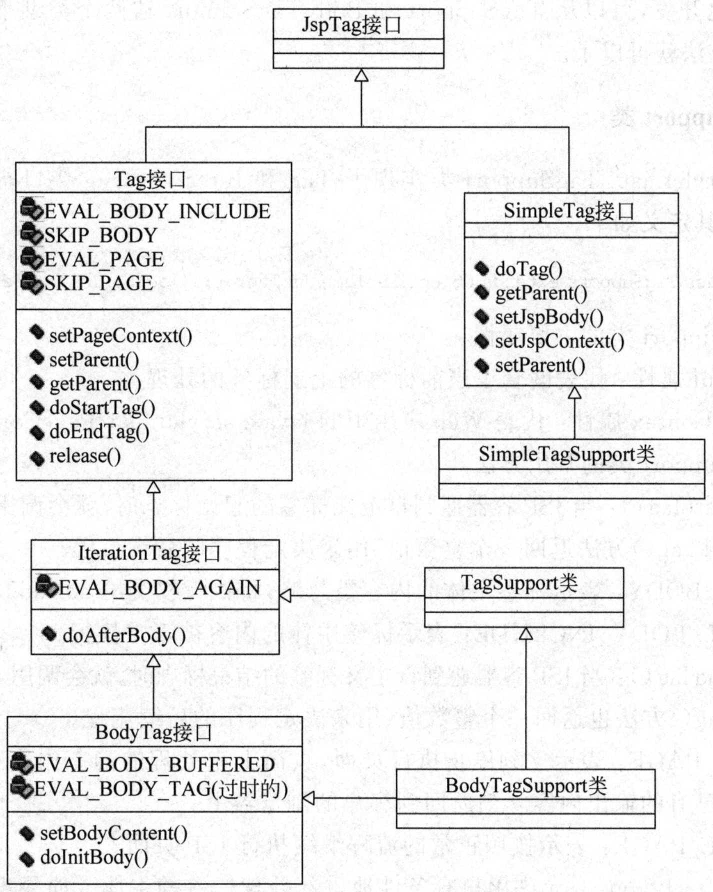

图 7-30 Tagext 包的接口及类的继承关系

(2) IterationTag：扩展了 Tag 接口，增加了一个控制重复处理标签主体内容的方法。IterationTag 接口的定义如下。

```
public interface IterationTag extends Tag {
  public static final int EVAL_BODY_AGAIN;//重复执行标签体内容,会再次调用 doAfterBody(),
                                          //直到出现 SKIP_BODY 为止
  public int doAfterBody() throws JspException;
}
```

(3) BodyTag 接口：扩展了 IterationTag 接口，根据需要对标签的主体内容进行访问并能够对其进行操作。

```
public interface BodyTag extends IterationTag {
  public static final int EVAL_BODY_BUFFERED; //表示标签体的内容应该被处理,所有的处理
                                              //结果都将保存在 BodyContent 类中
  public static final int EVAL_BODY_TAG;
  public void setBodyContent(BodyContent b)
  public void doinitBody() throws JspException;
}
```

为了简化开发，可以从 TagSupport 和 BodyTagSupport 这两个类继承，只需要重新定义所需的方法就可以了。

2. TagSupport 类

javax. servlet. jsp. TagSupport 类实现了 Tag 和 InterationTag 接口，是整个标签编程的核心类，其定义如下。

```
public class TagSupport extends Object implements IterationTag,Serializable
```

1）TagSupport 类的主要属性

（1）parent 属性：代表嵌套了当前标签的上层标签的处理类。

（2）pageContex 属性：代表 Web 应用中的 javax. servlet. jsp. PageContext 对象。

2）TagSupport 类的常用方法

（1）doStartTag()：当 JSP 容器遇到自定义标签的起始标志时，就会调用 doStartTag() 方法。doStartTag() 方法返回一个整数值，用来决定程序的后续流程。

① SKIP_BODY：表示标签主体的内容被忽略，将执行权交给 doEndTag() 方法。

② EVAL_BODY_INCLUDE：表示标签主体的内容被正常执行。

（2）doEndTag()：当 JSP 容器遇到自定义标签的结束标志时，就会调用 doEndTag() 方法。doEndTag() 方法也返回一个整数值，用来决定程序的后续流程。

① SKIP_PAGE：表示立刻停止执行页面，页面上未处理的静态内容和 JSP 程序均被忽略，任何已有的输出内容立刻返回到客户的浏览器上。

② EVAL_PAGE：表示按照正常的流程继续执行 JSP 页面。

（3）doAfterBody()：允许用户有条件地重新处理标签的主体。如果没有标签主体，则不会调用 doAfterBody() 方法。它有如下两种返回值。

① SKIP_BODY：表示标签主体的内容被忽略，将执行权交给 doEndTag() 方法。

② EVAL_BODY_AGAIN：表示重复执行标签体的内容，会重复调用 doAfterBody() 方法，一直循环执行下去，直到 doAferBody() 方法返回 SKIP_BODY 为止。

3）Tag 接口的执行流程

Tag 接口的执行流程如图 7-31 所示。

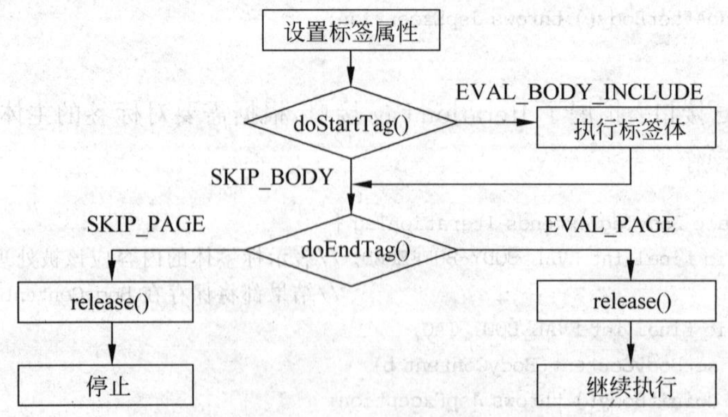

图 7-31　Tag 接口的执行流程

4) IterationTag 接口的执行流程

IterationTag 接口的执行流程如图 7-32 所示。

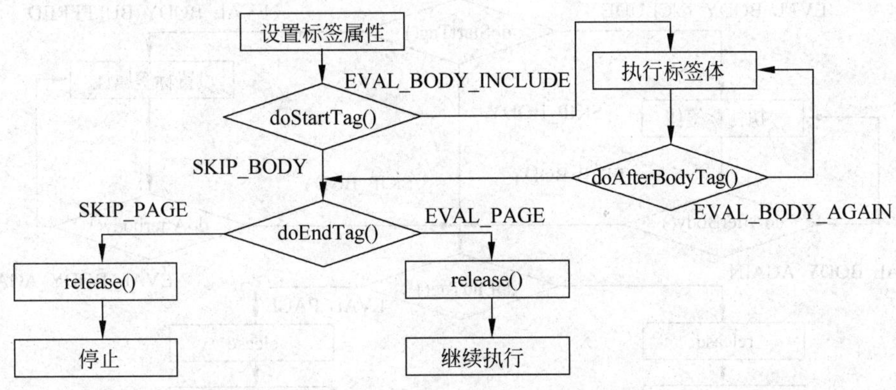

图 7-32 IterationTag 接口的执行流程

注意：JSP 容器在调用 doStartTag()或者 doEndTag()方法前，会先调用 setPageContext()和 setParent()方法设置 pageContext 和 parent。因此在标签处理类中可以直接访问 pageContext 变量。

3. BodyTagSupport 类

BodyTagSupport 是 TagSupport 类的子类，通过 BodyTagSupport 类实现的标签可以直接处理标签体内容的数据。

1) BodyTagSupport 主要属性

protected BodyContent bodyContent：存放处理结果。

2) BodyTagSupport 常用方法

public JspWriter getPreiousOut()：取得 JspWriter 的输出对象。

在 BodyTagSupport 类中定义了一个受保护的属性 bodyContent，而 bodyContent 是 BodyContent 类的对象，该类的定义如下。

```
public abstract class BodyContent extends JspWriter
```

该类主要有以下 3 个方法。

(1) public abstract Reader()：将所有的内容转换为 Reader 对象。

(2) public abstract String getString()：将所有的内容转换为 String 对象。

(3) public abstrract void writeOut(Writer out) throws IOException：将所有内容输出。

3) BodyTag 接口的执行流程

BodyTag 接口专门处理带标签体的标签，BodyTag 是 Tag 接口的子类，BodyTag 接口除了具有 Tag 接口的 doStartTag()、doEndTag()等方法外，还具有其他的方法。

BodyTag 接口的执行流程如图 7-33 所示。

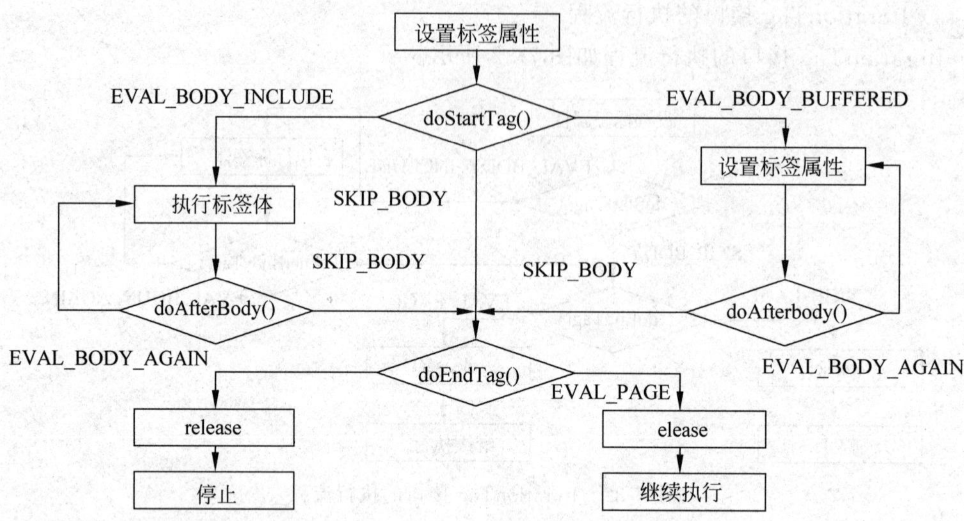

图 7-33 BodyTag 接口的执行流程

7.3.3 简单标签示例

下面通过一个简单的程序演示自定义标签编程。本程序的主要功能依然是在 JSP 页面上输出 Hello World! 的信息。

1. 定义一个标签操作类

任何一个标签都对应一个 Java 类，该类必须实现 Tag 接口。JSP 中遇到一个标签后，将通过一个 TLD 文件查找该标签的实现类，并运行该类的相关方法。

要实现一个标签，可以直接继承 javax. servet. jsp. tagext. TagSupport 类，如果要定义的标签内没有标签体，则直接覆写 TagSupport 类中的 doStartTa()方法即可。

【例 7-46】 定义标签类 HelloTag。

```
package org.myvtc.tag;
import javax.servlet.jsp.JspTagException;
import javax.servlet.jsp.JspWriter;
import javax.servlet.jsp.tagext.TagSupport;
public class Hellotag extends TagSupport {
  public int doStartTag()throws JspTagException{
      JspWriter out = super.pageContext.getOut();
      try{
        out.print("<h1>hello</h1>");
      }
      catch (Exception e){
        e.printStackTrace();
      }
      return this.SKIP_BODY;
  }
}
```

在 HelloTag 类中,首先继承 TagSupport 类。然后覆写 doStartTag()方法,之后通过 TagSupport 类中的 pageContext 属性取得当前页面的输出对象进行页面的输出。由于此时开发的标签没有任何的标签体,所以在程序的最后返回的是 SKIP_BODY 常量,表示不执行标签体的内容。

2. 标签库描述文件 TLD

Tag 实现类有了,还需要在 TLD 文件中配置标签文件的信息,以及标签与实现类的映射。在/WEB-INF/下新建 TLD 文件,其代码格式如下。

```
<taglib>
  <!-- 版本号 -->
  <tlib-version>1.0</tlib-version>
  <jsp-version>2.0</jsp-version>
  <short-name>cc</short-name>
  <uri>URI</uri>
    <tag>
        <!-- 指定标签名 -->
        <name>showUserInfo</name>
        <!-- 指定标签类文件的全路径 -->
        <tag-class>com.mytags.UserInfoTag</tag-class>
        <!-- 如果不需要标签体则设置为 empty,反之设置为 jsp -->
        <body-content>empty</body-content>
        <!-- 设定属性(如果有) -->
        <attribute>
        <!-- 指定标签名 -->
        <name>user</name>
        <!-- 是否必要,如果非必要没设置则为空 -->
        <required>false</required>
      <rtexprvalue>true</rtexprvalue><!—是否可在属性中使用表达式 -->
    </attribute>
  </tag>
<taglib>
```

在标签<taglib>内配置标签文件的基本信息包括 shortname、uri 等。shortname 就是推荐使用的 prefix,uri 就是引用这个标签库时使用的 URI。标签<tag>内配置的就是标签的信息。

(1) name:标签名。

(2) tag-class:实现类。

(3) bodyconten:标签体的限制,有 empty、JSP 与 tagdependent 3 种取值。具体区别如下。

① empty:不允许有标签体存在。如果有,执行时会抛出异常。

② JSP:允许有标签体存在。可以为 JSP 代码。

③ tagdependent:允许有标签体存在,但是标签体内的 JSP 代码不会被执行。

【例 7-47】 创建 HelloTag 的描述文件。

在 WEB-INF 目录下创建 mytaglib. tld 文件。

```
<?xml version = "1.0" encoding = "GBK" ?>
<!DOCTYPE taglib
PUBLIC " - //Sun Microsystems, Inc.//DTD JSP Tag Library 1.2//EN"
"http://java.sun.com/dtd/web - jsptaglibrary_1_2.dtd">
<taglib>
  <description>自定义标签库</description>
  <tlib - version>1.1</tlib - version>
  <jsp - version>1.2</jsp - version>
  <short - name>ct</short - name>
  <uri>http://www.myvtc.edu.cn</uri>
  <tag>
    <description>hello</description>
    <name>hello</name>
    <tag - class>org.myvtc.tag.hellotag</tag - class>
    <body - content>empty</body - content>
  </tag>
</taglib>
```

3. 配置 web. xml

如果 TLD 文件位于/WEB_ INF/下，Tomcat 会自动加载文件中的标签库。如果位于其他的位置，可以在 web. xml 中配置。web. xml 文件格式如下。

```
<jsp - config>
    <taglib>
    <!-- 标签库的 URI 路径，即 JSP 头文件中声明<%@ taglib uri = "/mytaglib"
      prefix = "cc" %>的 URI -->
        <taglib - uri>/mytaglib</taglib - uri>
          <!—TLD 文件所在的位置 -->
        <taglib - location>/WEB - INF/mytaglib.tld</taglib - location>
    </taglib>
</jsp - config>
```

【例 7-48】 配置标签文件。

在 web. xml 中，Hello 标签的配置如下。

```
<jsp - config>
    <taglib>
        <taglib - uri>/mytaglib</taglib - uri>
        <taglib - location>/WEB - INF/mytaglib.tld</taglib - location>
    </taglib>
</jsp - config>
```

或者在 JSP 中直接使用<%@ taglib uri＝"/WEB-INF/mytaglib. tld" prefix＝"taglib" %>导入该标签库。

4. 在 JSP 中引用标签

采用如下格式，就可以引用自定义标签。

```
<%@ taglib prefx = "标签前缀"uri = "TLD 文件路径" %>
```

prefix 表示标签使用时的前缀。
uri 表示的是此标签对应的 TLD 文件的路径。

【例 7-49】 编写 JSP 页面并调用标签 hellotag.jsp。

```
<%@ page language = "java" contentType = "text/html; charset = UTF - 8"
        pageEncoding = "UTF - 8" %>
<%@ taglib uri = "/mytaglib" prefix = "mytag" %>
<!DOCTYPE HTML PUBLIC " - //W3C//DTD HTML 4.01 Transitional//EN">
<html>
  <head><title>自定义标签</title></head>
    <body>
      <mytag:hello/>
    </body>
</html>
```

浏览页面，其效果如图 7-34 所示。

根据上面编程，可以发现要完成一个自定义标签的开发需要具有以下几个部分。

(1) 标签处理类：HelloTag.java。
(2) 标签描述文件：hellotag.tld。
(3) JSP 页面：通过<%taglib%>定义。
(4) 在 web.xml 文件中配置映射文件名(可选)。

标签库的执行流程如图 7-35 所示。

图 7-34 例 7-49 浏览效果

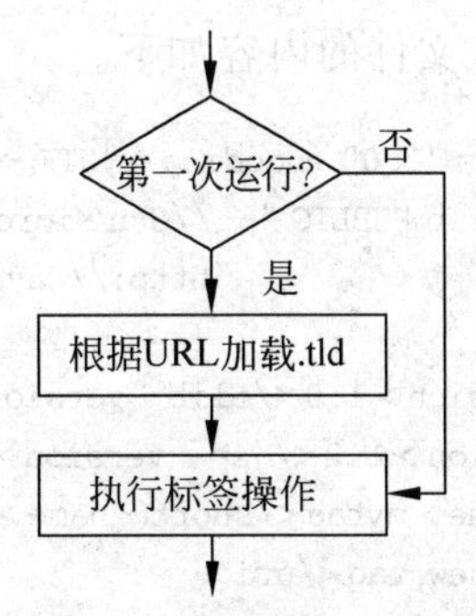

图 7-35 标签库的执行流程

7.3.4 定义带有属性的标签

标签的参数是通过 Tag 实现类的 setter 方法注释进去的，因此定义带参数的标签只需要给类加一个属性以及对应的 setter 方法就可以了，这样在使用自定义标签时可以自定义属性。

1. 定义一个带有属性标签的操作类

修改上一节中 helloTag 类，代码如下。

```
public class helloTag extends TagSupport
{
  private String name;
  public int doStartTag()throws JspTagException{
    JspWriter out = super.pageContext.getOut();
    try {
      out.print("<h1>hello " + name + "</h1>");
    }
    catch (Exception e) {
      e.printStackTrace();
    }
    return this.SKIP_BODY;
  }
  public String getName() {
      return name;
  }
  public void setName(String name) {
      this.name = name;
  }
}
```

在类中编写了 setter 和 getter 方法，当用户通过标签设置属性时，就会调用其中的 setter 方法完成属性的赋值。

2. 定义标签描述文件——/WEB-INF/helloTag.tld

helloTag.tld 文件的内容如下。

```
<?xml version="1.0" encoding="UTF-8"?>
<!DOCTYPE taglib PUBLIC "-//Sun Microsystems, Inc.//DTD JSP Tag Library 1.2//EN"
                        "http://java.sun.com/dtd/web-jsptaglibrary_1_2.dtd">
<taglib>
  <tlib-version>1.0</tlib-version>
  <jsp-version>1.2</jsp-version>
  <short-name>mytag</short-name>
  <uri>org.new.tag</uri>
  <tag>
    <name>hello</name>
    <tag-class>org.news.tag.helloTag</tag-class>
    <attribute>
      <name>name</name>
      <rtexprvalue>false</rtexprvalue>
    </attribute>
  </tag>
</taglib>
```

在标签中定义了一个 name 属性。在 TLD 文件中,其主要元素的作用如下。

(1) <attribute>表示定义的是标签中所具有的属性,一个<tag>元素中可以定义多个<attribute>元素。<name>为标签属性的名称。

(2) <required>:表示此属性是否必须设置,如果为 true,则表示必须设置;如果为 false,则表示可选。

(3) <rtexprvalue>:表示是否支持表达式输出,true 表示支持,false 表示不支持。

3. 配置 web.xml

web.xml 文件中的相关代码如下。

```
<taglib>
  <taglib-uri>mytag/taglib-uri>
  <taglib-location>/WEB-INF/helloTag.tld</taglib-location>
</taglib>
```

将映射名称定义为 mytag,所以在 JSP 中可直接使用此映射名称。

4. 在 JSP 中使用标签

使用标签的 JSP 代码如下。

```
<%@ page language="java" pageEncoding="GBK"%>
<%@ taglib uri="/WEB-INF/mytag.tld" prefix="mytag"%>
  <html>
    <head><title>带属性的自定义标签</title></head>
      <body>
    <mytag:hello name="张山"></mytag:hello>
    </body>
</html>
```

浏览页面,运行结果如图 7-36 所示。

图 7-36　运行结果

7.3.5　定义有标签体的标签库

在 TLD 文件中,只要标签的 bodycontent 属性不为 empty,标签就可以携带标签体。只要 Tag 实现类的 doStartTagO 方法返回 EVAL_ BODY_ INCLUDE,标签体就会输出。这是最简单的一种带标签体的标签,这种标签只能输出而得不到标签体的内容,更不能对标签体进行分析处理。

可以定义一个标签类,它继承自 BodyTagSupport 类,用于实现带有标签体的标签。设计一个包含方法的标签体,该标签带有一个 out 属性,如果为 true 则输出标签体的内容；如果取值为 false 则不输出标签体的内容。

1. 创建一个继承自 BodyTagSupport 的标签类

int Tag 类的代码如下。

```
package cn.edu.mypt;
import javax.servlet.jsp.JspException;
import javax.servlet.jsp.tagext.BodyTagSupport;
public class OutTag extends BodyTagSupport {
  private boolean out;
  public void setOut(boolean out){
    this.out = out;
  }
  public int doStartTag() throws JspException{
    if (out == true)
      return this.EVAL_BODY_INCLUDE;
    else
      return this.SKIP_BODY;
  }
}
```

在代码中,根据 doStartTag()方法中返回的两个参数的值,来决定是否显示标签体的内容。如果 out 属性取值为 true,则显示标签体的内容,并返回 EVAL_BODY_INCLUDE；并返回 SKIP_BODY,页面将不显示标签体的内容。

2. 定义标签描述文件

标签描述文件的内容如下。

```
<tag>
  <name>out</name>
  <tag-class>cn.edu.mypt.OutTag</tag-class>
<body-content>JSP</body-content>
  <attribute>
    <name>out</name>
    <required>true</required>
    <rtexprvalue>true</rtexprvalue>
  </attribute>
</tag>
```

<body-content>元素有 4 个属性：empty、JSP、tagdependent、scriptless。

(1) empty：表示没有标签体。

(2) JSP：表示标签体可以嵌套 JSP 语法,其 JSP 内容会先执行完再对标签进行处理。

(3) tagdependent：表示标签体中不能有任何脚本,但可以是 EL 或动作元素等。

(4) scriptless：表示标签体中不能有任何脚本,不做任何处理,完全传入标签。

3. 创建一个 JSP 文件，在 JSP 文件中调用标签

```
<%@ page language="java" pageEncoding="GBK" %>
<%@ taglib uri="/WEB-INF/mytag.tld" prefix="mytag" %>
  <html>
    <head><title>带标签体的自定义标签</title></head>
    <body>
      <mytag:out out="true">
        Hello!World
      </mytag:out>
    </body>
</html>
```

浏览页面，其效果如图 7-37 所示。

图 7-37　浏览页面效果

7.3.6　遍历标签

对于 Iterator 对象，在 JSP 的 Java 代码中需要用 while 循环或 for 循环来输出，难以维护，且可复用性不好，程序员总是在大量地做这样的工作，这时可以考虑用遍历标签来处理，需要输出数据时只须在 JSP 页面中声明标签即可。

开发遍历标签需要两个类：标签处理类和标签信息类。标签处理类继承自 BodyTagSupport 类，标签信息类要扩展 TagExtraInfo 类。TagExtraInfo 类用来提供标签运行的信息。

1. 定义一个继承自 BodyTagSupport 的类

iterateTag 类的代码如下。

```
iterateTag.java.
package cn.edu.mypt;
  //导入包,略
public class iterateTag extends BodyTagSupport {
  private String name;                    //遍历标签在 pageContext 中的名字
  private Collection collection;          //需要遍历的集合对象
  private Iterator it;                    //遍历器对象
  private String type;                    //遍历器中对象的类型
  public void setName(String name) {
     this.name = name;
  }
  public void setType(String type) {
```

```
      this.type = type;
    }
    //将页面中的集合对象传入程序
    public void setCollection(Collection collection) {
      this.collection = collection;
      if (collection.size() > 0)it = collection.iterator(); //生成遍历对象
    }
    //----标签结束时调用此方法-------
    public int doEndTag() throws JspException {
        if(bodyContent!= null){
          try {
            bodyContent.writeOut(bodyContent.getEnclosingWriter());
          } catch (IOException e) {
            throw new JspException("IO Error " + e.getMessage());
          }
        }
        return this.EVAL_PAGE;
    }
    //如果集合没有内容,不执行标签体
    public int doStartTag() throws JspException {
      if (it == null) return this.SKIP_BODY;
        //从遍历对象中取出数据放入 pageContext
        pageContext.setAttribute(name,it.next());
      return this.EVAL_BODY_INCLUDE;
    }
    //*在 doStartTag()方法后调用此方法,如果返回值是 EVAL_BODY_AGAIN 则反复调用此方法。
      直到返回值是 SKIP_BODY 时调用 doEndStart 方法*/
    public int doAfterBody() throws JspException {
          //从遍历对象中取出数据并放入 pageContext
      if(it.hasNext()) {
        pageContext.setAttribute(name,it.next());
        return this.EVAL_BODY_AGAIN;//此返回值将反复调用此方法
      }
      return this.SKIP_BODY;
    }
}
```

在标签实现类中有 3 个属性：name、type 和 it。其中，name 代表在 pageContext 中标识一个属性的名字；type 代表待遍历对象内容的数据类型；it 为要遍历对象的内容。在 doStartTag()方法中，如果 it 不为 null，就开始遍历，遍历时调用 continueNext()方法。

```
IterateTEI.java
package cn.edu.mypts;
//导入包 、略
  public class IterateTEI extends TagExtraInfo {
    public IterateTEI()
    { super();
    }
    public VariableInfo[] getVariableInfo(TagData data)
```

```
    {
      return new VariableInfo[]{
        new VariableInfo(data.getAttributeString("name"),
        data.getAttributeString("type"),true, VariableInfo.NESTED);
      }
    }
}
```

VariableInfo 类中有几个常量,具体的含义如下。

NESTED：表示标签中的参数在标签开始到标签结束之间是有效的。

AT_BEGIN：表示标签中的参数在标签开始到使用它的 JSP 页面结束是有效的。

AT_END：表示标签中的参数在标签的结束到使用它的 JSP 页面结束是有效的。

2. 编写标签描述 tld 文件

在 myTag.tld 文件中编写<taglib>与</taglib>之间的内容。

```
<!-- iterateTag -->
<taglib>
  <tag>
    <name>iterateTag</name>     <!-- 标签名称 -->
    <tag-class>cn.edu.mypt.iterateTag</tag-class>     <!-- 标签对应的处理类 -->
    <tag-class>cn.edu.mypt.IterateTEI</tag-class>
    <body-content>JSP</body-content>     <!-- 标签体内容,有标签体则设为 jsp -->
    <attribute>                  <!-- 标签的属性声明 -->
        <name>collection</name>
        <required>true</required>
        <rtexprvalue>true</rtexprvalue>
    </attribute>
    <attribute>
        <name>name</name>
        <required>true</required>
        <rtexprvalue>true</rtexprvalue>
      </attribute>
    <attribute>
        <name>type</name>
      <required>true</required>
        <rtexprvalue>true</rtexprvalue>
      </attribute>
    </tag>
</taglib>
```

3. 在 JSP 页面中声明并调用标签

代码如下。

```
<%@ page language="java"  pageEncoding="GBK" %>
<%@ taglib uri="/WEB-INF/mytag.tld" prefix="mytag" %>
<%@ page import="java.util.ArrayList" %>
```

```
<html>
    <head><title>迭代标签</title></head>
  <body>
    用户信息:<br>
    <%//----------设置一个 ArrayList 对象的初始值----------
      ArrayList UserList = new ArrayList();
      UserList.add("张上");
      UserList.add("李生");
      UserList.add("旺旺");
    %>
    <mytag:iterateTag name = "user" type = "String" collection = "<% = UserList %>">
      ${user}
    </mytag:iterateTag>
  </body>
</html>
```

浏览页面,其效果如图 7-38 所示。

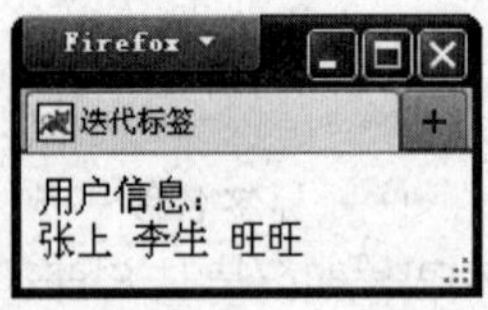

图 7-38 页面效果

7.3.7 自定义方法

EL 自定义方法可以扩展 EL 表达式的功能,让 EL 表达式完成普通 Java 程序代码所能完成的功能。

EL 自定义方法开发步骤如下。

(1) 编写 EL 自定义方法映射的 Java 类中的静态方法:这个 Java 类必须带有 public 修饰符,方法必须是这个类的带有 public 修饰符的静态方法。

(2) 编写标签库描述文件(TLD 文件),在 TLD 文件中描述自定义方法。

(3) 在 JSP 页面中导入和使用自定义方法。

例如,定义一个 Hello()方法。

```
package cn.edu.mypt.;
public class MyFunctions {
    public static String sayHello(String name) {
        return "Hello " + name;
    }
}
```

提供 TLD 文件,该文件必须放置于 WEB-INT 下或 WEB-INT 任意子目录下,其内容如下。

```
<?xml version = "1.0" encoding = "UTF - 8" ?>
<taglib xmlns = "http://java.sun.com/xml/ns/j2ee"
```

```
  xmlns:xsi = "http://www.w3.org/2001/XMLSchema - instance"
  xsi:schemaLocation = "http://java.sun.com/xml/ns/j2ee
  http://java.sun.com/xml/ns/j2ee/web - jsptaglibrary_2_0.xsd" version = "2.0">
  < description > my functions library </description >
  < display - name > my functions </display - name >
  < tlib - version > 1.0 </tlib - version >
  < short - name > my </short - name >
  < uri > http://www.mypt.edu.cn/functions </uri >
  < function >
    < name > sayHello </name >
    < function - class > cn.edu.mypt.MyFunctions </function - class >
    < function - signature >
      java.lang.String sayHello(java.lang.String)
    </function - signature >
  </function >
</taglib >
```

在 JSP 页面中调用自定义方法的方法如下。

```
<%@ taglib prefix = "my" uri = "http://www.mypt.edu.cn/functions" %>
${my:sayHello(name)
```

7.3.8 任务：自定义用户信息标签

1. 任务要求

自定义一个用表格显示用户信息的标签。

2. 任务实施

首先利用前面的案例，创建一个用户类 userBean，然后创建一个标签类，编写标签描述文件，最后编写一个 JSP 代码页面测试该标签。

（1）定义 JavaBean 用户类，参考 4.2 节中的任务。

（2）创建自定义标签类——usertag.java。

```
package cn.edu.mypt.tags;
import javax.servlet.jsp.JspException;
import javax.servlet.jsp.JspWriter;
import javax.servlet.jsp.tagext.TagSupport;
package cn.edu.mypt.bean.UserBean
public class usertag extends TagSupport {
  private UserBean user;
  public int doStartTag() throws JspException {
    try{
      JspWriter out = this.pageContext.getOut();
      if(user == null) {
        out.println("No UserInfo Found...");
        return SKIP_BODY;
      }
```

```
        out.println("<table width='300px' border='1' align='center'>");
        out.println("<tr>");
        out.println("<td width='30%'>编号:</td>");
        out.println("<td>" + user.getUserID() + "</td>");
        out.println("</tr>");
        out.println("<tr>");
        out.println("<td>姓名:</td>");
        out.println("<td>" + user.getUserName() + "</td>");
        out.println("</tr>");
        out.println("</table>");
      } catch(Exception e) {
        throw new JspException(e.getMessage());
      }
      return SKIP_BODY;
    }
    public int doEndTag() throws JspException {
      return EVAL_PAGE;
    }
    public UserBean getUser() {
      return user;
    }
    public void setUser(UserBean user) {
      this.user = user;
    }
}
```

(3) 在 WEB-INF 目录下创建标签库描述文件 mytag.tld。

```
<?xml version="1.0" encoding="UTF-8"?>
<!DOCTYPE taglib PUBLIC "-//Sun Microsystems, Inc.//DTD JSP Tag Library 1.2//EN"
                    "http://java.sun.com/dtd/web-jsptaglibrary_1_2.dtd">
<taglib>
  <tlib-version>1.0</tlib-version>
  <jsp-version>1.2</jsp-version>
  <short-name>mytag</short-name>
  <uri>cn.edu.mypt.tags</uri>
  <tag>
    <name>outUser</name>
    <tag-class>cn.edu.mypt.tags.usertag</tag-class>
    <attribute>
      <name>user</name>
      <required>true</required>
      <rtexprvalue>true</rtexprvalue>
    </attribute>
  </tag>
</taglib>
```

(4) 引用标签。

```
<%@ page language="java" import="java.util.*" pageEncoding="GBK"%>
```

```
<%@taglib uri = "cn.edu.mypt.tags" prefix = "mytag" %>
<%@ taglib prefix = "c" uri = "http://java.sun.com/jstl/core_rt" %>
<html>
  <head><title>显示用户信息</title></head>
    <body>
    <jsp:useBean id = "User" class = "org.news.bean.user"></jsp:useBean>
    <%
      User.setUserID("981101");
      User.setUserName("张山");
      pageContext.setAttribute("userinfo",User);
     %>
  <!-- 给标签设置 user 属性绑定要展现的 userinfo 对象 -->
    <mytag:outUser user = "${userinfo}"></mytag:outUser>
  </body>
</html>
```

项目 7　用户管理系统的视图层设计

1. 项目需求

获取所有用户信息并显示。

2. 项目实施

(1) 创建 Web 项目 testprj,利用项目 5 的业务逻辑层和数据访问层的代码,获取用户信息。

(2) 编写 Servlet,调取模型层的 getUser()方法,获取所有用户信息,保存到 request 属性中,跳转到显示页面。

GetUsersServlet.java 的代码如下。

```
package cn.edu.mypt.servlet;

//导入包(略)
import cn.edu.mypt.model.UserModel;
import cn.edu.mypt.model.impl.UserModelImple;
public class GetUsersServlet extends HttpServlet {
  //略
  public void doGet(HttpServletRequest request, HttpServletResponse response)
  throws ServletException, IOException {
    doPost(request,response);
  }
  public void doPost(HttpServletRequest request, HttpServletResponse response)
  throws ServletException, IOException {
    UserModel uM = new UserModelImple(); //参考项目 5 代码
    List ulist = uM.getUser();
    request.getSession().setAttribute("USERLIST", ulist);
    request.getRequestDispatcher("userList.jsp").forward(request, response);
```

```
    }
}
```

(3) 编写显示页面 userList.jsp 显示所有用户信息,代码如下。

```
<%@ page language="java" import="java.util.*" pageEncoding="GBK"%>
<%@ taglib uri="http://java.sun.com/jsp/jstl/core" prefix="c"%>
<!DOCTYPE HTML PUBLIC "-//W3C//DTD HTML 4.01 Transitional//EN">
<html>
  <head><title>用户列表</title></head>
    <body>
    <h2>用户信息</h2>
      <a href="index.jsp">首页</a>
    <table>
    <tr><td>用户 ID</td><td>用户名</td><td>密码</td><td>删除</td><td>修改</td>
        <td>更多</td></tr>
        <c:forEach var="user" items="${USERLIST}">
            <tr>
                <td>${user.userID}</td><td>${user.userName}</td>
                <td>${user.userPwd}</td>
        <c:url var="delurl" value="DeleteServlet">
        <c:param name="uID" value="${user.userID}"/>
        </c:url>
        <c:url var="modurl" value="ModifyServlet">
        <c:param name="uID" value="${user.userID}"/>
        </c:url>
        <c:url var="moreurl" value="DisplayServlet">
        <c:param name="uID" value="${user.userID}"/>
        </c:url>
          <td><a href="${delurl}">删除</a></td>
          <td><a href="${modurl}">修改</a></td>
          <td><a href="${moreurl}">更多</a></td>
        </tr>
      </c:forEach>
    </table>
    <jsp:include page="comInfo.jsp"/></jsp:include><!-- 导入公司信息页面 -->
  </body>
</html>
```

(4) 配置 web.xml。

```
<servlet>
  <servlet-name>GetUsersServlet</servlet-name>
  <servlet-class>cn.edu.mypt.servlet.GetUsersServlet</servlet-class>
</servlet
<servlet-mapping>
  <servlet-name>GetUsersServlet</servlet-name>
  <url-pattern>/GetUsersServlet</url-pattern>
</servlet-mapping>
```

(5) 启动服务器,部署并运行 http:127.0.0.1:8080/testprj/GetUsersServlet。

习 题 7

1. 选择题

(1) 自定义标签的配置文件放在(　　)目录下。

A. WebRoot　　B. lib　　C. classes　　D. WEB-INF

(2) EL 表达式＄{10 mod 3}的结果为(　　)。

A. 10 mod 3　　B. 1　　C. 3　　D. null

(3) 自定义标签的作用是(　　)。

A. 编写和使用方便

B. 规定是这样的,如果不用,别人会说我们不专业

C. 连数据库

D. 可以减少 JSP 中的 Java 代码,将代码与界面标签分离,简化前台开发

(4) 在 JSP 中,若要在 JSP 正确使用标签< x:getKing/>,则在 JSP 中声明的 taglib 指令为<%@tagliburi= "/WEB-INF/myTags.tld" prefix= "(　　)"%>。

A. x　　B. getKing

C. King　　D. myTags

(5) Login.jsp 为登录页面,表单代码如下。

```
<form?action = "index.jsp" method = "post">
  <input?type = "text&quot;?name = "name"/>
  <input?type = "submit" value = "login"/>
</form>
```

要在 index.jsp 中直接显示用户名,以下代码正确的是(　　)。

A. ＄{requestScope.name}　　B. <%=name%>

C. ＄{param.name}　　D. <%=param.name%>

(6) 以下代码的显示结果为(　　)。

```
<%
  int x = 5;
  request.setAttribute("a","123");
  session.setAttribute("a","456");
 %>
<c:out value = "${a}"/>
```

A. 5　　B. 123　　C. 456　　D. null

(7) 以下代码执行结果为(　　)。

```
<c:forEach var? = "i" begin = "1" end = "5" step = "2">
  <c:out value = "${i}"/>
</c:forEach>
```

A. 12345　　B. 135　　C. iii　　D. 15

(8) 编写自定义标签处理类后,需要编写一个(　　)文件去描述。

A. . tag　　B. . tld　　C. . dtd　　D. . xml

(9) 在 JSP 中可动态导入其他页面的标签是(　　)。

A. <%include/></textarea>

B. <%@ include%>

C. <jsp:import Page/>

D. <jsp:include/>

(10) (　　)方法不是 TagSupport 类的方法。

A. doPost()　　B. doStartTag()

C. doEndTag()　　D. doAfterBody()

2. 简答题

(1) 什么是标签库描述文件?其作用是什么?

(2) JSTL 的 5 种标签是什么,分别有什么作用?

(3) 设计一个遍历标签,用于对 java. until. Collection 类型数据的遍历。

第 8 章 JSP 应用开发

通过前面几章的学习,已经基本掌握了 JSP 基本编程技术,包括如何通过 JavaBean 实现业务层的数据处理、使用 JDBC 实现对数据库的访问、使用 Servlet 实现控制层的设计、使用 JSTL 与 EL 技术实现表示层的页面设计,通过这些知识能够开发出简单的 Web 应用程序。然而在实际工程应用中,Web 应用越来越丰富,有很多应用组件和框架技术为我们的开发提供了方便,提高了我们开发的效率。

本章介绍数据分页显示与文件上传、分层架构技术。

【技能目标】 掌握分页处理、文件上传/下载的编程方法;理解 MVC 编程。

【知识目标】 分页显示的原理;SmartUpload 组件的使用和 MVC 的基本概念。

【关键词】 模型(model) 分层(tier) 数据访问对象(DAO)
业务(business) 控制(control) 继承(extend)
接口(interface) 实现(implement) 抽象(abstract)
视图(view)

8.1 分页处理技术

基于 Internet 的 Web 应用变得越来越复杂,资源也越来越庞大,数据量也越来越大。在页面中常常以列表的形式显示数据,将数据按照指定格式显示,可以使布局清晰,不受信息数量的限制。但是当数据量很大时,受页面的限制用户必须滚动页面才能浏览更多的数据,而且页面也显得冗长,这个时候就要求分页显示数据。

8.1.1 常见的分页技术

在使用数据库的过程中,不可避免地需要使用到分页的功能,其处理的方法有多种。

1. 利用数据库实现分页

有一些数据库,如 SQL Server MySQL、Oracle 等有自己的分页方法,比如 MySQL 可以使用 limit 子句,Oracle 可以使用 ROWNUM 来限制结果集的大小和起始位置。

1) SQL Server 分页

第一种分页方法需用到的参数如下。

(1) pageSize:每页显示多少条数据。

(2) pageNumber:页码,从客户端传来。

(3) totalRecouds:表中的总记录数。SQL 语句如下。

```
select count( * ) from 表名
```

(4) totalPages：总页数。SQL 语句如下。

```
totalPages = totalRecouds % pageSize = = 0?totalRecouds/pageSize:totalRecouds/pageSize + 1
```

(5) pages：起始位置。SQL 语句如下。

```
pages = pageSize * (pageNumber - 1)
```

完成分页的 SQL 语句如下。

```
select top pageSize * from 表名
where id not in (select top pages id from 表名 order by id)
order by id
```

第二种分页方法需用到的参数如下。

(1) pageSize：每页显示多少条数据。

(2) pageNumber：页码，从客户端传来。

完成分页的 SQL 语句如下。

```
pages = pageSize * (pageNumber - 1) + 1
select top pageSize * from 表名
where id >= (select max(id) from (select top pages id from 表名
order by id asc ) t )
```

2) MySQL 分页

MySQL 分页需用到的参数如下。

(1) pageSize：每页显示多少条数据。

(2) pageNumber：页码，从客户端传来。

(3) totalRecouds：表中的总记录数。SQL 语句如下：

```
select count( * ) from 表名
```

(4) totalPages：总页数。SQL 语句如下。

```
totalPages = totalRecouds % pageSize == 0?totalRecouds/pageSize:totalRecouds/pageSize + 1
```

(5) pages：起始位置。

```
pages = pageSize * (pageNumber - 1)
```

完成分页的 SQL 语句如下。

```
select * from 表名 limit pages, pageSize;
```

3) Oracle 分页

Oracle 分页需用到的参数如下。

(1) pageSize：每页显示多少条数据。

(2) pageNumber：页码，从客户端传来。

(3) totalRecouds：表中的总记录数。SQL 语句如下。

```
select count( * ) from 表名
```

(4) totalPages：总页数。SQL 语句如下。

```
totalPages = totalRecouds % pageSize == 0?totalRecouds/pageSize:totalRecouds/pageSize + 1
```

(5) startPage 和 end Page：起始位置和结束位置。SQL 语句如下。

```
startPage =  pageSize * (pageNumber - 1) + 1
endPage = startPage + pageSize
```

完成分页的 SQL 语句如下。

```
select a. * from
( select rownum num ,t. * from 表名 t where 某列 = 某值 order by id asc ) a
  where a. num > = startPage and a. num < endPage
```

4) Access 分页

Access 分页需用到的参数如下。

(1) pageSize：每页显示多少条数据。

(2) pageNumber：页码，从客户端传来。

```
pages = pageSize * (pageNumber - 1) + 1
```

完成分页的 SQL 语句如下。

```
select top pageSize * from 表名 where id> = (select max(id) from (select top pages id from 表
名 order by id asc ) t )
```

【例 8-1】 在 MySQL 中获取分页数据，典型代码如下。

```
//计算总的记录条数
String SQL = "SELECT Count( * ) AS total " + this.QueryPart;
rs = db.executeQuery(SQL);
if (rs.next())
    Total = rs.getInt(1);
//设置当前页数和总页数
TPages = (int)Math.ceil((double)this.Total/this.MaxLine);
CPages = (int)Math.floor((double)Offset/this.MaxLine + 1);
//根据条件判断，取出所需记录
if (Total > 0) {
    SQL = Query + " LIMIT " + Offset + " , " + MaxLine;
    rs = db.executeQuery(SQL);
}
return rs;
```

2. 操作 ResultSet 实现分页

这种方法不使用任何封装，在需要分页的地方，直接操作 ResultSet 滚到相应的位置，再读取相应数量的记录即可。典型代码如下。

```
<%
  sqlStmt = sqlCon.createStatement(java.sql.ResultSet.TYPE_SCROLL_INSENSITIVE,
                                   java.sql.ResultSet.CONCUR_READ_ONLY);
  strSQL = "select name,age from test";
  //执行 SQL 语句并获取结果集
  sqlRst = sqlStmt.executeQuery(strSQL);
  //获取记录总数
  sqlRst.last();
  intRowCount = sqlRst.getRow();
  //记算总页数
  intPageCount = (intRowCount + intPageSize - 1) / intPageSize;
  //调整待显示的页码
  if(intPage > intPageCount) intPage = intPageCount;
%>
<table border = "1" cellspacing = "0" cellpadding = "0">
<tr><th>姓名</th><th>年龄</th></tr>
<%
  if(intPageCount > 0){
    //将记录指针定位到待显示页的第一条记录上
    sqlRst.absolute((intPage - 1) * intPageSize + 1);
    //显示数据
    i = 0;
    while(i < intPageSize && !sqlRst.isAfterLast()) {
%>
    <tr>
    <td><% = sqlRst.getString(1) %></td>
      <td><% = sqlRst.getString(2) %></td>
    </tr>
<%
      sqlRst.next();
      i++;
    }
  }
%>
</table>
```

很显然，这种方法没有考虑到代码复用的问题，不仅代码数量巨大，而且在代码需要修改的情况下将会无所适从。

3. 使用 Vector 进行分页

先将所有记录都选择出来，然后将 ResultSet 中的数据取出并存入 Vector 集合类中，再根据所需分页的大小、页数定位到相应的位置，读取数据。或者先使用前面提到的两种分页方法，取得所需的页面之后再存入 Vector。

这种方法代码执行的效率较低，代码的可移植性差，对字段类型有限制。

8.1.2 JSP+JavaBean 实现分页

此分页程序用到 3 个文件 test.jsp、Pagination.java 和 DBConnect.java 以及一个简

单数据库 test 中的表 test,测试用的 Web 发布服务器为 resin-2.1.6,其中 test 用于显示分页结果,DBConnect.java 用于连接 MySQL 数据库,Pagination.java 用于封装分页程序,而且 DBConnect.java 和 Pagination.java 放在 WEB-INF 目录下的 classes 目录中,数据库用的是 MySQL。

1. 创建数据库连接 DBConnect.java

DBConnect.java 的代码如下。

```
import java.sql.*;
public class DBConnect{
  String sDBDriver = "org.gjt.mm.mysql.Driver";
  //设置驱动变量(JDBC 的驱动程序放在 WEB-INF\lib 目录下)
  String sConnStr = "jdbc:mysql://127.0.0.1:3306/test?user = root&password = ";
  /*创建连接,数据库名为 test,连接 MySQL 的用户名是 root,密码为空(如果你的 MySQL 有用户
  名和密码则填上你的用户名和密码) */
  Connection conn = null;
  ResultSet rs = null;
  public DBConnect(){
    try{
      Class.forName(sDBDriver);            //创建数据库驱动
    }
    catch(java.lang.ClassNotFoundException e){
      System.out.println("Jdbc_conn():" + e.getMessage());
    }
  }
  //数据更新,这里只用到查询
  public void executeUpdate(String sql)throws Exception{
    sql = new String(sql.getBytes("GBK"),"ISO8859_1");
    try{
      conn = DriverManager.getConnection(sConnStr);
      Statement stmt = conn.createStatement();
      stmt.executeUpdate(sql);
      conn.close();
      stmt.close();
    }
    catch(SQLException ex){
      System.out.println("sql.executeUpdate:" + ex.getMessage());
    }
  }
  //查询
  public ResultSet executeQuery(String sql)throws Exception{
    rs = null;
    try{
      sql = new String(sql.getBytes("GBK"),"ISO8859_1"); //字符转换
      conn = DriverManager.getConnection(sConnStr);       //创建连接
      Statement stmt = conn.createStatement();            //数据操作对象
      rs = stmt.executeQuery(sql);                        //执行查询
      conn.close();                                       //关闭连接
```

```
        stmt.close();                                        //关闭对象
      }
      catch(SQLException ex){
        System.out.println("sql.executeQuery:" + ex.getMessage());
      }
      return rs;
    }
}
```

2. Pagination.java——封装分页的类

Pagination.java 的代码如下。

```
import java.sql.*;
import javax.servlet.*;                          //引入 servlet 包
import javax.servlet.http.*;
import java.math.*;
public class Pagination{
  private String strPage = null;                 //起始页码
  private int curPages;                          //当前页码
  private int m_rows;                            //设置每页显示的行数
  private int pages;                             //总页数
  /*取得 test.jsp 里的 test.jsp?page=<% =curPages-1%>或是 page=<% =curPages+1%>
的值给变量 strPage*/
  public String strPage(HttpServletRequest request, String page){
    try{
      strPage = request.getParameter(page); //
    }
    catch(Exception e){
      System.out.println("delcolumn" + e.getMessage());
    }
    return strPage;                              //返回这个值
  }
  //页面数
  public int curPages(String strPage){
    try{
      if(strPage == null){                       //默认没有就设置是第 1 页
        curPages = 1;
      }
      else{
        curPages = Integer.parseInt(strPage);    //取得 strPage 的整数值
        if(curPages < 1)                         //如果小于 1,同样返回是第 1 页
          curPages = 1;
      }
    }
    catch(Exception e){
      System.out.print("curPages");
    }
    return curPages;                             //返回页面数
```

```
    }
    //设置每页要显示的记录数
    public void setRows(int rows){
      m_rows = rows;
    }  //取得页数
    public int getPages(int rowcounts){
      int test;                                  //变量
      test = rowcounts % m_rows;                 //取得余数
      if(test == 0)
        pages = rowcounts/m_rows;                //每页显示的整数
        else
        pages = rowcounts/m_rows + 1;            //不是就加 1
      return pages;                              //返回页数
    }
    //返回结果集
    public ResultSet getPageSet(ResultSet rs, int curPages){
      if(curPages == 1){
        return rs;                               //如果是就 1 页就返回这个 rs
      }
      else{
        int i = 1;
        try{
          while(rs.next()){
            i = i + 1;
            if(i >((curPages - 1) * m_rows))
              break;                             //退出
          }
          return rs;                             //从退出开始将结果集返回
        }
        catch(Exception e){
          System.out.print(e.getMessage());
        }
      }
      return rs;
    }
}
```

3. test.jsp——显示页面

test.jsp 的代码如下。

```
<%@ page contentType = "text/html;charset = gb2312" %>
<%@ page import = "java.sql.*" %>
<jsp:useBean id = "m_pages" scope = "page" class = "Pagination"/>
<jsp:useBean id = "sql" scope = "page" class = "DBConnect"/>
<%
  int curPages = Pagination.curPages(Pagination.strPage(request,"page"));
  Pagination.strPage(request,"page")     //取 page 值传递给 curPages()方法
  Pagination.setRows(10);                //设置每页显示 10 条
```

```
%>
<%
  ResultSet rs_count = sql.executeQuery("select count( * ) as t from test");
  //传递进数据库处理的 JavaBean
  rs_count.next();
  int resultconts = rs_count.getInt("t");              //取得总的数据数
  int totalPages  =  Pagination.getPages(resultconts); //取得总页数
  ResultSet rs = m_pages.getPageSet(sql.executeQuery("select * from test"),curPages);
%>
<p>分类表</p>
<table border = "1">
<tr><td>1</td><td>2</td><td>3</td></tr>
<% int i = 1; %>
<% while (rt.next()){ %>
<tr>
<td><% = rt.getString("id") %>  </td>
<td><% = rt.getString("name") %>  </td>
</tr>
<%
  i = i + 1;
  if(i > 10) break;
  }
%>
</table>
<p align = "center">
<% if(curPages > 1){ %><a href = "testBean.jsp?page = <% = curPages - 1 %>" >
上一页</a><% } %>
<% if(curPages < totalPages){  %><a href = "testBean.jsp?page = <% = curPages + 1 %>">
下一页</a><% } %>
</p>
```

8.1.3 任务：实现用户信息的分页显示

1. 需求说明

分页显示所有用户信息。

2. 任务实施

本任务实现的基本思想是：建立一个分页模型，用于保存分页获取的用户信息，并根据数量和每页的记录数构造一个分页导航。采用分页 SQL 语句获取每页的数据信息。其实现的基本步骤如下。

(1) 新建 Web 项目 pageTest，利用项目 5 的业务逻辑层和数据访问层的代码获取用户信息。

(2) 定义分页模型，代码如下。

```
package cm.edu.mypt.comm;
//导入包(略
```

```
public class PageModel < T > implements Serializable {
    private static final long serialVersionUID = 1L;
    private int pageSize = 10; //页面大小
    private int pageNo = 1; //当前页编号
    private int recordCount = 0; //总记录数
    private int pageCount; //总页数
    private List < T > data; //分页数据
    private String pageNav; //导航
    //属性封装(略)
    //根据页面大小、当前页号、记录数构造分页模型
    public PageModel(int pageSize, int pageNo, int recordCount) {
      if(pageSize < 1){
        this.pageSize = 10;
      }else {
        this.pageSize = pageSize;
      }
      if(recordCount!= 0){
        this.pageCount = (int)((recordCount + this.pageSize - 1) / this.pageSize);
      }
      else{
          this.pageCount = 1;
      }
      if(pageNo < 1){
          this.pageNo = 1;
      }else if(pageNo > this.pageCount){
          this.pageNo = this.pageCount;
      }else{
          this.pageNo = pageNo;
      }
      this.recordCount = recordCount;
    }
    //构建导航栏
    public void setPageNav(String url){
      if(url.lastIndexOf('?') != - 1){
        url += "&";
      }else{
        url += "?";
      }
      StringBuilder sb = new StringBuilder();
      sb.append("共  ").append(recordCount).append(" 条   ");
      sb.append(pageSize).append(" 条/页   ");
      if (pageNo >= 2) {
        sb.append("< a href = '").append(url);
        sb.append("pageNo = ").append(1).append("&pageSize = ").append(pageSize);
        sb.append("'>首页</a >   ");
        sb.append("< a href = '").append(url);
        sb.append("pageNo = ").append(pageNo - 1).append("&pageSize = ");
        sb.append(pageSize).append("'>上一页</a >   ");
```

```
        } else {
            sb.append("首页   ");
            sb.append("上一页   ");
        }
        if (pageNo <= pageCount - 1 && pageCount != 0) {
            sb.append("<a href='").append(url);
            sb.append("pageNo=").append(pageNo+1).append("&pageSize=");
            sb.append(pageSize).append("'>下一页</a>   ");
            sb.append("<a href='").append(url);
            sb.append("pageNo=").append(pageCount).append("&pageSize=");
            sb.append(pageSize).append("'>尾页</a>   ");
        } else {
            sb.append("下一页   ");
            sb.append("尾页   ");
        }
        sb.append("跳转到第  ").append("<select>");
        for(int i = 1; i <= pageCount; i++){
            if(pageNo != i){
                sb.append("<option onclick=\"location.href='").append(url);
                sb.append("pageNo=").append(i);
                sb.append("&pageSize=").append(pageSize).append("';\">");
                sb.append(" ").append(i).append("/").append(pageCount);
                sb.append(" ").append("</option>");
            }else {
                sb.append("<option selected='selected' onclick=\"location.href='");
                sb.append(url).append("pageNo=").append(i);
                sb.append("&pageSize=").append(pageSize).append("';\">");
                sb.append(" ").append(i).append("/").append(pageCount);
                sb.append(" ").append("</option>");
            }
        }
        sb.append("</select>  页");
        this.pageNav = sb.toString();
    }
  }
}
```

(3) 定义数据库分页访问的 DAO,代码如下。

```
package cn.edu.mypt.dao.impl;
//导入包(略)
import com.cykj.db.ConnectionFactoryA;
import cn.edu.mypt.bean.UserBean;
public class UserPageDao {
    //分页查询记录
    //sql :查询表; key:关键字
    //pageSize:每页记录数; pageNo: 页码
    public List getUser(int pageSize, int pageNo,String sql,String key) {
        int pages;
        pages = pageSize * (pageNo - 1);
```

```
        ArrayList < UserBean > records = new ArrayList < UserBean >();
        Connection con = ConnectionFactoryA.getConnection();
        PreparedStatement pre = null;
        ResultSet rs = null;
        String pagWhere = "";
        pagWhere = " select top " + pageSize + " * " + sql + " where " +
          key + " not in (select top " + pages + " " + key + " " + sql + " order by " + key +
          ") order by " + key;
        try{
          pre = con.prepareStatement(pagWhere);
          rs = pre.executeQuery();
          while(rs.next()){
            UserBean record = new UserBean();
            record.setUserID(rs.getString(1)) ;
            record.setUserName(rs.getString(2));
            record.setUserPwd(rs.getString(3));
            records.add(record);
          }
        }catch (Exception e){
          e.printStackTrace();
        }finally{
          ConnectionFactoryA.Close(rs, pre, con);
        }
        return records;
      }
      //获取记录数,sql:查询表
      public int total(String sql) {
        int recordCount = 0;
        String countsql = "select count( * ) " + sql;
        Connection con = ConnectionFactoryA.getConnection();
        PreparedStatement pre = null;
        ResultSet rs = null;
        try {
          pre = con.prepareStatement(countsql);
          rs = pre.executeQuery();
          if (rs.next()) {
            recordCount = rs.getInt(1);
          }
        }catch (SQLException e) {
          e.printStackTrace();
        } finally {
          ConnectionFactoryA.Close(rs, pre, con);
        }
        return recordCount;
      }
    }
```

(4) 定义 Servlet,用于获取用户信息,并分页,代码如下。

```
package cn.edu.mypt.servlet;
```

```
//导入包(略)
import cm.edu.mypt.comm.PageModel;
import cn.edu.mypt.bean.UserBean;                    //参考项目5代码
import cn.edu.mypt.model.UserModel;                  //参考项目5代码
import cn.edu.mypt.model.impl.UserModelImple;        //参考项目5代码
public class UserPageListServlet extends HttpServlet {
  //...略
  public void doGet(HttpServletRequest request, HttpServletResponse response)
      throws ServletException, IOException {
    doPost(request,response);
  }
  public void doPost(HttpServletRequest request, HttpServletResponse response)
    throws ServletException, IOException {
    int pageSize,pageNo;
    try{
         pageSize = request.getParameter("pageSize") == null ? 10 : Integer.parseInt
(request.getParameter("pageSize").toString());
        pageNo = request.getParameter("pageNo") == null ? 1 : Integer.parseInt(request
.getParameter("pageNo").toString());
    }catch (NumberFormatException e) {
        pageSize = 10;
        pageNo = 1;
    }
    int num;
    UserPageDao uDao = new UserPageDao();
    //获取分页记录
    List<UserBean> users = uDao.getUser(pageSize,pageNo," from userTable","userid");
    //记录总数
    num = uDao.total("from userTable");
    构造分页模型
    PageModel<UserBean> pm = new PageModel<UserBean>(pageSize,pageNo,num);
    pm.setData(users);
    //设置导航
    pm.setPageNav(request.getRequestURI());
    request.setAttribute("pm", pm);
    request.getRequestDispatcher("userpageList.jsp").forward(request, response);
  }
}
```

(5) 设计分页显示用户信息页面 userpageList.jsp,代码如下。

```
<%@ page language="java" import="java.util.*" pageEncoding="GBK" %>
<%@ taglib uri="http://java.sun.com/jsp/jstl/core" prefix="c" %>
<!DOCTYPE HTML PUBLIC "-//W3C//DTD HTML 4.01 Transitional//EN">
<html>
  <head><title>用户信息</title></head>
    <body>
      <h3>用户信息</h3>
        <table class="dt" border="0" cellspacing="1">
          <tr><th>序号</th><th>用户名</th><th>用户名</th><th>密码</th></tr>
```

```
        <c:forEach items = "${requestScope.pm.data}" var = "record"
                   varStatus = "rows">
            <tr>
                <td>${(pm.pageNo - 1) * pm.pageSize + rows.count}</td>
                <td>${record.userID}</td>
                <td>${record.userName}</td>
                <td>${record.userPwd}</td>
            </tr>
            </c:forEach>
        </table>
        <div class = "pagenav">${pm.pageNav}</div>
    </body>
</html>
```

（6）配置 web.xml，关键代码如下。

```
<servlet>
  <servlet-name>UserPageListServlet</servlet-name>
  <servlet-class>cn.edu.mypt.servlet.UserPageListServlet</servlet-class>
</servlet>
<servlet-mapping>
  <servlet-name>UserPageListServlet</servlet-name>
  <url-pattern>/UserPageListServlet</url-pattern>
</servlet-mapping>
```

（7）启动服务器，部署并运行 http://127.0.0.1:8080/PageTest/UserPageListServlet。

8.2 文件的上传/下载

8.2.1 JSP SmartUpload 简介

JSP SmartUpload 是由 www.jspsmart.com 网站开发的一个可免费使用的全功能的文件上传/下载组件，适于嵌入执行上传下载操作的 JSP 文件中。该组件有以下 5 个特点。

（1）使用简单。在 JSP 文件中仅仅书写三五行 Java 代码就可以搞定文件的上传/下载，非常方便。

（2）能全程控制上传。利用 JSP SmartUpload 组件提供的对象及其方法，可以获得全部上传文件的信息（包括文件名、大小、类型、扩展名、文件数据等），方便存取。

（3）能对上传的文件在大小、类型等方面做出限制，如此可以滤掉不符合要求的文件。

（4）下载灵活。仅写两行代码，就能把 Web 服务器变成文件服务器。无论文件在 Web 服务器的目录下或在其他任何目录下，都可以利用 JSP SmartUpload 进行下载。

（5）能将文件上传到数据库中，也能将数据库中的数据下载下来。

JSP SmartUpload 组件可以从 www.jspsmart.com 网站上自由下载，压缩包的名字为 jspSmartUpload.zip。下载后，用 WinZip 或 WinRAR 将其解压到 Tomcat 的 webapps 目录下。

8.2.2 SmartUpload 组件常用方法

1. File 类

这个类包装了一个上传文件的所有信息。通过它，可以得到上传文件的文件名、文件大小、扩展名、文件数据等信息。

File 类主要提供以下方法。

1) saveAs()

作用：将文件换名另存。

原型：

```
public void saveAs(java.lang.String destFilePathName)
public void saveAs(java.lang.String destFilePathName,int optionSaveAs)
```

其中，destFilePathName 是另存的文件名。optionSaveAs 是另存的选项，该选项有三个值，分别是 SAVEAS_PHYSICAL、SAVEAS_VIRTUAL、SAVEAS_AUTO。SAVEAS_PHYSICAL 表示以操作系统的根目录为文件根目录另存文件，SAVEAS_VIRTUAL 表示以 Web 应用程序的根目录为文件根目录另存文件，SAVEAS_AUTO 则表示让组件决定，当 Web 应用程序的根目录存在另存文件的目录时，它会选择 SAVEAS_VIRTUAL，否则会选择 SAVEAS_PHYSICAL。

例如，saveAs("/upload/sample.zip",SAVEAS_PHYSICAL)执行后，若 Web 服务器安装在 C 盘，则另存的文件名实际是 c:\upload\sample.zip；saveAs("/upload/sample.zip",SAVEAS_VIRTUAL)执行后，若 Web 应用程序的根目录是 webapps/jspsmartupload，则另存的文件名实际是 webapps/jspsmartupload/upload/sample.zip；saveAs("/upload/sample.zip",SAVEAS_AUTO)执行后，若 Web 应用程序根目录下存在 upload 目录，则其效果同 saveAs("/upload/sample.zip",SAVEAS_VIRTUAL)，否则同 saveAs("/upload/sample.zip",SAVEAS_PHYSICAL)。

建议：对于 Web 程序的开发来说，最好使用 SAVEAS_VIRTUAL，以便移植。

2) isMissing()

作用：判断用户是否选择了文件，也即对应的表单项是否有值。选择了文件时，它返回 false。未选文件时，它返回 true。

原型：

```
public boolean isMissing()
```

3) getFieldName()

作用：取 HTML 表单中对应于此上传文件的表单项的名称。

原型：

```
public String getFieldName()
```

4) getFileName()

作用：取文件名(不含目录信息)。

原型：

```
public String getFileName()
```

5）getFilePathName()

作用：取文件全名(带目录)。

原型：

```
public String getFilePathName()
```

6）getFileExt()

作用：取文件扩展名。

原型：

```
public String getFileExt()
```

7）getSize()

作用：取文件长度(以字节计)。

原型：

```
public int getSize()
```

8）getBinaryData()

作用：取文件数据中指定位移处的一个字节，用于检测文件等处理。

原型：

```
public byte getBinaryData(int index)
```

其中，index 表示位移，其值为 0～getSize()－1。

2. Files 类

这个类表示所有上传文件的集合，通过它可以得到上传文件的数目、大小等信息。Files 类有以下方法。

1）getCount()

作用：取得上传文件的数目。

原型：

```
public int getCount()
```

2）getFile()

作用：取得指定位移处的文件对象 File(这是 com.jspsmart.upload.File，不是 java.io.File，注意区分)。

原型：

```
public File getFile(int index)
```

其中，index 为指定位移，其值为 0～getCount()－1。

3）getSize()

作用：取得上传文件的总长度，可用于限制一次性上传的数据量大小。

原型：

```
public long getSize()
```

4）getCollection()

作用：将所有上传文件对象以集合的形式返回，以便其他应用程序引用，浏览上传文件信息。

原型：

```
public Collection getCollection()
```

5）getEnumeration()

作用：将所有上传文件对象以枚举的形式返回，以便其他应用程序浏览上传文件信息。

原型：

```
public Enumeration getEnumeration()
```

3. Request 类

这个类的功能等同于 JSP 内置的对象 request。之所以提供这个类，是因为对于文件上传表单，通过 request 对象无法获得表单项的值，必须通过 jspSmartUpload 组件提供的 request 对象来获取。该类提供如下方法。

1）getParameter()

作用：获取指定参数的值。当参数不存在时，返回值为 null。

原型：

```
public String getParameter(String name)
```

其中，name 为参数的名称。

2）getParameterValues()

作用：当一个参数可以有多个值时，用此方法来取其值。它返回的是一个字符串数组。当参数不存在时，返回值为 null。

原型：

```
public String[] getParameterValues(String name)
```

其中，name 为参数的名称。

3）getParameterNames()

作用：取得 Request 对象中所有参数的名称，用于遍历所有参数。它返回的是一个枚举型的对象。

原型：

```
public Enumeration getParameterNames()
```

4. SmartUpload 类

这个类完成上传/下载工作,提供了以下方法。

1) initialize()

作用：执行上传下载的初始化工作,必须首先执行。

原型(有多个,主要使用下面这个)：

```
public final void initialize(javax.servlet.jsp.PageContext pageContext)
```

其中,pageContext 为 JSP 页面内置对象(页面上下文)。

2) upload()

作用：上传文件数据。对于上传操作,第一步执行 initialize 方法,第二步就要执行这个方法。

原型：

```
public void upload()
```

3) save()

作用：将全部上传文件保存到指定目录下,并返回保存的文件个数。

原型：

```
public int save(String destPathName)
public int save(String destPathName,int option)
```

其中,destPathName 为文件保存目录。option 为保存选项,它有三个值,分别是 SAVE_PHYSICAL、SAVE_VIRTUAL 和 SAVE_AUTO。同 File 类的 saveAs 方法的选项值类似,SAVE_PHYSICAL 指示组件将文件保存到以操作系统根目录为文件根目录的目录下,SAVE_VIRTUAL 指示组件将文件保存到以 Web 应用程序根目录为文件根目录的目录下,而 SAVE_AUTO 则表示由组件自动选择。

注：save(destPathName)作用等同于 save(destPathName,SAVE_AUTO)。

4) getSize()

作用：取得上传文件数据的总长度。

原型：

```
public int getSize()
```

5) getFiles()

作用：取得全部上传文件,以 Files 对象形式返回,可以利用 Files 类的操作方法来获得上传文件的数目等信息。

原型：

```
public Files getFiles()
```

6) getRequest()

作用：取得 request 对象,以便由此对象获得上传表单参数的值。

原型：

```
public Request getRequest()
```

7）setAllowedFilesList()

作用：设定允许上传带有指定扩展名的文件，当上传过程中有文件名不允许时，组件将抛出异常。

原型：

```
public void setAllowedFilesList(String allowedFilesList)
```

其中，allowedFilesList 为允许上传的文件扩展名列表，各个扩展名之间以逗号分隔。如果想允许上传没有扩展名的文件，可以用两个逗号表示。例如，setAllowedFilesList("doc,txt,,")将允许上传带.doc 和.txt 扩展名的文件以及没有扩展名的文件。

8）setDeniedFilesList()

作用：限制上传那些带有指定扩展名的文件。若有文件扩展名被限制，则上传时组件将抛出异常。

原型：

```
public void setDeniedFilesList(String deniedFilesList)
```

其中，deniedFilesList 为禁止上传的文件扩展名列表，各个扩展名之间以逗号分隔。如果想禁止上传那些没有扩展名的文件，可以用两个逗号来表示。例如，setDeniedFilesList("exe,bat,,")将禁止上传带.exe 和.bat 扩展名的文件以及没有扩展名的文件。

9）setMaxFileSize()

作用：设定每个文件允许上传的最大长度。

原型：

```
public void setMaxFileSize(long maxFileSize)
```

其中，maxFileSize 为每个文件允许上传的最大长度，当文件超出此长度时，将不被上传。

10）setTotalMaxFileSize()

作用：设定允许上传的文件的总长度，用于限制一次性上传的数据量大小。

原型：

```
public void setTotalMaxFileSize(long totalMaxFileSize)
```

其中，totalMaxFileSize 为允许上传的文件的总长度。

11）setContentDisposition()

作用：将数据追加到 MIME 文件头的 CONTENT-DISPOSITION 域。JSP SmartUpload 组件会在返回下载的信息时自动填写 MIME 文件头的 CONTENT-DISPOSITION 域，如果用户需要添加额外信息，请用此方法。

原型：

```
public void setContentDisposition(String contentDisposition)
```

其中,contentDisposition 为要添加的数据。如果 contentDisposition 为 null,则组件将自动添加"attachment:",以表明将下载的文件作为附件,结果是 IE 浏览器将会提示另存文件,而不是自动打开这个文件(IE 浏览器一般根据下载的文件扩展名决定执行什么操作,扩展名为. doc 的将用 Word 程序打开,扩展名为. pdf 的将用 Acrobat 程序打开等)。

12) downloadFile()

作用:下载文件。

原型:

```
public void downloadFile(String sourceFilePathName)
public void downloadFile(String sourceFilePathName,String contentType)
public void downloadFile(String sourceFilePathName,String contentType,String destFileName)
```

其中,sourceFilePathName 为要下载的文件名(带路径的文件全名),contentType 为内容类型(MIME 格式的文件类型信息,可被浏览器识别),destFileName 为下载后默认的另存文件名。

8.2.3 SmartUpload 组件的应用

1. 文件上传

1) 表单要求

对于上传文件的表单,有以下两个要求。

(1) method 应用 Post,即 method="post"。

(2) 增加属性:enctype="multipart/form-data"。

下面是一个用于上传文件的表单的例子。

```
<form method = "post" enctype = "multipart/form - data" action = "/jspSmartUpload/upload.jsp">
    <input type = "FILE" name = "fileName">
    <input type = "submit" value = "上传">
</form>
```

2) 上传页面 upload.html

本页面提供表单,让用户选择要上传的文件,单击"上传"按钮执行上传操作。页面源代码如下。

```
<meta http - equiv = "Content - Type" content = "text/html; charset = gb2312">
<html>
    <head> title>用 smartUpload 组件实现文件上传</title></head>
    <body>
    <p align = "center">用 smartUpload 组件实现文件上传</p>
      <p align = "center">请选择要上传的文件:</p>
    <FORM METHOD = "POST" ACTION = "do_upload.jsp" ENCTYPE = "multipart/form - data">
      <table width = "75 % " border = "1" align = "center">
        <tr><td>1 <input type = "FILE" name = "FILE1" size = "30"></td> </tr>
        <tr><td> 2 <input type = "FILE" name = "FILE2" size = "30"> </td></tr>
        <tr><td> 3 <input type = "FILE" name = "FILE3" size = "30"> </td></tr>
```

```
        <tr><td>4<input type="FILE" name="FILE4" size="30"></td></tr>
        <tr><td><input type="submit" name="Submit" value="上传"></td></tr>
      </table>
    </FORM>
    </body>
</html>
```

3）上传处理页面 do_upload.jsp

本页面执行文件上传操作，代码如下。

```
<%@ page contentType="text/html; charset=gb2312" language="java"
  import="java.util.*,com.jspsmart.upload.*" errorPage="" %>
<html>
  <head><title>文件上传处理页面</title></head>
  <body>
    <%
      SmartUpload su = new SmartUpload(); //新建一个 SmartUpload 对象
      su.initialize(pageContext); //上传初始化
      //设定上传限制
      su.setMaxFileSize(40000); //1.限制每个上传文件的最大长度.
      su.setTotalMaxFileSize(150000); //2.限制总上传数据的长度.
      su.setAllowedFilesList("doc,txt");
      //3.设定允许上传的文件(通过扩展名限制),仅允许 DOC、TXT 文件
      su.setDeniedFilesList("exe,bat,jsp,htm,html,,");
      //4.设定禁止上传的文件(通过扩展名限制),禁止上传带有.exe、.bat、.jsp、.htm、.html
      //扩展名的文件和没有扩展名的文件.
      su.upload(); //上传文件
      int count = su.save("/upload"); //将上传文件全部保存到指定目录
      out.println(count+"个文件上传成功!<br>");
      //逐一提取上传文件信息,同时可保存文件.
      for (int i=0;i<su.getFiles().getCount();i++) {
          com.jspsmart.upload.File file = su.getFiles().getFile(i);
          //若文件表单中的文件选项没有选择文件则继续
          if (file.isMissing()) continue;
          //显示当前文件信息
          out.println("<TABLE BORDER=1>");
          out.println("<TR><TD>表单项名(FieldName)</TD><TD>" +
            file.getFieldName()+"</TD></TR>");
          out.println("<TR><TD>文件长度(Size)</TD><TD>" +
            file.getSize()+"</TD></TR>");
          out.println("<TR><TD>文件名(FileName)</TD><TD>" +
            file.getFileName()+"</TD></TR>");
          out.println("<TR><TD>文件扩展名(FileExt)</TD><TD>" +
            file.getFileExt()+"</TD></TR>");
          out.println("<TR><TD>文件全名(FilePathName</TD><TD>" +
            file.getFilePathName()+"</TD></TR>");
          out.println("</TABLE><BR>");
      }
    %>
  </body>
</html>
```

2. 文件下载

(1) 下载链接页面 download.html 的代码如下。

```
<html>
  <head><title>文件下载</title></head>
  <body>
      <a href="do_download.jsp">点击下载</a>
    </body>
</html>
```

(2) 下载处理页面 do_download.jsp do_download.jsp 展示了如何利用 jspSmartUpload 组件来下载文件，源代码如下。

```
<%...@ page contentType="text/html;charset=gb2312"
import="com.jspsmart.upload.*" %>
<%
  SmartUpload su = new SmartUpload();                //新建一个 SmartUpload 对象
  su.initialize(pageContext);                        //初始化
  //设定 contentDisposition 为 null 以禁止浏览器自动打开文件，保证单击链接后是下载文件
  //若不设定，则下载的文件扩展名为 doc 时，浏览器将自动用 Word 打开它
  //扩展名为 pdf 时，浏览器将用 Acrobat 打开
  su.setContentDisposition(null);
  su.downloadFile("/upload/jspSmartUpload.jar");     //下载文件
 %>
```

8.2.4　任务：注册表的照片上传

1. 任务要求

当学生登录后，要求上传照片，利用 SmartUpload 组件实现。

2. 任务实施

假设当用户登录后，将用户名保存在 Session 中，根据用户名查找学生信息，照片的文件名采用用户的学号，上传照片结束后，将照片保留的文件名存储到数据库表中，有利用显示照片。其基本步骤如下。

(1) 定义学生类用户类。

```
public class Student {
  private String userName;                //用户名
  private String code;                    //学号
  private String name;                    //姓名
  private String phtolocate;              //照片保存位置
  //其他属性略
  //getter/settter 略
}
```

(2) 定义数据库访问接口 StudentDao。

```
public interface StudentDao {
  public Student FindByUserName(String useName);   //根据用户名获取学生信息
  public int UpdatePhto(String code,String phto);  //保存学生照片保留的文件名及位置
  //其他方法略
}
```

(3) 实现接口 StudentDao 的方法(略)。

(4) 设计 phtoupload.jsp 页面用于上传学生照片。

```
<%@ page language="java" import="java.util.*" pageEncoding="GBK"%>
<%@ taglib uri="http://java.sun.com/jsp/jstl/core" prefix="c"%>
<!DOCTYPE HTML PUBLIC "-//W3C//DTD HTML 4.01 Transitional//EN">
<html>
  <head>
    <c:set var="title" value="上传照片" />
    <title>${title}</title>
  </head>
  <body>
    <form action="UploadPhoto"
        method="post" name="upload" ENCTYPE="multipart/form-data">
        <table style="margin:0 auto;">
        <tr>
        <td> ${username}
          <div
                style="margin:0 auto;background:url(${requestScope.phtolocate}) no-
repeat center center;width:145px;height:175px;border:2px solid #999;">
          </div>
        </td>
        </tr>
        <tr>
              <td style="text-align:center;">
                <input type="file" name="phtofile"
                size="20" id="phtofile"
                style="border:1px solid #999;"/>
          </td>
        </tr>
        <tr/>
        <tr>
        <td style="text-align:center;"><input class="button"
            type="submit" name="submit" value="点击上传" id="submit" />
        <input class="button" type="button" value="照片预览"
            onclick="window.location.reload()" />
        </td>
        </tr>
        </table>
    </form>
      </body>
</html>
```

(5) 定义 Servlet,实现文件上传,并将学生照片位置信息保存到数据表中。

```
public class UploadPhoto extends HttpServlet {
  public UploadPhoto() {
    super();
  }
  public void destroy() {
    super.destroy();
  }
  public void doGet(HttpServletRequest request, HttpServletResponse response)
        throws ServletException, IOException {
    response.setContentType("text/html");
    doPost(request,response);
  }
  public void doPost(HttpServletRequest request, HttpServletResponse response)
  throws ServletException, IOException {
    HttpSession session = request.getSession();
    String username = session.getAttribute("username").toString();
    StudentDao DAO = new StudentDaoImpl();
    Student s = DAO.FindByUserName(username);
    String code,phto;
    code = s.getCode();
    SmartUpload upload = new SmartUpload();
    upload.initialize(this.getServletConfig(),request,response);
    upload.setAllowedFilesList("jpg,JPG");
    try {
        upload.upload();
        Files files = upload.getFiles();
        String ext = files.getFile(0).getFileExt();
        String filename = "./upload/" + java.io.File.separator + s.getCode() +
                          "." + ext;
        files.getFile(0).saveAs(filename, SmartUpload.SAVE_VIRTUAL);
    } catch (SmartUploadException e1) {
      e1.printStackTrace();
      }
    phto = "./upload/" + code + ".jpg";
    DAO.UpdatePhto(code, phto);
    request.getRequestDispatcher("/photo.jsp").forward(request, response);
  }
  public void init() throws ServletException {
  }
}
```

(6) 其他代码(略)。

8.3 分层架构开发(MVC 模式)

8.3.1 JSP 与分层模式

1. 为什么要使用三层架构

在软件开发的过程中,分层模式是常见的移植架构模式。如果不采用分层的架构,采用 Java 和 HTML 混合编程，造成看代码不方便。业务处理的代码与 JSP 代码混在一起，不易于阅读和代码维护。观察项目 3 的代码，可以发现，大量的数据库代码与 JSP 代码混杂在一起，如果程序发生变化，这个代码就无法使用，代码的可复用性、维护性差。对于一个简单的应用程序来说，代码量不是很多的情况下，一层架构或二层架构开发完全够用，没有必要将其复杂化。但对一个复杂的大型系统，设计为一层架构或二层架构开发，那么这样的设计存在很严重的缺陷。为了解决这个问题，可以采用分层设计的模式。

2. 什么是三层架构

分层模式是最常见的一种架构模式，它是很多架构模式的基础。分层是为了实现"高内聚、低耦合"。它采用"分而治之"的思想，把问题划分开来各个解决，易于控制，易于延展，易于分配资源。

所谓的三层开发就是将系统的整个业务应用划分为表示层、业务逻辑层、数据访问层，这样有利于系统的开发、维护、部署和扩展。

(1) 表示层：负责直接跟用户进行交互，一般是指系统的界面，用于数据录入、数据显示等。这意味着只做与外观显示相关的工作，不属于他的工作不用做。表示层一般为 Web 应用程序，以 JSP 文件、HTML 文件为主。

(2) 业务逻辑层：业务逻辑层的主要功能就是提供对业务逻辑处理的封装。在业务逻辑层中，通常会定义一些接口，表示层通过调用业务逻辑层的接口来实现各种操作，用于做一些有效性验证的工作，以更好地保证程序运行的健壮性，如完成数据添加、修改和查询业务，不允许指定的文本框中输入空字符串，数据格式是否正确及数据类型验证。业务逻辑层通常放在 Biz 包中。

(3) 数据访问层：顾名思义，其功能就是跟数据库进行交互、进行数据访问以及执行数据的添加、删除、修改和显示等。可以访问关系数据库、文本文件或 XML 文档等，数据访问层通常放在 DAO 包下。

3. 使用三层架构开发的优点

使用三层架构开发有以下优点。

(1) 从开发角度和应用角度来看，三层架构比二层架构或一层架构都有更大的优势。三层架构适合团队开发，每人可以有不同的分工，协同工作使效率倍增。开发二层或一层应用程序时，每个开发人员都应对系统有较深的理解，能力要求很高。开发三层应用程序时，则可以结合多方面的人才，只需少数人对系统全面了解即可，从一定程度降低了开发的难度。

(2) 三层架构可以更好地支持分布式计算环境。业务逻辑层的应用程序可以在多个计算机上运行，充分利用网络的计算功能。分布式计算的潜力巨大，远比升级 CPU 有效。

(3) 三层架构的最大优点是它的安全性。用户只能通过业务逻辑层来访问数据层，减少了入口，把很多危险的系统功能都屏蔽了。

4. 层与层之间的关系

在三层架构中,各层之间互相依赖。表示层依赖于业务逻辑层,业务逻辑层依赖数据访问层。各层之间的数据传递方向分为请求与响应两个方向,如图 8-1 所示。

表示层接收用户请求,根据用户请求通知业务逻辑层,业务逻辑层收到请求,首先对请求进行阅读审核,然后将请求通知数据访问层直接返回给表示层,数据访问层收到请求后便开始访问数据库。

数据访问层通过对数据库的访问得到请求结果,并把请求结果通知业务逻辑层;业务逻辑层收到请求结果,首先对结果进行审核,然后将请求结果通知表示层,表示层收到请求结果并展示给用户。

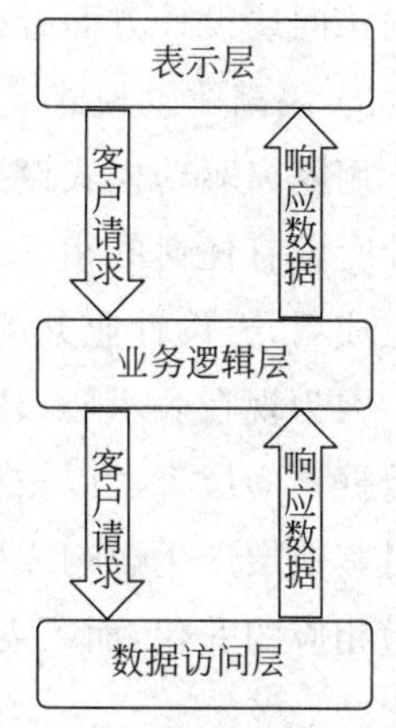

图 8-1　三层架构的数据传递方向

5. 三层架构的种类

目前,团队开发人员在开发项目时,大多都使用分层开发架构设计,目的是使各个层之间只能够被它相邻的层产生影响,但是这个限制常常在使用多层开发的时候被违反,这对系统的开发是有害的。三层架构按驱动模式可划分三种:数据访问层驱动模式、表示层驱动模式和隔离驱动模式,其中隔离驱动模式开发最为重要。下面通过三种模式的对比,介绍隔离驱动模式的重要性。

1）数据访问层驱动模式

所谓的数据层驱动模式,就是先设计数据访问层,表示层围绕数据访问层展开,一旦完成了数据层和表示层,业务逻辑层就围绕数据访问层展开。因为表示层是围绕数据访问层展开的,因此可能会使表示层中的约束不准确,并且限制了业务逻辑层的变更。由于业务逻辑层受到限制,一些简单变化可以通过 SQL 查询和存储过程来实现。

这种模式非常的普遍,它和传统的客户/服务器开发模式相似,并且是围绕已经存在的数据库设计的。由于表示层是围绕数据层设计的,它常常是凭直觉模仿数据访问层的实际结构。

常常存在一种额外的反馈循环在表示层到数据访问层之间,当设计表示层不容易实现的时候常常会去修改数据访问层,也就形成了这种反馈循环。开发者请求修改数据库以方便表示层的开发,但是对数据访问层的设计却是有害的。这种改变是人为的而没考虑到其他需求的限制,这种修改经常会违反至少损害数据的特有规则,导致不必要的数据冗余和数据的非标准化。

2）表示层驱动模式

表示层驱动模式是指数据访问层围绕表示层展开设计。业务逻辑层一般是通过简单的 SQL 查询和很少的变化完成的。由于数据库的设计是为了表示层的设计方便,而非从数据访问层设计方面考虑，所以数据库的设计在性能上通常很低。

3）隔离驱动模式

用隔离驱动模式设计，表示层和数据访问层被独立开发，常常是平行开发。这两层在设计时没有任何的相互干扰，所以不会存在人为的约束和有害的设计元素。当两层都设计完成后，再设计业务逻辑层。

因为现在表示层和数据访问层是完全独立的，当业务逻辑层需求改变的时候，表示层和数据访问层都可以做相应的修改而不影响对方。改变两个在物理上不相邻的层不会直接对其他层产生影响或发生冲突。这就允许数据访问层的调整或者表示层根据用户的需求做相应的变化，而不需要系统做大的调整或者修改。表 8-1 将对这三种驱动模式进行对比。

表 8-1　三种驱动模式对比

项　目	数据访问层驱动模式	表示层驱动模式	隔离驱动模式
数据库	(1) 很容易设计 (2) 产生负面影响 (3) 很难改变数据访问层，因为它和表示层紧密绑定	(1) 数据库设计不易 (2) 严重的不规范化设计 (3) 其他系统不易使用 (4) 很难改变数据访问层，由于它跟表示层紧密绑定	(1) 优化设计 (2) 集中设计数据库，表示层对它影响很小
业务需求	常常不能适应业务需求的变化	常常适应业务需求变化	适应需求变化
用户界面	围绕数据访问层而不是围绕用户，不易修改	适合用户扩展界面	适合用户扩展界面
可扩展性	通常可扩展，但是常常在用户界面做较多的工作以满足数据库的结构，同时数据库可能需要存储一些冗余的数据	完整性的扩展很难，常常只有通过“剪切”“粘贴”函数来实现	很容易扩展

综上所述，很容易看出隔离驱动模式的优点，隔离驱动模式设计可以极大地提高程序的可扩展性。

8.3.2　分层的实现

1 分层的原则

(1) 上层依赖其下层，依赖关系不跨层：表示层不能直接访问数据访问层；上层调用下层的结果，取决于下层的实现。

(2) 下层不能调用上层。

(3) 下层不依赖上层：上层的改变不会影响下层，下层的改变会影响上层得到的结果。

(4) 在上层中不能出现下层的概念，分工明确，各司其职。

2. 分层的特征

(1) 下层不知道上层的存在：仅完成自身的功能，不关心结果如何使用。

(2) 每一层仅知道其下层的存在，忽略其他层的存在。只关心结果的取得，不关心结果的实现过程。JSTL 通常会与 EL 表达式合作实现 JSP 页面的编码。

8.3.3　任务：利用三层架构实现用户管理系统

1. 任务要求

利用三层架构实现用户管理，主要实现用户的登录、注册、修改、删除、列表显示等功能。

2. 任务实施

实现的基本步骤和方法参考项目 4～项目 7，将其结合在一起，即可实现其功能，创建一个项目 UserPrj，下面简要说明其基本步骤。

1) 模型层的设计

(1) 定义实体类 UserBean，参考项目 4 定义用户类 user.java。

```
package cn.edu.mypt.bean;
public class UserBean {
    private String userID, userName,userPwd, userType;
    //略
}
```

(2) 定义接口 UserModel。

```
package cn.edu.mypt.model;
import java.util.List;
import cn.edu.mypt.bean.*;
public interface UserModel {
  public boolean userReg(UserBean userr); //用户注册,若注册成功,返回 true,否则返回 false
  public UserBean userLogin(String userId,String userPwd); //用户登录,登录成功,返回用户
                                                            //信息,否则返回 null
  public boolean UserModify(UserBean user); //修改用户信息,若修改成功,返回 true,否则
                                              //返回 false
  public boolean UserDelete(String userID); //删除用户,若删除成功,返回 true,否则返回 false
  public UserBean findById(String userID); //根据用户 ID 查找用户,找到返回用户信息,否则
                                            //返回 null
  public List<UserBean> findByUType(String utype); //获取指定类型的用户信息
  public List<UserBean> getUser(); //获取所有用户信息
  public boolean ModifyPwd(String userID,String oldPwd,String newPwd);
  //根据用户 ID 和旧密码修改用户密码,若修改成功,返回 true,否则返回 false
}
```

(3) 创建数据访问接口，实现对数据库数据的增、删、改、查的功能，参考 5.4 节中的任务。

```
package cn.edu.mypt.dao;
import java.util.List;
import cn.edu.mypt.bean.UserBean;
public interface UserDao {
  public int add(UserBean user);                //增加用户信息
  public int delele(String userID);             //删除用户指定用户信息
  public int updata(UserBean user);             //修改指定用户信息
  public UserBean findByID(String userID);      //查找指定用户
  public List findByType(String uType);         //查找指定类型的用户
  public List getUsers();                       //获取所有用户信息
}
```

(4) 参考 5.2 节或 5.3 节中的任务，创建 ConnectionFactory 类。

(5) 参考 5.4 中的任务，定义 UserDaoimpl 类，实现数据库访问接口。

(6) 参考项目 5，定义类 UserModelImple，它继承自 UserModel，实现用户管理的功能。

2) 控制层的设计

利用 Servelt 实现控制层的设计。

(1) 获取所有用户信息。定义 UserMangerServlet 类，其关键代码如下。

```
package cn.edu.mypt.servlet;
   //导入包(略)
public class UserMangerServlet extends HttpServlet {
  public void doGet(HttpServletRequest request, HttpServletResponse response)
  throws ServletException, IOException {
    doPost(request,response);
  }
  public void doPost(HttpServletRequest request, HttpServletResponse response)
  throws ServletException, IOException {
    UserModel uModel = new UserModelImple();
    List ulist = uModel.getUsers();
    request.getSession().setAttribute("USERLIST", ulist);
    request.getRequestDispatcher("userList.jsp").forward(request, response);
  }
}
```

(2) 增加用户类 AddUserServlet 的关键代码如下。

```
package cn.edu.mypt.servlet;
//导入包 (略)
public class AddUserServlet extends HttpServlet {
    //略
    public void doGet(HttpServletRequest request, HttpServletResponse response)
    throws ServletException, IOException {
      doPost(request,response);
    }
    public void doPost(HttpServletRequest request, HttpServletResponse response)
    throws ServletException, IOException {
```

```
        String uId = request.getParameter("userID");
        String uPwd = request.getParameter("userPwd");
        String uName = request.getParameter("userName");
        UserModel uModel = new UserModelImple();
        UserBean user = uModel.findById(uId);
        String result = "userList.jsp";
        if (user!= null) {
            request.setAttribute("message", "用户 ID 已经存在");
            result = "adduser.jsp";
        }
        else {
          user = new UserBean();
          user.setUserID(uId);
          user.setUserName(uName);
          user.setUserPwd(uPwd);
          if (!uModel.userReg(user) ) {
              request.setAttribute("message", "增加用户失败");
              result = "Error.jsp";
          }
        }
        request.getRequestDispatcher(result).forward(request, response);
    }
}
```

(3) 删除用户类 AddUserServlet 的关键代码如下。

```
package cn.edu.mypt.servlet;
//导入包（略）
public class DelUserServlet extends HttpServlet {
    //略
    public void doGet(HttpServletRequest request, HttpServletResponse response)
    throws ServletException, IOException {
      doPost(request,response);
    }
    public void doPost(HttpServletRequest request, HttpServletResponse response)
    throws ServletException, IOException {
      String uId = request.getParameter("userID");
      UserModel uModel = new UserModelImple();
      UserBean user = uModel.findById(uId);
      String result = "userList.jsp";
      if (user = null) {
          request.setAttribute("message", "用户不存在");
      }
      else {
          if (!uModel.UserDelete(uId) ) {
              result = "Error.jsp";
              request.setAttribute("message", "用户删除失败");
          }
      }
      request.getRequestDispatcher(result).forward(request, response);
```

```
    }
}
```

(4) 修改用户信息类 ModifyUserServlet 的关键代码如下。

```
package cn.edu.mypt.servlet;
//导入包（略）
public class ModifyUserServlet extends HttpServlet {
    //略
    public void doGet(HttpServletRequest request, HttpServletResponse response)
    throws ServletException, IOException {
      doPost(request,response);
    }
    public void doPost(HttpServletRequest request, HttpServletResponse response)
    throws ServletException, IOException {
        String uId = request.getParameter("userID");
        String uPwd = request.getParameter("userPwd");
        String uName = request.getParameter("userName");
        UserBean user = new UserBean(uId,uName,uPwd);
        UserModel uModel = new UserModelImple();
        String result = "userList.jsp";
        if (!uModel.UserModify(uId) ) {
            result = "Error.jsp";
            request.setAttribute("message", "用户修改失败");
        }
        request.getRequestDispatcher(result).forward(request, response);
    }
}
```

(5) 获取指定用户信息类 GetUserServlet 的关键代码如下。

```
package cn.edu.mypt.servlet;
//导入包（略）
public class GetUserServlet extends HttpServlet {
    //略
    public void doGet(HttpServletRequest request, HttpServletResponse response)
    throws ServletException, IOException {
      doPost(request,response);
    }
    public void doPost(HttpServletRequest request, HttpServletResponse response)
    throws ServletException, IOException {
        String uId = request.getParameter("userID");
        String action = request.getParameter("action");
        String result = "userList.jsp";
        UserModel uModel = new UserModelImple();
        UserBean user = uModel.findByID(uId);
        if (user == null ) {
          result = "Error.jsp";
```

```
            request.setAttribute("message", "用户不存在");
        }
        else {
            request.setAttribute("USER", user);
            if (action.equals("find") result = "displayUser.jsp";    //查找用户
            else return = "modifyUser.jsp";                          //修改用户
        }
        request.getRequestDispatcher(result).forward(request, response);
    }
}
```

3）视图层的设计

(1) 显示所有用户信息页面 userList.jsp 的代码如下。

```
<%@ page language = "java" import = "java.util.*" pageEncoding = "GBK" %>
<%@ taglib uri = "http://java.sun.com/jsp/jstl/core" prefix = "c" %>
<!DOCTYPE HTML PUBLIC " - //W3C//DTD HTML 4.01 Transitional//EN">
<html>
  <head><title>用户列表 </title></head>
    <body>
      <h2>用户信息</h2>
        <a href = "index.jsp">首页</a>
      <h3>${message}</h3>
    <table>
    <tr><td>用户 ID </td><td>用户名</td><td>密码</td><td>删除</td>
        <td>修改</td><td>更多</td></tr>
    <c:forEach var = "user" items = "${USERLIST}">
    <tr><td>${user.userID}</td><td>
    ${user.userName}</td><td>${user.userPwd}</td>
    <c:url var = "delurl" value = "DeleteServlet"><c:param name = "uID" value = "${user
.userID}"/></c:url>
    <c:url var = "modurl" value = "ModifyServlet">
        <c:param name = "uID" value = "${user.userID}"/>
        <c:param name = "action" value = "modify"/>
    </c:url>
    <c:url var = "findurl" value = "DisplayServlet">
        <c:param name = "uID" value = "${user.userID}"/>
        <c:param name = "action" value = "find"/>
    </c:url>
        <td><a href = "${delurl}">删除</a></td>
        <td><a href = "${modurl}">修改</a></td>
        <td><a href = "${findurl}">更多</a></td>
        </tr>
      </c:forEach>
      </table>
        <a href = "userAdd.jsp">增加</a>
    </body>
</html>
```

(2) 增加用户页面 userAdd.jsp 的代码如下。

```
<%@ page language = "java" import = "java.util.*" pageEncoding = "GBK" %>
<!DOCTYPE HTML PUBLIC " - //W3C//DTD HTML 4.01 Transitional//EN">
<html>
  <head><title>用户注册</title></head>
  <body>
    <h2>增加用户</h2>
    <hr><a href = "userList.jspt">返回</a><hr>
    <h3>${message}</h3>
      <form action = "UserAddServlet" method = "post">
          用户 ID:<input type = "text" name = "userID"/><br>
          用户名:<input type = "text" name = "userName"/><br>
          <input type = "hidden" name = "userPwd"/><br>
        <input type = "submit" value = "增加"/>
      </form>
    </body>
</html>
```

(3) 修改用户页面 modifyUser.jsp 的代码如下。

```
<%@ page language = "java" import = "java.util.*" pageEncoding = "GBK" %>
<!DOCTYPE HTML PUBLIC " - //W3C//DTD HTML 4.01 Transitional//EN">
<html>
  <head><title>用户修改</title></head>
  <body>
  <h3>修改用户信息</h3><hr>
    <a href = "userList.jsp">返回</a>
     <hr><h3>${message}</h3>
     <form action = "ModifyUserServlet" method = "post">
          用户 ID: ${USER.userID}<br>
          用户名:<input type = "text" name = "userName" value = "${USER.userName}"/><br>
          密码:<input type = "text" name = "userPwd" value = "${USER.userPwd}"/><br>
        <input type = "hidden" name = "userID" value = "${USER.userID}"/>
        <input type = "submit" value = "修改"/>
    </form>
  </body>
</html>
```

(4) 显示用户信息页面 displayUser.jsp 的代码如下。

```
<%@ page language = "java" import = "java.util.*" pageEncoding = "GBK" %>
<!DOCTYPE HTML PUBLIC " - //W3C//DTD HTML 4.01 Transitional//EN">
<html>
  <head><title>用户信息</title></head>
    <body>
    <h3>用户信息</h3>
    <hr><a href = "userList.jsp">返回</a><hr>
      用户 ID: ${USER.userID } <br>
      用户名: ${USER.userName }<br>
      密码: ${USER.userPwd }
```

```
  </body>
</html>
```

4）页面编码转换

创建一个过滤器 EncodingFilter，实现对所有页面的编码转换。

```
package cn.edu.mypt.filter;
//导入包（略）
public class EncodingFilter implements Filter {
    //略
     public void doFilter(ServletRequest arg0, ServletResponse arg1, FilterChain arg2)
     throws IOException, ServletException {
         arg0.setCharacterEncoding("GBK");
         arg2.doFilter(arg0, arg1);
    }
}
```

5）配置 web.xml

代码如下。

```
<?xml version="1.0" encoding="UTF-8"?>
<web-app version="3.0"
    xmlns="http://java.sun.com/xml/ns/javaee"
    xmlns:xsi="http://www.w3.org/2001/XMLSchema-instance"
    xsi:schemaLocation="http://java.sun.com/xml/ns/javaee
    http://java.sun.com/xml/ns/javaee/web-app_3_0.xsd">
    <servlet>
      <servlet-name>UserMangerServlet</servlet-name>
    <servlet-class>cn.edu.mypt.servlet.UserMangerServlet</servlet-class>
    </servlet>
    <servlet>
      <servlet-name>GetUserServlet</servlet-name>
      <servlet-class>cn.edu.mypt.servlet.GetUserServlet</servlet-class>
    </servlet>
    <servlet>
      <servlet-name>DelUserServlet</servlet-name>
      <servlet-class>cn.edu.mypt.servlet.DelUserServlet</servlet-class>
    </servlet>
  <servlet>
      <servlet-name>AddUserServlet</servlet-name>
      <servlet-class>cn.edu.mypt.servlet.AddUserServlet</servlet-class>
  </servlet>
  <servlet>
    <servlet-name>ModifyUserServlet</servlet-name>
    <servlet-class>cn.edu.mypt.servlet.ModifyUserServlet</servlet-class>
  </servlet>
  <servlet-mapping>
    <servlet-name>UserMangerServlet</servlet-name>
    <url-pattern>/UserMangerServlet</url-pattern>
  </servlet-mapping>
```

```
    <servlet-mapping>
      <servlet-name>GetUserServlet</servlet-name>
      <url-pattern>/GetUserServlet</url-pattern>
    </servlet-mapping>
    <servlet-mapping>
      <servlet-name>DelUserServlet</servlet-name>
      <url-pattern>/DelUserServlet</url-pattern>
    </servlet-mapping>
    <servlet-mapping>
      <servlet-name>ModifyUserServlet</servlet-name>
      <url-pattern>/ModifyUserServlet</url-pattern>
    </servlet-mapping>
    <servlet-mapping>
      <servlet-name>AddUserServlet</servlet-name>
      <url-pattern>/AddUserServlet</url-pattern>
    </servlet-mapping>
    <welcome-file-list>
      <welcome-file>index.jsp</welcome-file>
    </welcome-file-list>
    <filter>
      <filter-name>EncodingFilter</filter-name>
      <filter-class>cn.edu.mypt.filter.EncodingFilter</filter-class>
    </filter>
    <filter-mapping>
      <filter-name>EncodingFilter</filter-name>
      <url-pattern>/*</url-pattern>
    </filter-mapping>
</web-app>
```

6）部署并运行

启动服务器，部署项目并运行 http://127.0.0.1:8080/userPrj/UserMangerServlet。

项目 8 消息管理系统

1. 项目需求

在很多的平台，如 BBS、办公管理等，都会针对用户实现消息留言与管理，其基本功能主要有以下 4 项。

(1) 用户登录后，浏览消息。

(2) 用户给指定的人员发送消息。

(3) 用户回复消息。

(4) 用户处理消息，如删除、收藏、标记消息等。

根据需求，系统必须提供用户注册、登录、登出、权限控制等基本功能。用户登录短信息系统，登录成功系统自动跳转至短信息列表页面，用户可以查看个人短信息、发送短信息、回复短信息、删除短信息，登录失败则跳转至登录页面。

2. 项目实施

(1) 根据用户需求创建数据库 MSG_DB,数据表包括用户表和消息表。

用户表 userTable 的结构如下。

{#用户名(username char(20)),密码(password char(20)),邮箱(email char(50))}

消息表 msgTable 的结构如下。

{#消息 ID(msgid number),发送方(usernsme char(20)),接收方(sendto char(20)),
标题(titl char(60)),内容(content char(255)),发送时间(senddate datetime),
消息状态(status int)}

(2) 编写实体类和数据库操作类。

① 用户信息类 UserInfo.java。

```
public class UserInfo{
  private String username = null;                    //用户名
  private String password = null;                    //密码
  private String email = null;                       //电子邮箱
  //geter 、setter 略
}
```

② 消息类 BBSMessage.java。

```
public class BBSMessage{
  private String msgid = null;                       //ID
  private String username = null;                    //发送方
  private String title = null;                       //标题
  private String content = null;                     //内容
  private int state = 0;                             //消息状态
  private String sendto = null;                      //收件方
  private Date senddate = null;                      //发送时间
    //geter 、setter 略
}
```

③ 数据库操作类 BaseDao.java。

```
//导入包,略
public class BaseDao {
  protected Connection con = null;
  protected PreparedStatement ps = null;
  protected ResultSet rs = null;
  static protected String dbDriver;
  static protected String dbUrl;
  static protected String dbuName;
  static protected String dbuPwd;
  public BaseDao()
  {
    dbDriver = "com.microsoft.sqlserver.jdbc.SQLServerDriver";
```

```
    dbUrl = "jdbc:sqlserver://127.0.0.1:1433;DatabaseName = DB_MSG";
    dbuName = "sa";
    dbuPwd = "123456";
  }
  //打开链接
  protected void openConnection(){
      try {
        Class.forName(dbDriver);
        con = DriverManager.getConnection(dbUrl,dbuName,dbuPwd);
      } catch (ClassNotFoundException e) {
        e.printStackTrace();
      } catch (SQLException e) {
        e.printStackTrace();
      }
  }
  //更新数据库
  public int executeUpdata(String sql, List < Object > list){
    openConnection();
    try {
      ps = con.prepareStatement(sql);
      if(list == null) return ps.executeUpdate();
      int i = 1;
      for(Object obj:list){
        ps.setObject(i, obj);
        i++;
      }
      return ps.executeUpdate();
    } catch (SQLException e) {
      e.printStackTrace();
    }finally{
      closeResource();
    }
    return 0;
  }
  //关闭流
  protected void closeResource(){
    try {
      if(rs != null) rs.close();
      if(ps != null) ps.close();
      if(con != null) con.close();
    } catch (SQLException e) {
      e.printStackTrace();
    }
  }
}
```

(3) 定义数据访问接口。

① 用户管理接口 UserinfoDao.java,实现按照条件查找用户和添加更新用户。

```
public interface UserinfoDao {
```

```
    //查找用户
    public List<UserInfo> getUsers(Map<String,String> contitions);
    //添加或更新用户
    public int executeUpdate(String operate, UserInfo userinfo);
}
```

② 消息管理接口 BBSMessageDao .java,实现按照条件查询消息、添加或更新消息、获取用户的消息条数。

```
public interface BBSMessageDao {
    //分页查询消息
    public List<BBSMessage> getMessagesByPage(Map<String,Object> elements);
    //更新消息
    public int executeUpdate(String operate, Map<String,Object> elements,
                             Map<String,Object> conditions);
    //获取消息条数
    public int getMsgsCount(String username);
}
```

(4) 数据访问接口的实现。

① 访问消息数据库类 BBSMessageDaoImpl.java,实现对消息数据表的增、删、改、查功能。

```
public class BBSMessageDaoImpl extends BaseDao implements BBSMessageDao {
    String sql = null;
    String msgid = null ,username = null,title = null,msgcontent = null,state = null,
           sendto = null;
    Date msg_create_date = null;
    List<Object> list = null;
    @SuppressWarnings("unused")
    //解析更新字段
    private void getUpdateParm(Map<String,Object> elements) {
        if(elements!= null && elements.size()> 0){
            msgid = (String)elements.get("msgid");
            username = (String)elements.get("username");
            title = (String)elements.get("title");
            msgcontent = (String)elements.get("msgcontent");
            state = (String)elements.get("state");
            sendto = (String)elements.get("sendto");
            msg_create_date = (Date)elements.get("msg_create_date");
        }
    }
    //插入
    private void Insert(Map<String,Object> elements) {
        sql = "insert into msgTable values(?,?,?,?,?,?,?)";
        getUpdateParm(elements);
        list = new ArrayList<Object>();
        list.add(msgid);
        list.add(username);
```

```
        list.add(title);
        list.add(msgcontent);
        list.add(state);
        list.add(sendto);
        list.add(new Timestamp(msg_create_date.getTime()));
    }
    //删除
    private void Delete(Map < String,Object > conditions)
    {
        sql = "delete from msgTable where msgid = ?";
        list = new ArrayList < Object >();
        msgid = (String)conditions.get("msgid");
        list.add(msgid);
    }
    //更新
    private void Update(Map < String,Object > elements, Map < String,Object > conditions)
    {
        sql = "update msgTable set ";
        //获取条件元素
        String ctn_msgid = (String)conditions.get("msgid");
        String ctn_username = (String)conditions.get("username");
        String ctn_title = (String)conditions.get("title");
        String ctn_msgcontent = (String)conditions.get("msgcontent");
        String ctn_state = (String)conditions.get("state");
        String ctn_sendto = (String)conditions.get("sendto");
        Date ctn_msg_create_date = (Date)conditions.get("msg_create_date");
        //获取更新值
        getUpdateParm(elements);
        //写出 SQL 更新语句
        String and = "";
        sql += msgid!= null ? and + " msgid = ? ":"";
        and = msgid!= null ? " and ":"";
        sql += username!= null ? and + " username = ? ":"";
        and = username!= null ? " and ":"";
        sql += title!= null ? and + " title = '" + title + "' ":"";
        and = title!= null ? " and ":"";
        sql += msgcontent!= null ? and + " msgcontent = ? ":"";
        and = msgcontent!= null ? " and ":"";
        sql += sendto!= null ? and + " sendto = ? ":"";
        and = sendto!= null ? " and ":"";
        sql += state!= null ? and + " state = ? ":"";
        and = state!= null ? " and ":"";
        sql += msg_create_date != null ? and + " senddate = ? ":"";
        //写出 SQL 条件语句
        sql = sql + " where 1 = 1 ";
        sql += ctn_msgid != null ? " and msgid = ? ":"";
        sql += ctn_username != null ? " and username = ? ":"";
        sql += ctn_title != null ? " and title = '" + title + "' ":"";
        sql += ctn_msgcontent != null ? " and msgcontent = ? ":"";
```

```
    sql += ctn_sendto != null ? " and sendto = ? ":"";
    sql += ctn_state != null ? " and state = ? ":"";
    sql += ctn_msg_create_date != null ? " and senddate = ? ":"";
    list = new ArrayList<Object>();
    //设置更新参数
    if(msgid != null) list.add(msgid);
    if(username != null) list.add(username);
    if(title != null) list.add(title);
    if(msgcontent != null) list.add(msgcontent);
    if(sendto != null) list.add(sendto);
    if(state != null) list.add(state);
    if(msg_create_date != null)
        list.add(new Timestamp(msg_create_date.getTime()));
    //设置条件参数
    if(ctn_msgid != null) list.add(ctn_msgid);
    if(ctn_username != null) list.add(ctn_username);
    if(ctn_title != null) list.add(ctn_title);
    if(ctn_msgcontent != null) list.add(ctn_msgcontent);
    if(ctn_sendto != null) list.add(ctn_sendto);
    if(ctn_state != null) list.add(ctn_state);
    if(ctn_msg_create_date != null)
        list.add(new Timestamp(ctn_msg_create_date.getTime()));
}
/* 执行 update 操作
 * @param
 * operate:执行插入、更新还是删除
 * elements:需要插入或者更新的值集合,key 对应 column,value 是插入或者更新的值
 * conditions:update 执行的条件集合,key 对应 column,value 是条件值
 */
public int executeUpdate(String operate, Map<String,Object> elements,
                         Map<String,Object> conditions) {
    //更新操作
    if(operate.equals("update")) Update(elements,conditions);
    //添加操作
    else if(operate.equals("insert")) Insert(elements);
    //删除操作
    else if(operate.equals("delete")) Delete(conditions);
    return executeUpdata(sql, list);
}
//执行查询操作 elements:查询条件集合
public List<BBSMessage> getMessagesByPage(Map<String,Object> elements) {
    //获取查询语句中的条件参数
    Integer msgs_in_one_page = (Integer)elements.get("msgs_in_one_page");
    Integer page_no = (Integer)elements.get("page_no");
    String username = (String)elements.get("username");
    String msgid = (String)elements.get("msgid");
    String title = (String)elements.get("title");
    String msgcontent = (String)elements.get("msgcontent");
    String sendto = (String)elements.get("sendto");
```

```
        Integer state = (Integer)elements.get("state");
        Date msg_create_date = (Date)elements.get("msg_create_date");
        SimpleDateFormat sdf = new SimpleDateFormat("yyyy-MM-dd HH:mm");
        //编写 SQL 语句
        String sql = "select * from msg where 1 = 1 ";
        //编写条件语句
        sql += msgid!= null ? " and msgid = '" + msgid + "'":"";
        sql += username != null?" and username = '" + username + "'":"";
        sql += title!= null ? " and title = '" + title + "'":"";
        sql += msgcontent!= null ? " and msgcontent = '" + msgcontent + "'":"";
        sql += sendto!= null ? " and sendto = '" + sendto + "'":"";
        sql += state!= null ? " and state = " + state + " ":"";
        sql += msg_create_date!= null ? " and msg_create_date = '" +
            sdf.format(msg_create_date) + "'":"";
        //不同的数据库分页 SQL 语句有差异
        if(page_no != null){
            sql += " LIMIT " + msgs_in_one_page * (page_no - 1) + ", " +
                msgs_in_one_page;
        }
        //返回结果
        List<BBSMessage> list = null;
        openConnection();
        list = new ArrayList<BBSMessage>();
        try {
            ps = con.prepareStatement(sql);
            rs = ps.executeQuery();
            while(rs.next()){
                BBSMessage bms = new BBSMessage();
                bms.setMsgid(rs.getString("msgid"));
                bms.setTitle(rs.getString("title"));
                bms.setContent(rs.getString("msgcontent"));
                bms.setSendto(rs.getString("sendto"));
                bms.setState(rs.getInt("state"));
                bms.setDatetime(rs.getDate("msg_create_date"));
                bms.setUsername(rs.getString("username"));
                list.add(bms);
            }
        } catch (SQLException e) {
            e.printStackTrace();
        }finally{
            closeResource();
        }
        return list;
    }
    public int getMsgsCount(String username) {
        String always_true = " where 1 = 1 ";
        int count = 0;
        if(username == null)
```

```
            username = "";else username = " and username = '" + username + "'";
        String sql = "select count( * ) from msg " + always_true + username;
        openConnection();
        try {
            ps = con.prepareStatement(sql);
            rs = ps.executeQuery();
            while(rs.next())
                count = rs.getInt(1);
        } catch (SQLException e) {
            e.printStackTrace();
        }
        return count;
    }
}
```

② 用户信息的数据访问类 UserinfoDaoImpl.java，实现对用户信息的增、删、改、查的处理。

```
public class UserinfoDaoImpl extends BaseDao implements UserinfoDao {
  //添加或更新用户
  public int executeUpdate(String operate, UserInfo userinfo) {
    List<Object> list = null;
    String sql = null;
    if(operate.equals("update")){
      sql = "update userTable set password = ? where username = '" +
            userinfo.getUsername() + "'";
      list = new ArrayList<Object>();
      list.add(userinfo.getPassword());
    }else if(operate.equals("insert")){
      sql = "insert into userTable values(?,?,?)";
      list = new ArrayList<Object>();
      list.add(userinfo.getUsername());
      list.add(userinfo.getPassword());
      list.add(userinfo.getEmail());
    }
    return executeUpdata(sql, list);
  }
  //查找用户
  public List<UserInfo> getUsers(Map<String,String> contitions) {
    String username = (String) contitions.get("username");
    String pwd = (String) contitions.get("pwd");
    String operate = (String)contitions.get("operate");
    String sql = "select * from userTable where 1 = 1 ";
    if(operate != null && operate.equals("except")){
      sql += username != null ? " and username != '" + username + "'":"";
    }else{
      sql += username != null ? " and username = '" + username + "'":"";
    }
    sql += pwd != null ? " and password = '" + pwd + "'":"";
    openConnection();
```

```
    List < UserInfo > users = new ArrayList < UserInfo >();
    String uName,uPwd,em;
    try {
      ps = con.prepareStatement(sql);
      rs = ps.executeQuery();
      while(rs.next()){
        UserInfo uf = new UserInfo();
        uName = rs.getString("username");
        uPwd = rs.getString("password");
        em = rs.getString("email");
        uf.setUsername(uName);
        uf.setPassword(uPwd);
        uf.setEmail(em);
        users.add(uf);
      }
    } catch (SQLException e) {
      e.printStackTrace();
    }finally{
      closeResource();
    }
    return users;
  }
}
```

(5) 定义业务逻辑接口。

① 用户管理业务逻辑接口 UserInfoBiz .java。

```
public interface UserInfoBiz {
  //查找用户
  public List < UserInfo > findUser(Map < String,String > elements);
  //添加或更新用户
  public int doInsertOrUpdateUser(String operate, UserInfo userinfo);
}
```

② 消息处理业务逻辑接口 BBSMessageBiz .java。

```
public interface BBSMessageBiz {
  //查询短信息
  public List < BBSMessage > findMessages(Map < String,Object > elements);
  //更新或删除短信息
  public int doInsertOrUpdateMsg ( String operate, Map < String, Object > elements, Map
< String,Object > conditions);
}
```

(6) 业务逻辑的实现。

① 用户管理业务逻辑类 UserInfoBizImpl.java。

```
public class UserInfoBizImpl implements UserInfoBiz {
  private UserinfoDao userinfoDao = new UserinfoDaoImpl();
```

```
  //添加或更新用户
  public int doInsertOrUpdateUser(String operate, UserInfo userinfo) {
    return userinfoDao.executeUpdate(operate, userinfo);
  }
  //查找用户
  public List<UserInfo> findUser(Map<String,String> elements) {
    return userinfoDao.getUsers(elements);
  }
}
```

② 消息处理业务逻辑类 BBSMessageBizImpl.java。

```
public class BBSMessageBizImpl implements BBSMessageBiz {
  private BBSMessageDao bmd = new BBSMessageDaoImpl();
  //更新或删除短信息
  public int doInsertOrUpdateMsg(String operate, Map<String,Object> elements,
                                 Map<String,Object> conditions) {
    return bmd.executeUpdate(operate, elements, conditions);
  }
  //查询短信息
  public List<BBSMessage> findMessages(Map<String,Object> elements) {
    return bmd.getMessagesByPage(elements);
  }
}
```

(7) 控制层采用 Servelet 实现。

① 用户管理 Servlet：UserServlet.java。

```
public class UserServlet extends HttpServlet {
  HttpServletRequest request;
  HttpServletResponse response;
  UserInfoBiz userInfoBiz = new UserInfoBizImpl();
  private static final long serialVersionUID = 1L;
  public void doGet(HttpServletRequest request, HttpServletResponse response)
  throws ServletException, IOException {
    doPost(request, response);
  }
  public void doPost(HttpServletRequest req, HttpServletResponse res)
  throws ServletException, IOException {
    request = req;
    response = res;
    String action = request.getParameter("action");
    if ("login".equals(action)) {                    //登录
        if (login().equals("error"))
          request.getRequestDispatcher("index.jsp").forward(request, response);
        else
          response.sendRedirect("MsgServlet?action=list");
    } else if("regist".equals(action)) {
        //注册
```

```
            if (register().equals("error"))
              request.getRequestDispatcher("register.jsp").forward(request,response);
            else
              request.getRequestDispatcher("index.jsp").forward(request,response);
        }
        else if("logout".equals(action)){                           //退出
          request.getSession().removeAttribute("loginuser");
          response.sendRedirect("index.jsp");
        }else if("findUsers".equals(action)){                       //查找其他所有的用户
          Find();
          request.getRequestDispatcher("sendMsg.jsp").forward(request,response);
        }
    }
    private String login() {
        String username = request.getParameter("username");
        String password = request.getParameter("password");
        Map<String,String> contitions = new HashMap<String,String>();
        contitions.put("username", username);
        contitions.put("pwd", password);
        List<UserInfo> list = userInfoBiz.findUser(contitions);
        if(list == null || list.size() != 1) {
          request.setAttribute("error", "用户名或密码错误!");
          return "error";
        }
        else {
          request.getSession().setAttribute("loginuser", username);
          return "success";
        }
    }
    private String register() {
        String username = request.getParameter("username");
        String password = request.getParameter("password");
        String email = request.getParameter("email");
        Map<String,String> conditions = new HashMap<String,String>();
        conditions.put("username", username);
        List<UserInfo> list = userInfoBiz.findUser(conditions);
        if(list.size() > 0){
          request.setAttribute("error", "此用户名已被注册");
          return "error";
        }else{
          UserInfo uf = new UserInfo();
          uf.setUsername(username);
          uf.setPassword(password);
          uf.setEmail(email);
          int isRun = userInfoBiz.doInsertOrUpdateUser("insert", uf);
          if(isRun > 0){
            request.setAttribute("error", "已经成功注册!");
            return "success";
          }
```

```
        else {
          request.setAttribute("error", "注册失败!");
          return "error";
        }
      }
    }
    private String Find() {
      String username = (String)request.getSession().getAttribute("loginuser");
      Map<String,String> contitions = new HashMap<String,String>();
      contitions.put("username", username);
      contitions.put("operate","except");
      List<UserInfo> list = userInfoBiz.findUser(contitions);
      request.setAttribute("users", list);
      return "success";
    }
}
```

② 消息处理 Servlet：MsgServlet.java。

```
public class MsgServlet extends HttpServlet {
  HttpServletRequest request;
  HttpServletResponse response;
  BBSMessageBiz bmb = new BBSMessageBizImpl();
  private static final long serialVersionUID = 1L;
  public void doGet(HttpServletRequest request, HttpServletResponse response)
  throws ServletException, IOException {
    doPost(request, response);
  }
  public void doPost(HttpServletRequest req, HttpServletResponse res)
  throws ServletException, IOException {
    request = req;
    response = res;
    String action = request.getParameter("action");
    if(action.equals("del")){
      //删除短信息
      delMsg();
      request.getRequestDispatcher("MsgServlet?action = list").forward(request,response);
    }else if(action.equals("send")) {
      //发送短信息
      sendMsg();
      response.sendRedirect("MsgServlet?action = list");
    }else if(action.equals("list")){      //显示当前用户的收件箱
        listMsg();
      request.getRequestDispatcher("main.jsp").forward(request,response);
    }else if(action.equals("read")){      //读取某条短信息
      readMsg();
      request.getRequestDispatcher("readMsg.jsp").forward(request,response);
    }
  }
  private String delMsg() {
```

```
    String msgid = request.getParameter("msgid");
    Map<String,Object> conditions = new HashMap<String,Object>();
    conditions.put("msgid", msgid);
    int isSuccess = bmb.doInsertOrUpdateMsg("delete",new HashMap<String,Object>(),
                                            conditions);
    if(isSuccess > 0){
      request.setAttribute("error", "删除成功!");
      return "success";
    }else{
      request.setAttribute("error", "删除属于非法操作,请确认当前用户权限!");
      return "error";
    }
  }
  private String sendMsg() {
    Random r = new Random();
    String msgid = String.valueOf(r.nextInt());
    String fromUser = (String)request.getSession().getAttribute("loginuser");
    String toUser = request.getParameter("toUser");
    String title = request.getParameter("title");
    String content = request.getParameter("content");
    String state = "0";
    Timestamp ts = new Timestamp((new Date()).getTime());
    Map<String,Object> map = new HashMap<String,Object>();
    map.put("msgid", msgid);
    map.put("sendto", toUser);
    map.put("title", title);
    map.put("msgcontent",content);
    map.put("state", state);
    map.put("username", fromUser);
    map.put("msg_create_date", ts);
    bmb.doInsertOrUpdateMsg("insert",map,null);
    return "success";
  }
  private String readMsg() {
    Map<String,Object> elements = null;
    Map<String,Object> conditions = null;
    String msgid = request.getParameter("msgid");
    String state = request.getParameter("state");
    if(state.equals("0")){
      elements = new HashMap<String,Object>();
      conditions = new HashMap<String,Object>();
      elements.put("state", "1");
      conditions.put("msgid", msgid);
      bmb.doInsertOrUpdateMsg("update", elements, conditions);
    }
    if(conditions == null){
      conditions = new HashMap<String,Object> ();
      conditions.put("msgid", msgid);
    }
```

```
    List < BBSMessage > list = bmb.findMessages(conditions);
    request.setAttribute("msg", list);
    return "success";
  }
  private String listMsg()
  {
    String username = (String)request.getSession().getAttribute("loginuser");
    Map < String,Object > map = new HashMap < String,Object >();
    map.put("sendto", username);
    List < BBSMessage > list = bmb.findMessages(map);
    request.setAttribute("msgs", list);
    return "success";
  }
}
```

(8) 编写过滤器。

① 编码过滤器类 EncodeFilter.java，用于设置编码。

```
public class EncodeFilter implements Filter {
  private String encode = null;
  public void destroy() {
    encode = null;
  }
  public void doFilter(ServletRequest request, ServletResponse response,
  FilterChain chain) throws IOException, ServletException {
    if (null == request.getCharacterEncoding()) {
      request.setCharacterEncoding(encode);
    }
    chain.doFilter(request, response);
  }
  public void init(FilterConfig filterConfig) throws ServletException {
    String encode = filterConfig.getInitParameter("encode");
    if (this.encode == null) {
      this.encode = encode;
    }
  }
}
```

② 强制登录类 UserLoginFilter.java。

```
public class UserLoginFilter extends HttpServlet implements Filter {
  //略
  public void doFilter(ServletRequest request, ServletResponse response,
  FilterChain filterChain) throws ServletException, IOException {
    HttpSession session = ((HttpServletRequest)request).getSession();
    if(session.getAttribute("loginuser") == null){
      PrintWriter out = response.getWriter();
      out.print("< script language = javascript > alert('您还没有登录!');window.location
.href = '../index.jsp';</script >");
      request.getRequestDispatcher("index.jsp").forward(request, response);
```

```
        }else{
            filterChain.doFilter(request, response);
        }
    }
}
```

(9) 配置 web.xml。

```
<servlet>
    <servlet-name>UserServlet</servlet-name>
    <servlet-class>cn.edu.mypt.web.UserServlet</servlet-class>
</servlet>
<servlet>
    <servlet-name>MsgServlet</servlet-name>
    <servlet-class>> cn.edu.mypt.web.MsgServlet</servlet-class>
</servlet>
<servlet-mapping>
    <servlet-name>UserServlet</servlet-name>
    <url-pattern>/UserServlet</url-pattern>
</servlet-mapping>
<servlet-mapping>
    <servlet-name>MsgServlet</servlet-name>
    <url-pattern>/MsgServlet</url-pattern>
</servlet-mapping>
<welcome-file-list>
    <welcome-file>index.jsp</welcome-file>
</welcome-file-list>
<filter>
    <filter-name>EncodeFilter</filter-name>
    <filter-class>> cn.edu.mypt.filter.EncodeFilter</filter-class>
    <init-param>
        <param-name>encode</param-name>
        <param-value>UTF-8</param-value>
    </init-param>
</filter>
<filter-mapping>
    <filter-name>EncodeFilter</filter-name>
    <url-pattern>/*</url-pattern>
</filter-mapping>
<filter>
    <filter-name>userlogin</filter-name>
    <filter-class>cn.edu.mypt.filter.UserLoginFilter</filter-class>
</filter>
<filter-mapping>
    <filter-name>userlogin</filter-name>
    <url-pattern>/jsp/*</url-pattern>
</filter-mapping>
```

(10) 视图层。

① 用户注册页面 register.jsp。

```
<%@ page language="java" import="java.util.*" pageEncoding="UTF-8"%>
<html>
```

```
<head><title>在线消息管理平台</title></head>
<body>
    <div id = "regTitle" class = "png"></div>
    <div id = "regForm" class = "userForm png">
    <form action = "UserServlet?action = regist" method = "post">
      <dl>
        <div id = "error"> ${error }</div>
        <dt>用 户 名:</dt><dd><input type = "text" name = "username" /></dd>
        <dt>密      码:</dt><dd><input type = "password" name = "password" /></dd>
        <dt>确认密码:</dt><dd><input type = "password" name = "affirm" /></dd>
        <dt>邮      箱:</dt><dd><input type = "text" name = "email" /></dd>
      </dl>
      <div class = "buttons">
        <input class = "btn - reg png" type = "submit" name = "register" value = " " />
        <input class = "btn - reset png" type = "reset" name = "reset" value = " " />
      </div>
      <div class = "goback"><a href = "index.jsp" class = "png">返回登录页</a></div>
    </form>
  </div>
</body>
</html>
```

② 用户登录页面(首页)index.jsp。

```
<%@ page language = "java" import = "java.util.*" pageEncoding = "UTF - 8" %>
<head>
<title>在线消息管理平台</title>
<body>
<div id = "loginTitle" class = "png"></div>
<div id = "loginForm" class = "userForm png">
  <form method = "post" name = "loginform" action = "UserServlet?action = login" >
    <dl>
      <div id = "error"> ${error}</div>
      <dt>用户名:</dt><dd><input type = "text" name = "username" /></dd>
      <dt>密   码:</dt><dd><input type = "password" name = "password" /></dd>
    </dl>
    <div class = "buttons">
      <input class = "btn - login png" type = "submit" name = "submit" value = " " />
      <input class = "btn - reg png" type = "button" name = "register" value = " " />
    </div>
  </form>
</div>
</body>
</html>
```

③ 消息发送页面 sendMsg.jsp。

```
<%@ page language = "java" import = "java.util.*" pageEncoding = "UTF - 8" %>
<html>
<body>
```

```
<form action = "MsgServlet?action = send" method = "post" >
  < div id = "main">
    < div class = "mainbox">
      < div class = "menu">
        < span >当前用户:< a href = "main.jsp"> $ {sessionScope.loginuser}</a ></span >
        < span >< a href = "UserServlet?action = findUsers">发短消息</a ></span >
        < span >< a href = "UserServlet?action = logout">退出</a ></span >
      </div >
      < div class = "content">
          < div class = "message">
            < div class = "tmenu">
              < ul class = "clearfix">
                < li >
                  发送给:< %
                  String to = request.getParameter("sendto");
                  System.out.print(" -- to -- " + to);
                  request.setAttribute("sendto",to );
                  % >
                  < select name = "toUser">
                    < c:forEach var = "user" items = " $ {users}">
                      < option > $ {user.username}</option >
                    </c:forEach >
                  </select >
              </li >
              < li>标题:< input type = "text" name = "title" id = "title"/></li >
            </ul >
          </div >
        < div class = "view">
            < textarea name = "content" id = "content"></textarea >
            < div class = "send">
              < input type = "submit" name = "submit" value = " " /></div >
        </div >
      </div >
    </div >
  </div >
 </div >
</form >
</body >
</html >
```

④ 消息显示页面 main.jsp。

```
< % @ page language = "java" import = "java.util. * " pageEncoding = "UTF - 8" % >
< html >
< body >
< div id = "main">
  < div class = "mainbox">
    < div class = "title myMessage png"></div >
    < div class = "menu">
      < span >当前用户:< a href = "main.jsp"> $ {sessionScope.loginuser}</a ></span >
```

```
      <span><a href = "UserServlet?action = findUsers">发短消息</a></span>
      <span><a href = "UserServlet?action = logout">退出</a></span>
    </div>
    <div id = "error">${error}</div>
    <div class = "content messageList">
      <ul>
        <c:forEach var = "msg" items = "${msgs}">
          <c:if test = "${msg.state == 0}">
            <li class = "unReaded">
          </c:if>
          <c:if test = "${msg.state == 1}">
            <li>
          </c:if>
          <em><c:out value = "${msg.datetime}"/></em>
          <em><a href = ${"UserServlet?action = findUsers&sendto = "}${msg.username}
${"&state = "}${msg.state}>回信</a></em>
          <em><a href = ${"MsgServlet?action = del&msgid = "}${msg.msgid}>删除</a></em>
          <p>
            <strong><a href = MsgServlet?action = read&msgid = ${msg.msgid}&state =
${msg.state}>
            <c:out value = "${msg.title}"/></a></strong>
            <c:if test = "${fn:length(msg.content) > 8}">
              <c:out value = "${fn:substring(msg.content,0,7)}"/>....
            </c:if>
            <c:if test = "${fn:length(msg.content) <= 8}">
              <c:out value = "${msg.content}"/>
            </c:if>
          </p>
          </li>
        </c:forEach>
      </ul>
    </div>
  </div>
</div>
</body>
</html>
```

(11) CSS 代码(略)。

习　题　8

1. 选择题

(1) MVC 中的 MVC 分别用(　　)表示。

A. JSP、Servlet、JavaBean　　B. HTML、JavaBean、JSP

C. JavaBean、JSP、Servlet　　D. Servlet、HTML、JSP

(2) 关于 MVC 模式说法正确的是(　　)。

A. 用来将代码分开

B. 将显示、流程控制、业务逻辑分开，提高维护性和分离复杂度

C. 视图模型控制器模型

D. 只用 Java 才有的模型

(3) MVC 架构中使用 HTML 界面、JSP 界面实现用户界面，使用(　　)实现控制逻辑。

A. JSP　　B. Servle　　C. JavaScript　　D. EJB

(4) 下面关于 MVC 的说法不正确的是(　　)。

A. M 表示 Model 层，是存储数据的地方

B. View 表示视图层，负责向用户显示外观

C. Controller 是控制层，负责控制流程

D. 在 MVC 架构中 JSP 通常做控制层

(5) 在 JSP MVC 设计模式体系结构中，(　　)是实现控制器的首选方案。

A. JSP　　B. Servlet　　C. JavaBean　　D. HTML

2. 简答题

(1) 什么是软件系统开发的三层架构，使用三层开发有什么优点？

(2) 在三层架构中，其分层的基本原则是什么？

参 考 文 献

[1] 刘京华. Java Web 整合开发王者归来[M]. 北京：清华大学出版社，2010.

[2] 李兴华，王月清. Java Web 开发实战经典基础篇[M]. 北京：清华大学出版社，2013.

[3] 覃国蓉. 基于工作任务的 Java Web 应用教程[M]. 北京：电子工业出版社，2009.

[4] 明日科技. Java Web 开发之道[M]. 北京：电子工业出版社，2011.

[5] 孙卫琴. Tomcat 与 Java Web 开发技术详解[M]. 3 版. 北京：电子工业出版社，2019.